GEOGRAPHY
FOR EDEXCEL

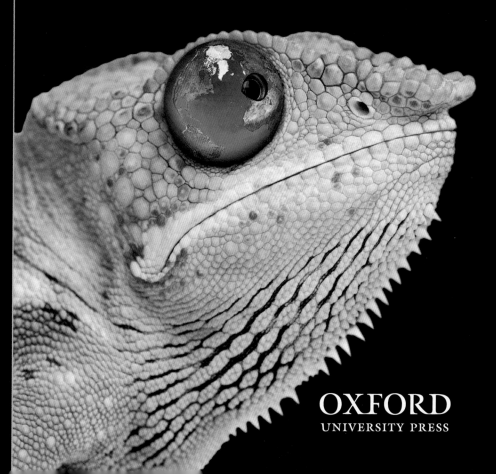

A LEVEL YEAR 1
AND AS LEVEL

Series editor Bob Digby

**Catherine Hurst   Lynn Adams   Russell Chapman
Simon Ross   Jane Ferretti**

T0202382

OXFORD
UNIVERSITY PRESS

OXFORD
UNIVERSITY PRESS

Great Clarendon Street, Oxford, OX2 6DP, United Kingdom

Oxford University Press is a department of the University of Oxford. It furthers the University's objective of excellence in research, scholarship, and education by publishing worldwide. Oxford is a registered trade mark of Oxford University Press in the UK and in certain other countries

Series editor: Bob Digby

Authors: Catherine Hurst, Lynn Adams, Russell Chapman, Simon Ross, Jane Ferretti

The moral rights of the authors have been asserted

Database right of Oxford University Press (maker) 2016

First published in 2016

British Library Cataloguing in Publication Data
Data available

ISBN 978-0-19-836645-4

11

The manufacturing process conforms to the environmental regulations of the country of origin.

Printed in Great Britain by Bell and Bain Ltd, Glasgow

Acknowledgements

Cover: Eric Isselee/Shutterstock; **Cover:** Earth Imaging/Getty Images; **p7:** Roni Chowdhury/Barcroft India via Getty Images; **p7:** Daniel Berehulak/Getty Images Reportage; **p14:** Than WIN/AFP/Getty Images; **p14:** Reuters; **p14:** AP Photo/PA Archive; **p15:** Fairfax Media/Getty Images; **p16:** NASA's MODIS Rapid Response Team; **p17:** Ena Media Hawaii/Getty Images; **p17:** Y. T Haryono/Anadolu Agency/Getty Images; **p17:** Jacques Langevin/Sygma/Corbis; **p18:** Kevin West/AP photo/PA Archive; **p18:** DOMINIQUE CHOMEREAU-LAMOTTE/AFP/Getty Images; **p19:** USGS; **p20:** David Rydevik; **p21:** YOMIURI SHIMBUN/AFP/Getty Images; **p23:** Digital Globe; **p23:** Digital Globe; **p24:** Florain Kopp/imageBROKER/Alamy Stock Photo; **p24:** Ulet Ifansasti/Getty Images; **p28:** U.S. Navy via Getty Images; **p28:** David Gilkey/NPR; **p30:** Chien-min Chung/Getty Images; **p31:** China Photos/Getty Images; **p31:** Evens Lee/Color China Photo/AP Images/PA; **p32:** KeystoneUSA-ZUMA/REX/Shutterstock; **p32:** YOSHIKAZU TSUNO/AFP/Getty Images; **p37:** REUTERS/Romeo Ranoco; **p39:** PATRICE COPPEE/AFP/Getty Images; **p39:** THONY BELIZAIRE/AFP/Getty Images; **p42:** Apichart Weerawong/AP Photo/PA Archive; **p43:** Dr Xiaoming Wang, GNS Science; **p43:** www.buildchange.org; **p45:** Courtesy of SOS Children,Äôs Villages; **p45:** AAMIR QURESHI/AFP/Getty Images; **p48:** Erlend Bjrtvedt (CC-BY-SA); **p49:** ArcticPhoto.com; **p49:** stuart thomson/Alamy Stock Photo; **p53:** The Print Collector/Alamy; **p54:** ISS006-E-41533.JPG: Image courtesy of the Earth Science and Remote Sen; **p56:** The University of Texas at Austin; **p57:** Canadian Parks Agency; **p57:** Isabelle Laurion; **p57:** Simon Ross; **p58:** Lynda Dedge/ESS Photo Collection; **p59:** Steve Garufi; **p60:** EnviroFoto; **p60:** H. Mark Weidman Photography/Alamy Stock Photo; **p61:** Eric Baccega/age fotostock/Alamy Stock Photo; **p62:** eric bargis/Fotolia; **p64:** Patrick J. Endres/Corbis Documentary/Getty Images; **p68:** Simon Ross; **p68:** Simon Ross; **p70:** Photograph courtesy from www.markhewittphotography; **p71:** © Crown copyright and database rights (2016) OS (100000249); **p72:** © Prisma Bildagentur AG/Alamy Stock Photo; **p72:** © Jo Katanigra/Alamy Stock Photo; **p72:** Gregory Perkins/Alamy Stock Photo; **p73:** Simon Ross; **p74:** funkyfood London - Paul Williams/Alamy Stock Photo; **p76:** © david speight/Alamy Stock Photo; **p77:** © Crown copyright and database rights (2016) OS (100000249); **p78:** © Naturfoto-Online/Alamy Stock Photo; **p78:** © Tom Bean/Alamy Stock Photo; **p79:** Tony Waltham Geophotos; **p80:** © robertharding/Alamy Stock Photo; **p82:** Courtesy of Lake District National Park; **p82:** Courtesy of Lake District National Park; **p83:** Bob Digby; **p84:** Jeff J Mitchell/Getty Images; **p84:** Courtesy of Lake District National Park; **p86:** Namgyal Sherpa/AFP/Getty Images; **p87:** Christian Kober/AWL Images/Getty Images; **p89:** David Woodfall/Photoshot License Ltd/Alamy; **p90:** Jason Edwards/National Geographic/Getty Images; **p90:** Image/photo courtesy of the National Snow and Ice Data Center, University of Colorado, Boulder. **p92:** JamesTung/iStockphoto; **p97:** Martin Zwick/Getty; **p97:** Dave Porter Peterborough UK/Getty; **p97:** Nigel Gladwell/Alamy Stock Photo; **p97:** esp_imaging/iStockphoto; **p98:** Neil McAllister/Alamy Stock Photo; **p100:** Christopher Nicholson/Alamy Stock Photo; **p102:** David Goddard/Getty Images; **p102:** © LatitudeStock/Alamy †Stock Photo; **p103:** NASA Earth Observatory image; **p104:** © Crown copyright and database rights (2016) OS (100000249); **p104:** Skyscan.co.uk/J Farmar; **p108:** Matt Cardy/Getty Images; **p109:** Bob Digby; **p109:** Bob Digby; **p110:** Robyn Mackenzie/Shutterstock;

p110: David Waugh; **p110:** Simon Ross; **p112:** Christophe D. Assogba; **p113:** MARK BRAZIER/Alamy Stock Photo; **p115:** Dan Burton Photo/Alamy Stock Photo; **p115:** David Robertson/Alamy Stock Photo; **p115:** Strutt & Parker; **p155:** Les Gibbon/Alamy Stock Photo; **p118:** Dorset Media Service/Alamy Stock Photo; **p118:** Dave Caulkin/Associated Press; **p119:** Doru Cristache/Shutterstock; **p121:** Bob Digby; **p122:** Richard Vogel/AP/Press Association Images; **p124:** DigitalGlobe/Getty Images; **p124:** DigitalGlobe/Getty Images; **p124:** GEORGE BERNARD/SCIENCE PHOTO LIBRARY; **p125:** Catherine Hurst; **p125:** frans lemmens/Alamy Stock Photo; **p126:** Les Gibbon/Alamy Stock Photo; **p126:** Rolf Richardson/Alamy Stock Photo; **p127:** Les Gibbon/Alamy Stock Photo; **p129:** David Lauberts; **p130:** Richard Middleton/University of Hull Department of Geography/National Library of Scotland/Ordnance Survey/© Bluesky International Limited; **p130:** Powered by Light/Alan Spencer/Alamy Stock Photo; **p132:** Sgt. Ezekiel R. Kitandwe/U.S. Marine Corps; **p133:** Joerg Boethling/Alamy Stock Photo; **p133:** Jesse Allen/Earth Observatory/NASA; **p139:** Alan Curtis/Alamy Stock Photo; **p139:** © geogphotos/Alamy Stock Photo; **p140:** Andrew Bira/Reuters; **p141:** Ashley Cooper/Alamy Stock Photo; **p142:** Ahmad Masood/ Reuters; **p144:** Ian Murray/ Geography Photos; **p148:** Albanpix Ltd/REX/Shutterstock; **p152:** Warwick Page/Panos Pictures; **p152:** INTERFOTO/Alamy Stock Photo; **p153:** Nikolas Georgiou/Alamy Stock Photo; **p156:** Ulrich Doering/ Alamy Stock Photo; **p157:** Chau Doan/LightRocket via Getty Images; **p161:** TED ALJIBE/AFP/Getty Images; **p164:** Zute Lightfoot/Alamy Stock Photo; **p165:** Joerg Boethling/Alamy Stock Photo; **p167:** Henry Westheim Photography/Alamy Stock Photo; **p168:** epa european pressphoto agency b.v./Alamy Stock Photo; **p169:** Tim Graham/Alamy Stock Photo; **p169:** Xinhua/Alamy Stock Photo; **p170:** Andrew Melbourne/Alamy Stock Photo; **p174:** Jon Wilson/Alamy Stock Photo; **p174:** MARWAN NAAMANI/AFP/Getty Images; **p176:** Noah Friedman-Rudovsky/Bloomberg via Getty Images; **p177:** Joe Giddens/PA Archive/Press Association Images; **p178:** epa european pressphoto agency b.v./Alamy Stock Photo; **p179:** Dutourdumonde/Alamy Stock Photo; **p184:** Mark Mercer/Alamy Stock Photo; **p184:** Janine Wiedel Photolibrary/Alamy Stock Photo; p186: Zhao jian kang/Shutterstock; **p189:** Courtesy of Transition Network/Artist: Jennifer Johnson; **p189:** Bristol City Council/Courtesy of TransitionNetwork; **p194:** Rick Potter; **p195:** AC Manley/Shutterstock; **p196:** Mark Richardson/Alamy Stock Photo; **p197:** Sabena Jane Blackbird/Alamy Stock Photo; **p200:** Greg Balfour Evans/Alamy Stock Photo; **p200:** christopher Pillitz/Alamy Stock Photo; **p201:** Bob Digby; **p202:** Bob Digby; **p204:** Frederick Wilfred; **p205:** Photoplan; **p205:** UniversalImagesGroup/Getty Images; **p208:** Bob Digby; **p209:** Bob Digby; **p209:** Bob Digby; **p211:** Sean Smith/Guardian Syndication; **p212:** Bob Digby; **p212:** Bob Digby; **p213:** Bob Digby; **p215:** Bob Digby; **p216:** Anthony Palmer/Alamy Stock Photo; **p218:** Bob Digby; **p218:** Produced by AECOM on behalf of Linkcity; **p220:** Frances Wilmot; **p221:** © Crown copyright; **p222:** NurPhoto/Getty Images; **p223:** A.P.S. (UK)/Alamy Stock Photo; **p223:** Matt Cardy/Getty Images; **p224:** Travel England - Paul White/Alamy Stock Photo; **p225:** Extreme Academy, Watergate Bay/Kirstin Prisk Photography; **p227:** Aerohub at Cornwall Airport Newquay; **p228:** Jeff Gilbert/Alamy Stock Photo; **p229:** Andy Buchanan/Alamy Stock Photo; **p229:** Stuart Robertson/Alamy Stock Photo; **p230:** Copyright unknown/supplied by Stephen Braggs retrowow66@googlemail.co.uk; **p231:** Courtesy of Urban Splash; **p232:** Bob Digby; **p236:** Bob Digby; **p236:** Bob Digby; **p237:** Loop Images Ltd/Alamy Stock Photo; **p239:** Trevor Llewelyn; **p241:** Bob Digby; **p242:** Bob Digby; **p242:** © geogphotos/Alamy Stock Photo; **p242:** Superfast Cornwall; **p243:** Simon Burt/Apex News & Pictures; **p243:** Bob Digby; **p246:** © A.P.S. (UK)/Alamy†Stock Photo; **p247:** © Justin Kase zsixz/Alamy†Stock Photo; **p247:** © Greg Balfour Evans/Alamy†Stock Photo; **p254:** © Jeffrey Blackler/Alamy†Stock Photo; **p255:** © Robert Stainforth/Alamy†Stock Photo; **p255:** © roger parkes/Alamy†Stock Photo; **p258:** Leeds Library and Information Service, David Atkinson Archive www.leodis.net; **p259:** © A.P.S. (UK)/Alamy†Stock Photo; **p259:** © Leo Rosser/Alamy†Stock Photo; **p261:** © Leo Rosser/Alamy†Stock Photo; **p262:** Courtesy of Manchester Libraries, Information and Archives, Manchester; **p262:** © Mike Robinson/Alamy†Stock Photo; **p263:** © Joanne Moyes/Alamy†Stock Photo; **p266:** Bob Digby; **p267:** Bob Digby; **p271:** Bob Digby; **p277:** © Homer Sykes Archive/Alamy†Stock Photo; **p278:** © John Warburton-Lee Photography/Alamy†Stock Photo; **p278:** © Percy Ryall/Alamy†Stock Photo; **p279:** © Cath Harries/Alamy†Stock Photo; **p281:** © Mark Dunn/Alamy†Stock Photo; **p281:** © LH Images/Alamy†Stock Photo; **p281:** © Stephen Chung/Alamy Stock Photo; **p282:** © Mark Thomas/Alamy Stock Photo; **p282:** Bob Digby; **p283:** No Airport in Newham, 1983. Peter Dunn and Loraine Leeson, Docklands Community Poster Project. A2 poster.; **p283:** © Loop Images Ltd/Alamy Stock Photo; **p285:** © Liam White/Alamy†Stock Photo; **p286:** Bob Digby; **p289:** © Stephen Chung/Alamy†Stock Photo; **p290:** © Roger Sedres/Alamy†Stock Photo; **p291:** © Carl Court/Alamy†Stock Photo; **p292:** © Stan Kujawa/Alamy Stock Photo; **p294:** Martin Cooke/Alamy Stock Photo; **p294:** Bob Digby; **p295:** Bob Digby; **p295:** Bob Digby; **p299:** SFL Choice/Alamy Stock Photo; **p301:** sasaperic/Shutterstock; **p303:** Sabena Jane Blackbird/Alamy Stock Photo; **p317:** Bob Digby; **p317:** Bob Digby; **p319:** © Ashley Cooper/Alamy Stock Photo; **p320:** Bob Digby; **p321:** Bob Digby; **p322:** Bob Digby.

Artwork by Barking Dog Art, Dave Russell Illustration, Kamae Design, Lovell Johns, Mike Connor, and Simon Tegg.

Design/page layout: Kamae Design.

Bob Digby would like to thank Adam Jameson for helping with research, particularly 'Books to read', 'Music to listen to' and 'Films to see' for each chapter.

Third party website addresses referred to in this publication are provided by Oxford University Press in good faith and for information only and Oxford University Press disclaims any responsibility for the material contained therein

Every effort has been made to contact copyright holders of material reproduced in this book. Any omissions will be rectified in subsequent printings if notice is given to the publisher.

How to use this book

This is the first of two books in this series, written for the Edexcel GCE in Geography. This particular book (Book 1) has been written to meet the content requirements of the A Level course, but can equally well be used for the separate AS course.

About the questions in this book:

- ◆ 'Over to you' questions are designed for collaborative work in pairs or larger groups.

- ◆ 'On your own' questions provide opportunities for independent, private study.

- ◆ Skills questions (indicated by the ✦ icon) are aimed at meeting the geographical and statistical skills requirements for both AS and A Level.

- ◆ Exam-style questions have been included for both AS and A Level, with marks allocated.

- ◆ At appropriate points, chapters focus on the synoptic themes – Players, Attitudes, and Futures. These, plus associated questions, will help to prepare you for synoptic questions at AS, and for Paper 3 in the A Level examination.

Chapter overview – introducing the topic

This chapter studies primary tectonic hazards (earthquakes and volcanic eruptions) and secondary hazards (e.g. tsunami), as well as the risks posed by both.

In the Specification, this topic has been framed around three Enquiry Questions:

1 Why are some locations more at risk from tectonic hazards?

2 Why do some tectonic hazards develop into disasters?

3 How successful is the management of tectonic hazards and disasters?

The sections in this chapter provide the content and concepts needed to help you answer these questions.

Synoptic themes

Underlying the content of every topic are three synoptic themes that 'bind' or glue the whole Specification together:

1 Players

2 Attitudes and Actions

3 Futures and Uncertainties

Both 'Players' and 'Attitudes and Actions' are discussed below. You can find further information about 'Futures and Uncertainties' on the chapter summary page (page 46).

1 Players

Players are individuals, groups and organisations involved in making decisions that affect both people and places. They can be national or international individuals and organisations (e.g. inter-governmental organisations like the UN), national and local governments, businesses (from small operations to the largest TNCs), plus pressure groups and non-governmental organisations.

Players that you'll study in this topic include:

- Sections 1.6 and 1.7 – How **local and national governments** have affected vulnerability and the resilience of communities to withstand and cope with hazards in Haiti, China and Japan.

- Sections 1.3, 1.4, 1.10 and 1.11 – How hazard prediction and forecasting involves **scientists** and **emergency planners**.

- Section 1.11 – How **planners** and **engineers** attempt to modify hazard events through hazard-resistant design and engineering defences.

- Section 1.11 – How **insurers** can modify losses that result from hazards, as well as the parts played by **communities**.

2 Attitudes and Actions

Actions are the means by which players try to achieve what they want. For example, an engineer working in hazard prevention would seek approval for building techniques and building regulations to ensure that the world becomes a safer place.

Actions could pose the following questions:

1 **Strategic** – Faced with the likelihood of future hazard events, have governments worked with engineers, planners and insurance companies to make sure that future damage is limited?

2 **Implementation** – Faced with a future tectonic disaster, would it have similar impacts to those of previous disasters, or would lessons have been learned?

3 **Social justice** – Do those who are most vulnerable in hazardous environments face a better or safer future?

In groups, consider and discuss two tectonic disasters from this chapter, where actions taken by government, planners, scientists, or community groups were found to be **(a)** effective, and **(b)** ineffective in the face of disaster. Explain your views.

In this section, you'll explore natural hazards and why they can sometimes turn into disasters.

Nepal – a disaster waiting to happen

When a 7.8 magnitude earthquake struck Nepal in April 2015, scientists weren't surprised. Just a week earlier, earthquake experts had gathered in Nepal's capital, Kathmandu, to discuss how the country could prepare for 'the big one' they knew was eventually coming. What they didn't know was that the big one would occur the following week!

Nepal is situated in south Asia – sandwiched between India and China (see Figure 1). As the map shows, it sits right on the boundary between the Indian and Eurasian **tectonic plates** (see Section 1.3). As these plates push against each other, pressure builds up due to friction. The 25 April 2015 earthquake was the result of a sudden release of this pressure. Its epicentre was located 80 km northwest of Kathmandu (at a depth of 15 km), and the initial earthquake and its aftershocks killed a total of 8633, injured 21 000 – and led to three million people being made homeless.

CHINA

EURASIAN PLATE

Himalayas

NEPAL

Mt. Everest

Kathmandu

INDIA

0   100 km

N

Major aftershock
1. **Date:** April 25, 2015, **Magnitude:** 6.6
2. **Date:** April 26, 2015, **Magnitude:** 6.7
3. **Date:** May 12, 2015, **Magnitude:** 7.3

INDIAN PLATE

Figure 1   The epicentre and affected area of the April 2015 earthquake. Nepal was the most severely affected country, but the earthquake also affected China and India to a lesser degree.

Key
Perceived shaking
■ severe
■ very strong
☐ strong
☐ moderate
◎ epicentre
◈ major aftershock
---- fault line
▲ represents the subduction of one plate under the other

Natural hazards

The earthquake that caused so much damage, injury and loss of life in Nepal is a natural process – caused by the movement of the Earth's tectonic plates (see Section 1.2). But a natural process like this becomes a **natural hazard** when it affects people. An earthquake occurring deep under the sea, away from humans, may harm nobody (although there is a risk of it causing a tsunami). However, if that earthquake were to strike the middle of a busy city, then it becomes a natural hazard.

Disasters

Under certain conditions, a natural hazard can become a **disaster**. Exactly when this happens is hard to define (different governments and organisations use different definitions) but, generally, a disaster happens when a natural hazard strikes a vulnerable population that can't cope using its own resources (see Figure 2). The greater the scale of the natural hazard, and the more **vulnerable** the population, the greater the disaster will be.

Key word

Natural hazard – A naturally occurring process or event that has the potential to affect people.

Natural disaster – A major natural hazard that causes significant social, environmental and economic damage.

Vulnerability – The ability to anticipate, cope with, resist and recover from a natural hazard.

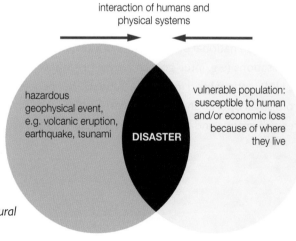

interaction of humans and physical systems

hazardous geophysical event, e.g. volcanic eruption, earthquake, tsunami

vulnerable population: susceptible to human and/or economic loss because of where they live

DISASTER

Figure 2   Degg's disaster model shows how the overlap of a natural hazard and vulnerability can cause a disaster

Nepal's vulnerability and its effects

The 2015 Nepal earthquake was a major disaster with potentially long-lasting effects:

◆ The earthquake occurred at midday, when most people were at work (many tending their fields). This helped to reduce the death toll considerably, although many did return to damaged or destroyed homes.

◆ Most of Nepal has a fairly low population density. Although the earthquake did affect the densely populated capital, Kathmandu, the worst damage was in rural areas.

◆ Nepal is a vulnerable country. It's one of the world's poorest (in 2016, it was 197th in a world ranking of GDP per capita, out of 229 countries), so the Nepalese were unprepared for the earthquake.

▲ **Figure 3** *Damage in the capital, Kathmandu, caused by the 2015 earthquake*

◆ The weak infrastructure (such as roads, bridges and safe water supplies) was severely damaged or destroyed in the area of the earthquake.

◆ Most of Kathmandu's buildings were not built to withstand earthquakes, and many collapsed (see Figure 3).

◆ After the initial earthquake, there were many large aftershocks (and nearly 100 smaller ones), which caused further destruction and deaths – and made the rescue work very dangerous.

◆ Nepal is in the Himalayas, so it's very mountainous (see Figure 4). The earthquake created landslides, which devastated many rural areas and cut them off. This made rescue and aid efforts very difficult.

◆ Nepal's emergency services were not able to cope with the level of destruction, and relied on international aid agencies for help.

▲ **Figure 4** *Barpak – one of many remote mountain settlements in Nepal destroyed by the 2015 earthquake (fewer than 10 houses out of 1200 remained standing)*

◆ Tourism, an important part of the Nepalese economy, fell significantly after the earthquake – causing the loss of much-needed income.

◆ Nepal's economy is estimated to have lost US$5 billion – about 25% of its Gross Domestic Product (GDP).

◆ US$6.6 billion was needed for rebuilding work (almost all from foreign aid).

Yet this isn't the end of earthquakes for Nepal. The country is located in one of the most earthquake-prone areas in the world. Scientists believe that the April 2015 earthquake, despite its size, did not release all of the built-up pressure between the two tectonic plates – making the risk of another big earthquake highly likely.

Over to you

1 Devise a fact file for the Nepal earthquake to summarise its nature and its impacts.

2 In pairs, design a spider diagram or mind map to show (**a**) why the earthquake happened, (**b**) why casualties were actually lower than could have been predicted.

3 Explain what makes Nepal 'vulnerable' as far as natural hazards are concerned.

On your own

4 **a** Distinguish between a natural hazard and a disaster.
   b Define 'vulnerability' and how the term applies to hazard studies.

5 Suggest why Degg's model in Figure 2 is widely used in hazard studies.

6 **a** Classify the impacts of the Nepal earthquake into economic, social and environmental.
   b Referring to Figures 3 and 4, suggest how actions taken now could reduce the impacts of any future earthquakes

In this section, you'll look at plate-tectonic theory and how it links to the movement of the Earth's tectonic plates.

The structure of the Earth

Learning about the structure of the Earth is difficult. It's about 6500 km to the centre of the core (see Figure 1), and its heat (some believe over 6000°C!) means that scientists can't drill into it. Instead, they map it using evidence from **seismic waves** (shock waves released by tectonic movements).

Plate-tectonic theory

The lithosphere is broken up into seven major and several minor parts – called **tectonic plates** – that move relative to each other over the asthenosphere. It is this movement that causes earthquakes and volcanic eruptions. Tectonic plates are large, irregularly shaped slabs of solid rock that vary greatly in size and move slowly (about 2 cm to 15 cm a year) – the speed at which your fingernails grow! The study of these plates is called **plate-tectonic theory**, and their movement is driven by a number of different processes (outlined below):

1 Mantle convection

Mantle convection has long been thought responsible for plate movement, but this argument is now less accepted. In mantle convection, heat produced by the decay of radioactive elements in the Earth's core heats the lower mantle – creating **convection currents**. These hot, liquid magma currents are thought to move in circles in the asthenosphere – thus, causing the plates to move.

2 Slab pull

Today, **slab pull** is increasingly being seen as a major driving force for plate movement. Newly formed oceanic crust at **mid-ocean ridges** (see 4 below) becomes denser and thicker as it cools. This causes it to sink into the mantle under its own weight – pulling the rest of the plate further down with it.

3 Subduction

Seafloor spreading (see below) raises a question: if new crust is being created, the Earth should be expanding – but it isn't. This is because, as new crust is being created in one place, it's being destroyed in another – by **subduction**. As two oceanic plates (or an oceanic plate and a continental plate) move towards each other, one slides under the other into the mantle – where it melts in an area known as a **subduction zone**.

4 Seafloor spreading

In the middle of many oceans are huge mid-ocean ridges, or underwater mountain ranges. These are formed when hot **magma** (molten rock) is forced up from the asthenosphere and hardens – forming new oceanic crust. This new crust pushes the tectonic plates apart in a process called **seafloor spreading**.

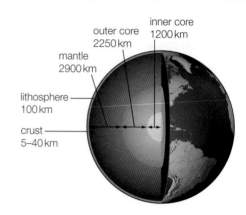

inner core
1200 km
outer core
2250 km
mantle
2900 km
lithosphere
100 km
crust
5–40 km

🔺 **Figure 1** A cut-away of the Earth to show its internal structure (see below for more details)

The Earth has three main layers:

1 **The core** – at the centre. This consists of two parts:

- **The inner core** – at the very centre of the Earth, and the hottest part (about 6000°C). It is solid and mostly consists of iron. In 2015, seismic-wave data suggested that the inner core has another, distinct, area at its centre (now called the 'inner inner core').
- **The outer core** – which is semi-molten and mostly consists of liquid iron and nickel. Temperatures there range between 4500-6000°C.

2 **The mantle** – surrounds the core, and is the widest layer making up the Earth. The upper part is solid, but below it the rock is semi-molten – forming the **asthenosphere** (on which tectonic plates 'float').

3 **The crust** – forms the outer shell of the Earth. There are two types of crust:

- **Oceanic** – a thin, dense layer (6-10 km thick), which lines the ocean floors.
- **Continental** – an older, thicker layer (usually 45-50 km thick), which makes up the Earth's landmasses. It is less dense than oceanic crust.

Between them, the crust and upper mantle make up the **lithosphere** – the solid layer from which tectonic plates are formed.

In the 1950s, studies of **palaeomagnetism** confirmed that the seafloor was spreading. Every 400 000 years or so, the Earth's magnetic fields change direction – causing the magnetic north and south poles to swap. When lava cools and becomes rock, minerals inside the rock line up with the Earth's magnetic direction (polarity) at the time (see Figure 2). Scientists studying mid-ocean ridges found the same pattern of magnetic direction on either side of the ridges (something that could only happen if new rock was being formed at the same time on both sides).

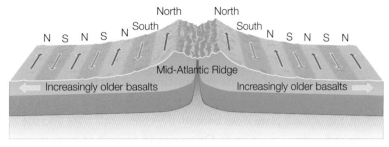

⊘ Figure 2 *Evidence for seafloor spreading (note how the seafloor gets progressively older, the further it gets from the mid-ocean ridge)*

Key word

Palaeomagnetism – The study of past changes in the Earth's magnetic field (determined from rocks, sediment or archaeological records).

Tectonic plate boundaries

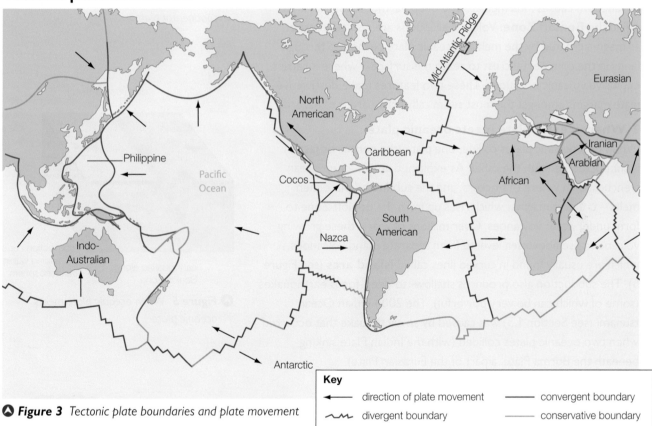

Key

— direction of plate movement   —— convergent boundary
᠕᠕ divergent boundary   —— conservative boundary

⊘ Figure 3 *Tectonic plate boundaries and plate movement*

When two tectonic plates meet, they form a **plate boundary** (see Figure 3) – each type creating distinct processes and landforms. It is at these plate boundaries that most tectonic activity (mountain building, volcanoes, earthquakes) occurs. The areas adjacent to plate boundaries are called **plate margins**, and includes areas either side of the boundary that may be affected by movement, e.g. an earthquake focus that is quite far underground may be some distance away from the surface boundary.

The three types of boundary are:

- convergent – where two plates collide, also known as **destructive margins**.

- divergent – where two plates move apart, also known as **constructive margins**.

- conservative – where two plates slide past each other, also known as **transform margins**.

Destructive plate margins (convergent boundaries)

At destructive plate margins, the plates move towards each other (converge). There are three types of destructive plate margin:

1 When oceanic plate meets continental plate

Oceanic plate is denser than continental plate, so, when the plates collide, the oceanic plate slides beneath the continental plate into the mantle and melts (see Figure 4). **Deep ocean trenches** mark the place where the oceanic plate starts to sink beneath the continental plate.

This subduction also leads to the formation of **fold mountains** (as the two plates collide, the continental plate is folded and slowly pushed up, forming chains of fold mountains). Since the plates are constantly moving towards each other, most fold mountains will continue to grow.

The friction created between the colliding plates (and the resultant subduction) causes intermediate and deep earthquakes in an area called the **Benioff Zone**. Volcanic eruptions are also generated as magma created by the melting oceanic plate pushes up through faults in the continental crust to reach the surface – where it causes explosive volcanic eruptions. These two features make destructive plate margins amongst the most seismically active areas of hazard.

2 When oceanic plate meets oceanic plate

When two oceanic plates collide, one plate (the denser or faster) is subducted beneath the other. As explained above, deep ocean trenches form where this occurs, and the subducted plate then melts – creating magma, which rises up from the Benioff Zone to form underwater volcanoes. Over millions of years, these growing volcanoes rise above sea level to form separate island volcanoes, which are usually found in curved lines called **island arcs** (see Figure 5). The subduction also produces shallow- to deep-focus earthquakes (some of which can be very powerful). The 2004 Indian Ocean tsunami (see Section 1.5) was caused by an earthquake that occurred when two oceanic plates collided (with the Indian Plate sinking beneath the Burma Plate, a part of the Eurasian Plate).

3 When continental plate meets continental plate

When two continental plates meet, a **collision margin** occurs. As both plates have about the same density, and are less dense than the asthenosphere beneath them, neither plate is actually subducted. Instead, they collide and sediments between them are crumpled and forced up to form high fold mountains, like the Himalayas (see Figure 6). However, inevitably there can be some subduction (such as the Nepal earthquake in Figure 1, Section 1.1), caused when the compressed (and therefore denser) sediments result in plate subduction beneath them. There is no volcanic activity, but any earthquakes are likely to have a shallow focus – increasing their severity.

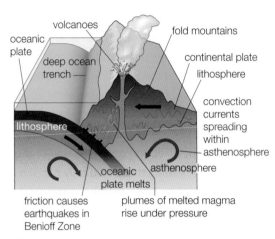

▲ *Figure 4 When oceanic plate meets continental plate*

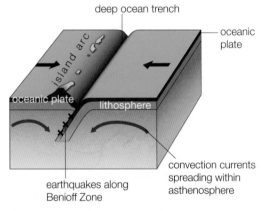

▲ *Figure 5 When oceanic plate meets oceanic plate*

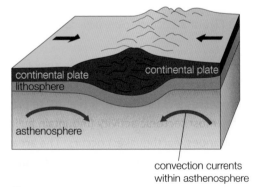

▲ *Figure 6 When continental plate meets continental plate*

Constructive plate margins (divergent boundaries)

At constructive plate margins, two plates are moving apart (diverging) – which leads to the formation of new crust. In oceans, this divergence forms mid-ocean ridges, and on continents it forms rift valleys.

Mid-ocean ridges

Mid-ocean ridges of underwater mountains extend for over 60 000 km across the world's ocean floors! Regular breaks (called **transform faults** – see below) cut across these ridges as they spread at different rates. Shallow-focus earthquakes (at a depth of less than 70 km) occur, but they pose little risk to humans because the shocks are minor and occur underwater. Regular volcanic eruptions also create **submarine volcanoes** along these mid-ocean ridges, some of which grow above sea level to create new islands such as Iceland on the Mid-Atlantic Ridge (see Figure 7).

Rift valleys

When plates move apart on continents, the crust stretches and breaks into sets of parallel cracks (faults). The land between these faults then collapses, forming steep-sided valleys called **rift valleys**.

Conservative plate margins

Along some boundaries, two plates slide past each other, forming a **conservative plate margin** (see Figure 8). This results in a major break in the crust between them as they move. The break itself is called a **fault,** and where it occurs on a large scale is known as a **transform fault**, which affects a wider area. Although no crust is made or destroyed here (and there is no volcanic activity), this type of plate margin is tectonically very active – and can be associated with powerful earthquakes. The two plates sometimes stick as they move past each other, causing stress and pressure to build up, which is suddenly released as a strong shallow-focus earthquake. One of the most famous conservative plate margins is the San Andreas Fault in California, which has generated significant earthquakes.

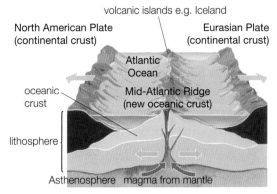

Figure 7 *A cross-section of the Mid-Atlantic Ridge, which forms the boundary between the Eurasian and North American Plates*

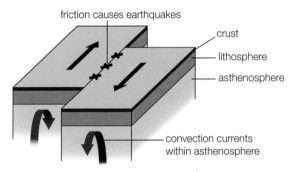

Figure 8 *A conservative plate margin*

Exam-style questions

AS 1 Explain the reasons why volcanoes are more likely along some plate margins than others. *(6 marks)*

A 2 Assess the contribution of plate-tectonic theory to our knowledge of the Earth's structure. *(12 marks)*

Over to you

 1 Using the text, copy and annotate with further detail the block diagrams in Figures 5 and 6 to show key features found at different plate boundaries/margins.

2 Plan an eight-slide presentation for a Year 10 class who have never studied plate tectonics before. Create one slide of explanation and one photo slide of features to explain to Year 10s what each of the following four terms means: (a) destructive margin, (b) collision margin (c) constructive margin, (d) conservative margin.

On your own

3 Distinguish between the following pairs of terms: (a) inner and outer core of the Earth, (b) mantle and crust, (c) oceanic and continental crust, (d) asthenosphere and lithosphere, (e) subduction and slab pull, (f) plate boundary and plate margin, (g) mid-ocean ridges and rift valleys.

4 Explain why (a) the most violent earthquakes are on destructive and conservative margins – rarely on constructive margins, (b) volcanoes rarely occur along collision or conservative margins.

In this section, you'll look at earthquakes: what causes them, how they're measured, and their primary and secondary effects.

Every day, hundreds of earthquakes occur around the world. Most are so small that they aren't felt at all, or they occur in remote areas and cause no damage. Earthquakes are a hazard only when people are involved – and then they can cause a lot of death and destruction. On average, about 10 000 people die every year as a result of earthquakes, but this number can vary enormously.

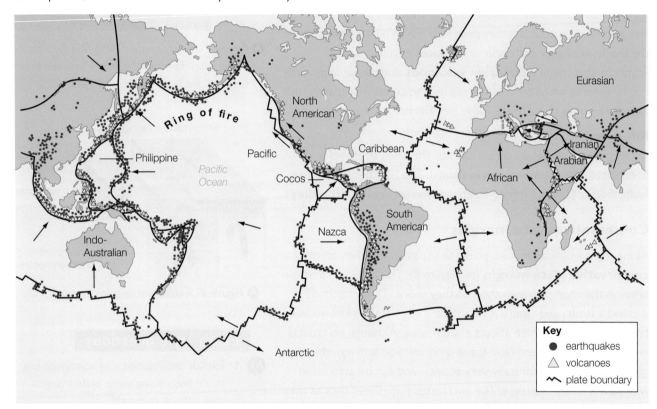

Key
- ● earthquakes
- △ volcanoes
- ∿ plate boundary

⬆ **Figure 1** *The relationship between tectonic plate boundaries, earthquakes and volcanoes*

95% of earthquakes occur along tectonic plate boundaries (see Figure 1). When the plates move against each other, they sometimes stick – causing huge amounts of pressure to build up. When the pressure becomes too much, the rock fractures along cracks called **faults** – and energy is suddenly released as **seismic waves** (causing the ground to shake). The point inside the crust from which the pressure is released is called the **focus** or **hypocentre** (see Figure 2). The point on the surface directly above that is called the **epicentre.** This is where most shaking and damage occurs.

Seismic waves

An earthquake's energy is released as **seismic waves**. These waves radiate out from the focus, like the ripples when a stone is thrown into a pond. There are three main types of wave, each travelling at different speeds (see Figures 3 and 4). Primary and Secondary waves are called **body waves**, because they travel through the Earth's body; Love waves are **surface waves**, because they travel along the Earth's surface.

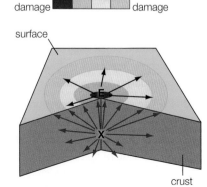

most damage ▮▯▯▯ least damage

surface

crust

⬆ **Figure 2** *The focus (hypocentre) H, and epicentre (E) of an earthquake*

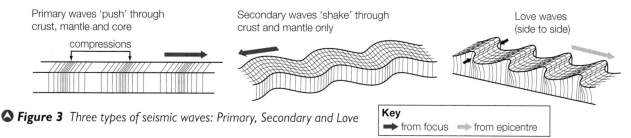

Primary waves 'push' through crust, mantle and core

compressions

Secondary waves 'shake' through crust and mantle only

Love waves (side to side)

🅐 **Figure 3** *Three types of seismic waves: Primary, Secondary and Love*

Key
➡ from focus ➡ from epicentre

Seismic waves are measured using a **seismometer**, which detects and measures ground movement (see Figure 5). By knowing the speed of different waves, Earth scientists use the results from several seismographs to calculate the time, location and magnitude of an earthquake.

P waves (Primary or pressure waves) – these are the fastest and first to reach the surface. They travel through both solids and liquids (and shake in a backwards and forwards motion). They are only damaging in the most powerful earthquakes.

S waves (Secondary or shear waves) – these are slower (60% of the speed of P waves). They only travel through solids (and move with a sideways motion, shaking at right angles to the direction of travel). They do more damage than P waves.

L waves (surface Love waves) – these are the slowest (last to arrive), but they cause the most damage (shaking the ground from side to side). They are larger and focus all of their energy on the Earth's surface.

Measuring earthquakes

Earth scientists use two characteristics – **magnitude** and **intensity** – to measure earthquakes.

Earthquake magnitude

Magnitude measures the amount of energy released at the epicentre. Several scales are used to measure magnitude, but the **Moment Magnitude Scale** (MMS) is generally preferred, because it's accurate and better at measuring large earthquakes. It measures the total energy released by an earthquake at the moment it occurs (called the **seismic moment**), using the:

◆ size of the seismic waves

◆ amount of slippage or rock movement

◆ area of the fault surface broken by the earthquake

◆ resistance of the affected rocks.

The scale goes from 1 (smallest) and is infinite – but it generally stops at 10, since the largest earthquake ever recorded was a magnitude 9.5 (in Chile in 1960). The scale is logarithmic (each number is ten times the magnitude of the number before), so a magnitude 5 earthquake is ten times more powerful than a magnitude 4.

Earthquake intensity

An earthquake's effect on people, structures and the natural environment is called its **intensity**. One scale used to measure this is the **Modified Mercalli Intensity Scale**, which takes observations from people who experienced the earthquake and rates them on a scale from I (hardly noticed) to X (catastrophic).

The effects of an earthquake

An earthquake's impact depends on a range of factors, both physical (e.g. magnitude and depth, and distance from the epicentre) and human (e.g. a country's level of development, its population, its level of preparedness, the effectiveness of emergency responses, and the impact of indirect hazards, e.g. fires, landslides). An earthquake's effects are classified as either **primary** or **secondary**.

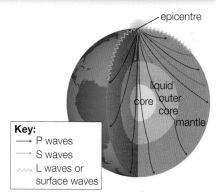

Key:
→ P waves
→ S waves
〜 L waves or surface waves

🅐 **Figure 4** *The paths of P waves, S waves and L waves*

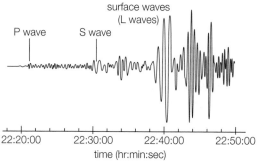

🅐 **Figure 5** *A seismometer recording (seismogram) of seismic waves in the UK from a distant earthquake. Note the time at which each type of wave was recorded.*

Primary effects

Primary effects are those that happen as a **direct** result of an earthquake, such as:

◆ **ground shaking**, which causes buildings, bridges, roads and infrastructure to collapse – killing or injuring those nearby

◆ **crustal fracturing**, when energy released during an earthquake causes the Earth's crust to crack – leaving gaps like those in Figure 6.

Secondary effects

In many earthquakes, secondary effects cause as much or more damage than the initial shaking. These include:

◆ **liquefaction** – The violent shaking during an earthquake causes surface rocks to lose strength and become more liquid than solid. The subsoil loses its ability to support building foundations, so buildings and roads tilt or sink. It can make rescue efforts more difficult, and also disrupt underground power and gas lines (leading to fires).

◆ **landslides** and **avalanches** – The ground shaking places stress on slopes, so that they fail (resulting in landslides, rock slides, mudslides and avalanches). Many of these effects (such as the one pictured in Figure 7) account for a large proportion of the damage and injuries caused by an earthquake.

◆ **tsunami** – Some underwater earthquakes generate tsunami that cause major problems for coastal areas (see Section 1.5).

▲ **Figure 6** *Crustal fracturing caused by an earthquake in Myanmar (formerly Burma)*

▲ **Figure 7** *This huge landslide – caused by an earthquake in El Salvador, Central America – suddenly sent thousands of tons of mud down a hillside into a town (destroying homes and burying many people)*

The 1989 Loma Prieta earthquake

The geology of an area can have an enormous impact on the effect of an earthquake. At 5:04 on 17 October 1989, a magnitude 6.9 earthquake struck near San Francisco. The city's Marina District suffered some of the worst of the damage. Built on man-made landfill, the area's soft, sandy soil amplified the ground shaking, which increased the damage experienced by buildings and other structures. The sandy soil also liquefied, which caused buildings to collapse.

A few miles away, part of the two-level Cypress freeway also collapsed (see Figure 8) – causing 42 of the 67 earthquake-related deaths (many drivers on the lower level were crushed by the collapsing upper level). As the map and seismogram in Figure 9 show, the part of the freeway that collapsed was built on soft mud; whereas adjoining parts of the freeway, built on firmer ground, remained standing.

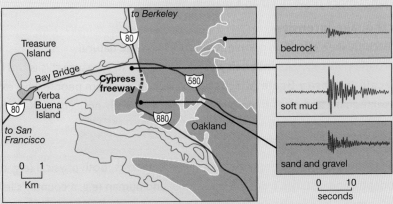

▲ **Figure 8** *The collapsed section of the Cypress freeway built on soft mud*

▲ **Figure 9** *The dashed red line on the map shows the part of the Cypress freeway that collapsed. The solid red line shows the part that remained standing.*

Aftershocks

For weeks, months or even years after an earthquake, other smaller earthquakes, called **aftershocks**, may follow. Aftershocks occur in the general area of the original earthquake, and are a result of the Earth 'settling down' or readjusting along the part of the fault that slipped at the time of the main earthquake. In general, the larger the earthquake, the larger and more numerous the aftershocks.

Aftershocks can cause additional damage (for example, structures weakened by the initial earthquake may collapse, injuring or killing people and hampering rescue efforts). Some aftershocks are very dangerous in their own right. The 2011 magnitude 6.3 aftershock that struck Christchurch (New Zealand) caused more damage and loss of life than the initial 2010 earthquake (see Figure 10). Unlike the 2010 earthquake, which was centred in a rural area, the 2011 aftershock was very shallow focused, so it caused more ground shaking – and it was also centred closer to the city itself.

Figure 10 Damage caused by the February 2011 aftershock in Christchurch, New Zealand

BACKGROUND

Intra-plate earthquakes

Not all earthquakes happen at plate margins. Some, called **intra-plate earthquakes**, occur in the middle of plates – far from the margins e.g. three large earthquakes in 1811–12 which occured in central USA (New Madrid, Missouri), with a magnitude of about 7.5. Scientists aren't certain what causes them, but some think that they occur when stresses build up in ancient faults – causing them to become active again. Because they don't occur in well-defined patterns along plate margins, it's harder to predict them.

Predicting the next 'big one'

Currently, there is no method to accurately predict when or where an earthquake will strike. But an understanding of plate tectonics allows scientists to know which areas are most at risk.

Geologists know that most earthquakes happen along plate boundaries – and areas that have had one big earthquake are likely to have another. This information allows them to **forecast** that an earthquake is likely to happen in an area, but not exactly when. For example, the US Geological Survey forecasts a 67% chance of another 'serious' earthquake striking San Francisco in the next 30 years.

Research today focuses on identifying 'warning signs', called **precursors**, that may suggest a major earthquake is about to happen. Various precursors are being looked at, such as foreshocks (small earthquakes that happen before a larger one), but none has proved to be a reliable sign that an earthquake is about to happen.

Over to you

1 **a** Use Figure 1 to identify (**i**) the geographical distribution of earthquakes, (**ii**) the distribution of volcanoes.
  b Explain to what extent (**i**) earthquakes and volcanoes always coincide, (**ii**) all plate boundaries produce earthquakes and/or volcanoes.

2 In pairs, analyse the 1989 Loma Prieta earthquake in terms of its primary and secondary effects. Which were more significant, and why?

3 In pairs, classify the different impacts of each hazard in this section into economic, social, and environmental. Which seem to be most significant?

On your own

4 Distinguish between the following sets of terms: (**a**) P, S and L waves, (**b**) earthquake magnitude and intensity, (**c**) primary and secondary effects of a hazard.

5 Compare the Christchurch earthquakes in 2010 and 2011 with that in San Francisco in 1989 in terms of (**a**) cause, (**b**) severity, (**c**) impacts.

Exam-style questions

AS 1 Explain the causes of one earthquake. *(6 marks)*

A 2 Assess the relative importance of the hazards associated with destructive plate margins. *(12 marks)*

In this section, you'll look at the hazards associated with volcanic eruptions.

Eyjafjallajökull

In 2010, flights over large parts of Europe were cancelled for a week, as a spreading ash cloud from the eruption of Eyjafjallajökull in Iceland (see Figures 1 and 2) threatened to clog aircraft engines and stop them working.

This eruption had social and economic impacts across Europe – and also around the world:

◆ 100 000 commercial flights were cancelled worldwide.

◆ Over 10 million passengers around the world were stranded or unable to board flights travelling either to or from Europe (or even via Europe).

◆ Worldwide, airlines lost US$1.7 billion in revenue.

◆ 30% of global airline capacity was cut – with European capacity cut by 75%.

◆ The European economy lost US$5 billion as a result of the disruption.

The economic effects of this Icelandic eruption were felt as far away as Kenya! Twenty percent of Kenya's economy is based on the export of green vegetables and flowers – mainly to Europe. When flights into Europe were cancelled, Kenyan businesses were forced to dump tonnes of fresh vegetables and flowers (costing US$1.3 million a day in lost revenue).

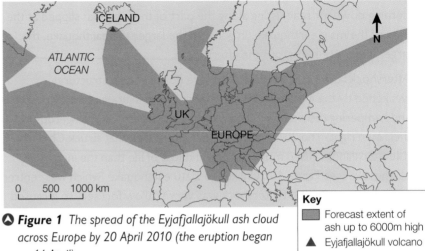

Figure 1 *The spread of the Eyjafjallajökull ash cloud across Europe by 20 April 2010 (the eruption began on 14 April)*

Key
- ■ Forecast extent of ash up to 6000m high
- ▲ Eyjafjallajökull volcano

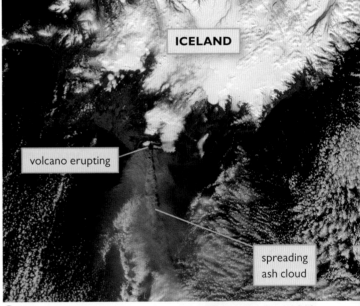

Figure 2 *Thick ash pours from Eyjafjallajökull in April 2010*

Volcanoes

Volcanoes are openings in the Earth's crust through which lava, ash and gases erupt. Like earthquakes, they are closely associated with plate margins (see page 12). As tectonic plates move, pressure builds and hot magma and gases push up from the mantle to the Earth's crust – and erupt. When the magma reaches the Earth's surface, it's called **lava**. When lava cools, it forms rock. So, as a volcano continues to erupt over time, it gets bigger.

More than 500 million people worldwide are at risk from hazards caused by volcanoes. Over the past 300 years, approximately 260 000 people have died as a result of volcanic eruptions. Today, about 1900 volcanoes are considered to be active – which means that they show some activity and are likely to erupt at some point.

Like earthquakes, volcanoes have a range of primary and secondary hazards – many of which can be deadly.

Primary hazards

Lava flows

Lava flows are streams of lava that have erupted from a volcano onto the Earth's surface. They are very hot (reaching up to 1170°C) and can take years to cool completely. However, despite this, lava flows are generally not a threat to humans, because most of them move so slowly that people can easily get out of their way. Nevertheless, lava flows do still cause a problem for people, because they destroy everything in their path (see Figure 3).

Pyroclastic flows

Pyroclastic flows are one of the greatest volcanic hazards (see Figure 4). These are a mixture of dense hot rock, lava, ash and gases ejected from a volcano, which move very quickly across the Earth's surface. Like lava flows, they destroy everything they touch – but, because they move so quickly, they are far more dangerous. They are extremely hot (up to 700°C), and travel at around 100 km per hour – so, unlike lava flows, you can't outrun one!

Tephra and ash falls

Tephra are pieces of volcanic rock and ash that blast into the air during volcanic eruptions. The larger pieces tend to fall near the volcano, where they can cause injury or death (as well as damaging structures). The smaller pieces (ash) can travel for thousands of kilometres.

Ash falls are very disruptive. Where the ash lands, it covers everything – causing poor visibility and slippery roads. Roofs may collapse under the weight, and engines may get clogged up and stop working (as was the fear with the Eyjafjallajökull ash cloud).

Gas eruptions

Magma contains dissolved gases that are released into the atmosphere during a volcanic eruption (some of which are a potential hazard to people, animals and structures). The volcanic gases include water vapour (about 80%), carbon dioxide and sulphur dioxide. Once in the air, the gases can travel for thousands of kilometres.

Secondary hazards

Lahars

Lahars are masses of rock, mud and water that travel quickly down the sides of a volcano. They vary in size and speed. The largest can be hundreds of metres wide and can flow at tens of metres per second – again, too fast for people to outrun. They are caused when an eruption quickly melts snow and ice. Or, alternatively, heavy rainfall during or after an eruption erodes loose rock and soil – causing it to surge downslope.

Jökulhlaup

The heat of a volcanic eruption can melt the snow and ice in a glacier – causing heavy and sudden floods called **jökulhlaups** (or glacial outburst floods). These floods can be very dangerous, because they suddenly release large amounts of water, rock, gravel and ice that can catch people unawares, and flood and damage land and structures.

▲ *Figure 3   A slowly moving stream of lava from the eruption of Kilauea volcano in Hawaii destroys farmland and sets a house on fire in November 2014 (while people stand safely nearby and watch)*

▲ *Figure 4   A pyroclastic flow caused by an eruption of Mount Sinabung (on the island of Sumatra, Indonesia) in July 2015*

▲ *Figure 5   A large lahar sweeping down from the eruption of Nevado del Ruiz destroys the town of Armero, Colombia, in November 1985*

Hot spots

While most volcanoes are located along plate margins, there are some exceptions. Volcanoes can also form in the middle of a plate, where **plumes** of hot magma rise upwards and erupt onto the sea floor (at what is called a **hot spot**).

As a tectonic plate moves over a hot spot, the volcano is carried away with it, and a new one forms. Eventually, this will create a chain of volcanic islands (such as the Hawaiian Islands).

Social and economic impacts

Volcanic eruptions and their secondary hazards can have enormous effects on a population – sometimes changing lives permanently (as the example below of Montserrat shows).

Montserrat

Montserrat is part of an island arc in the Caribbean Sea (formed where the Atlantic Plate subducts beneath the Caribbean Plate). Only 16 km long and 10 km wide, the island consists almost entirely of volcanic rock.

On 18 July 1995, the Soufriere Hills Volcano in the south of the island began to erupt huge clouds of ash (see Figure 6). Over the next five years, these eruptions continued – with pyroclastic flows also affecting much of the island.

The regular eruptions had huge social, economic and environmental consequences:

◆ Dozens of people lost their lives, and more than 7000 moved to other countries (over half of the original 11 000 residents).

◆ The capital, Plymouth, was destroyed (see Figure 7). As the capital, it contained all of the island's main services (such as government offices and hospitals).

◆ Two-thirds of all houses, and three-quarters of the infrastructure, were destroyed.

◆ Unemployment rose as the island's tourist industry collapsed.

◆ A lot of farmland was destroyed or abandoned, because it was too close to the volcano – severely affecting agriculture.

◆ A top-heavy population pyramid was created, as younger people no longer saw an economic future on the island and moved elsewhere (leaving the older residents behind).

Today, the volcano is still active and two-thirds of the island is still uninhabitable. A volcanic observatory has been built in the south to monitor the volcano. New infrastructure, such as roads and a new airport, have been built in the safer northern part of the island, and Montserrat is now trying to rebuild its tourism industry.

Figure 6 *Ash clouds erupting from the Soufriere Hills Volcano in 1997*

Figure 7 *Montserrat's capital, Plymouth, was covered by up to three metres of ash. The town had to be evacuated and abandoned, because it was too dangerous to live there.*

Measuring volcanic eruptions

The size and force of volcanic eruptions can vary – from gentle eruptions, producing small amounts of lava, to huge explosions that eject enough ash to block out the Sun! Scientists use the **Volcanic Explosivity Index (VEI)** to describe and compare the size or magnitude of volcanic eruptions. The VEI uses a scale from 0 (non-explosive) to 8 (extremely large). It's a logarithmic scale – with each number increasing by a factor of ten (see Figure 8).

The VEI uses several factors to assign a number, including:

- the amount and height of the volcanic material ejected (tephra and ash fall, etc.)
- how long the eruption lasts
- qualitative descriptive terms (such as 'gentle', 'explosive', etc.).

Predicting eruptions

Unlike with earthquakes, scientists can often predict volcanic eruptions with some accuracy. Using equipment placed on a volcano, as well as remote equipment (such as GPS and satellite-based radar), scientists can monitor a volcano for signs that it might erupt. These signs include:

- small earthquakes – as the magma rises to the surface, it breaks the rock, causing small earthquakes which scientists can detect on seismograms
- changes to the surface of the volcano – as it pushes upwards, the magma builds pressure (causing the surface of the volcano to swell)
- changes to the 'tilt' of the volcano – as the magma moves inside the volcano, it changes the slope angle or 'tilt' of the volcano (see Figure 9).

But, these predictions aren't 100% accurate, and not all volcanoes around the world are monitored.

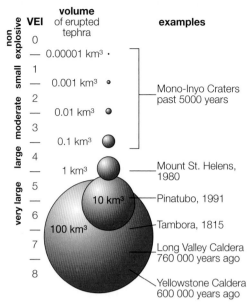

▲ **Figure 8** *The Volcanic Explosivity Index. There have been no known eruptions with a VEI greater than eight.*

▲ **Figure 9** *A tiltmeter (in the foreground) measures the tilt of the ground slope*

Over to you

1 In pairs, decide whether the eruption of Eyjafjallajökull was a hazard, a disaster – or just an inconvenience!

2 Also in pairs, (**a**) devise a set of criteria for assessing the severity of each of the primary and secondary hazards associated with volcanic eruptions, (**b**) using your criteria, assess each of the primary and secondary hazards, and (**c**) explain your findings.

3 What makes volcanoes potentially (**a**) more, (**b**) less hazardous than earthquakes?

On your own

4 Distinguish between the following pairs of terms: (**a**) lava and pyroclastic flows, (**b**) ash and tephra, (**c**) lahars and Jökulhlaup, (**d**) plume and hotspot.

5 How useful is the Volcanic Explosivity Index? Explain.

6 Compare the economic, social and environmental impacts of the eruption of Eyjafjallajökull with that of Montserrat. Which was the bigger disaster?

Exam-style questions

AS 1 Explain the hazards caused by one volcanic eruption. *(6 marks)*

A 2 Assess the range of hazards caused by explosive volcanic eruptions. *(12 marks)*

In this section, you'll look at what causes a tsunami, as well as studying the 2004 Indian Ocean tsunami in more detail.

How tsunami are formed

Boxing Day 2004, and countries around the Indian Ocean were devastated by a tsunami (see Figure 1). How do tsunami form?

Most are caused by large underwater earthquakes along subduction zones. Energy released during the earthquake causes the sea floor to uplift – displacing the **water column** above. This displaced water forms tsunami waves (see Figure 2). Most tsunami waves are no higher than 3 m, but the largest waves can reach heights of up to 30 m.

A tsunami wave moves fast – up to 805 km per hour (about the speed of a jet aircraft). When the wave's crest reaches the shore, it first produces a vacuum effect – sucking the water back out to sea and exposing a large amount of sea floor. The suddenly retreating water and exposed sea floor is a key early warning sign of an approaching tsunami. The water then returns rapidly in the form of a series of large waves (see Figure 2).

🔺 **Figure 1** *This dramatic photo shows a tsunami wave hitting Ao Nang, Thailand, during the 2004 Indian Ocean tsunami*

A **tsunami** is a series of larger-than-normal waves, which are usually caused by volcanic eruptions or underwater earthquakes. Because they're usually linked to tectonic events, they tend to occur along plate boundaries – particularly the Pacific Basin's 'Ring of Fire' (see page 12). Less frequently, tsunami are caused by underwater landslides, or by meteor or asteroid strikes, which suddenly displace large amounts of seawater. Smaller tsunami occur almost every day, with little effect, but large tsunami can cause many deaths and widespread destruction.

Key word

Water column – The area of seawater from the surface to the sea floor.

🔽 **Figure 2** *How a tsunami wave forms*

1 Generation of a tsunami in deep ocean
Tsunami are difficult to detect by ships due to small wave height and long wave length

2 Tsunami run-up
Nature of the waves will depend upon:
- cause of the wave, e.g. eruption or earthquake
- distance travelled from source, as energy is lost as they travel
- water depth over route affects energy loss through friction
- offshore topography and coastline orientation

3 Landfall
Death and destruction will depend upon land uses, population density and any warning given, as well as the physical geography or relief of coastal areas

Ocean is displaced. Waves radiate from the source in all directions

wave length 150–250m

water column

wave height 0.5–5m

wave period 10–60 minutes

Displacement of a large area of the sea floor

As water shallows, waves slow down and increase in height to produce onshore waves of up to 30m high

Wave energy is concentrated into a smaller volume of water

Waves which are 1m in height in the open ocean may reach 30m

Uplifted fault block. Seismic activity (i.e. earthquake)

Sea-floor irregularities reflect some wave energy so that less energy reaches the coast.

A tsunami event consists of a sequence of waves which may last for several hours. The fourth or fifth wave is often the largest. Forty per cent of wave energy is scattered back to sea, and sixty per cent is expended at or near the coast

The impacts of tsunami

Tsunami can be very destructive. The tall and rapidly advancing tsunami waves form walls of water that destroy everything in their paths (see Figure 1).

◆ Large tsunami can travel inland for several miles – sweeping away buildings, trees, bridges and people (see Figure 3).

◆ They also wash away the soil – undermining the foundations of buildings, bridges and roads, uprooting trees and destroying farmland.

◆ Tsunami can completely change the landscape (see Figure 7 on page 23). Small islands hit by a tsunami are often totally destroyed.

Most tsunami-related deaths are from drowning, but many people are also killed or injured by collapsing buildings or being hit by large debris, such as floating vehicles or trees. The flooding inland can also contaminate food and water supplies with salt, raw sewage (and unburied bodies), which can lead to serious illnesses such as cholera.

Figure 3 *Large amounts of debris washed inland as a result of the 2011 tsunami in Japan, which was caused by a magnitude 9.0 earthquake offshore (see Section 1.7)*

BACKGROUND

Predicting tsunami

Since most tsunami are caused by underwater earthquakes, the lack of a way to predict earthquakes also means that there is no way to predict tsunami before they occur – although it is possible to give people some early warning before an activated tsunami actually reaches the coast and threatens lives.

Tsunami **early warning systems** are now in place in both the Pacific and Indian Oceans. These systems use seismic sensors to detect undersea earthquakes. Yet, because not all undersea earthquakes cause tsunami, scientists use additional scientific equipment to gather more information. This includes a system called DART (Deep-ocean Assessment and Reporting of Tsunami).

The DART system (see Figure 4) uses seabed sensors and surface buoys to monitor changes in sea level and pressure. When tsunami waves are detected, the system sends the information via a satellite to tsunami warning stations. These stations review the transmitted information and use computer modelling to estimate the size and direction of the tsunami, before informing the areas at risk. Depending on where the tsunami originates, some people might receive a warning hours in advance of any threat – giving them time to move to higher ground further inland.

In 2004, the Indian Ocean didn't have an early warning system. The Boxing Day tsunami took 2 hours to reach Sri Lanka, where over 31 000 people died. Had an early warning system been in place, many people might have

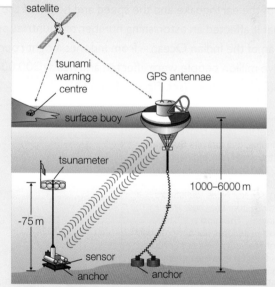

Figure 4 *The DART II system*

had time to evacuate. Therefore, because of the terrible impacts of the 2004 tsunami (see the next page), an Indian Ocean early warning system was developed and began operating in 2006.

Japan has the most extensive earthquake and tsunami warning systems in the world. When the magnitude 9.0 earthquake struck in 2011, the Japan Meteorological Agency issued a major tsunami warning within three minutes of the earthquake. But the system, in part, failed. The size of the earthquake was underestimated, which led to the size and power of the tsunami also being underestimated. As a result, some people thought the tsunami was going to be small, so they didn't take steps to prepare or evacuate.

The 2004 Indian Ocean tsunami

Most adults remember when they first saw pictures of the 2004 Indian Ocean tsunami, which showed bemused holidaymakers and locals watching the sea suddenly retreat an abnormal distance and then return rapidly as a wall of water – many realising far too late how huge the returning waves were (see Figure 1). It rates as one of the world's worst human disasters (on a scale that occurs about once every 100 years).

The earthquake off the coast of Sumatra (Indonesia) that caused the tsunami was estimated at between magnitude 9.0 and 9.3. Its thrust heaved the floor of the Indian Ocean towards Indonesia by about 15 metres, and – in so doing – sent out shock waves. Once started, these radiated out in a series of 'ripples', which moved almost unnoticed until they hit land. The waves that struck the shallow coastline near Banda Ache (only 15 minutes from their origin) were nearly 17 metres high on impact. By contrast, islands in the Maldives experienced a four-metre-high sea swell, rather than a crashing wall of water.

The size of the earthquake and the speed and height of the tsunami meant that it affected an astonishing number of countries around the entire span of the Indian Ocean – from Indonesia right round to South Africa. Five million people were affected and nearly 300 000 died, with 1.7 million left homeless (see Figures 5 and 6).

Affected country	Dead and missing
Indonesia	236 169
Sri Lanka	31 147
India	16 513
Thailand	5395
Somalia	150
The Maldives	82
Malaysia	68
Burma (Myanmar)	61
Tanzania	10
The Seychelles	3
Bangladesh	2
Kenya	1
Total	**289 601**

▲ **Figure 5** *The distribution of the Indian Ocean dead and missing by country. In addition to these casualties from Indian Ocean countries, 9000 tourists from all over the world were killed (especially in Thailand).*

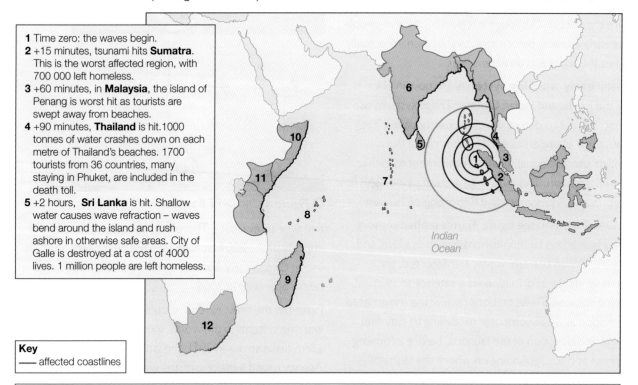

1 Time zero: the waves begin.
2 +15 minutes, tsunami hits **Sumatra**. This is the worst affected region, with 700 000 left homeless.
3 +60 minutes, in **Malaysia**, the island of Penang is worst hit as tourists are swept away from beaches.
4 +90 minutes, **Thailand** is hit. 1000 tonnes of water crashes down on each metre of Thailand's beaches. 1700 tourists from 36 countries, many staying in Phuket, are included in the death toll.
5 +2 hours, **Sri Lanka** is hit. Shallow water causes wave refraction – waves bend around the island and rush ashore in otherwise safe areas. City of Galle is destroyed at a cost of 4000 lives. 1 million people are left homeless.

Key
— affected coastlines

6 **India** is hit. Tsunami reaches up to 3km inland. Isolated Andaman and Nicobar islands suffered severe damage. 376 000 people are left homeless.
7 The **Maldives** are hit. 20 of 199 inhabited islands are totally destroyed. Low-lying islands are saved further losses because:
 • surrounding coral reefs break up the wave's energy
 • an absence of a continental shelf keeps the wave height down

8 The **Seychelles** are hit.
9 **Madagascar** is hit.
10 In **Somalia** 2000 structures are damaged.
11 The east coast of Africa is hit. In Kenya, warnings issued in time allow beaches to be evacuated and reduces the loss of human life.
12 **South Africa** is hit.

▲ **Figure 6** *Indian Ocean countries affected by the 2004 tsunami*

The nature of the tsunami

The 2004 Indian Ocean tsunami was so destructive for a number of reasons:

- The earthquake that caused the tsunami was especially large.
- The epicentre was close to some densely populated coastal communities, which had no time to react to the tsunami's rapid arrival within minutes.
- The low-lying coastlines of many Indian Ocean countries and islands meant that the tsunami waves were able to travel several kilometres inland.
- There was no early warning system in place in the Indian Ocean.
- Many of the countries in the region are lower-income countries. They did not have the resources to spend on tsunami protection.
- In some areas, such as Sri Lanka (see page 133), mangrove forests along the coast had been destroyed to allow for tourist developments – resulting in less natural protection when the waves struck.

The impacts

In addition to the many deaths and injuries, the tsunami had an enormous social, economic and environmental impact:

- Coastal settlements were devastated.
  - In some coastal villages, 70% of the villagers were killed.
  - In Sumatra, 1500 villages were completely destroyed (see Figure 7).
- Much infrastructure was destroyed. For example, the Andaman and Nicobar Islands were all but cut off as all jetties were washed away.
- Economies were devastated, especially fishing, tourism and agriculture.
  - In Sri Lanka, more than 60% of the fishing fleet and industrial infrastructure was destroyed.
  - In Thailand, the tourism industry lost about US$25 million a month, and 120 000 workers lost their jobs.
- There was also an enormous environmental impact.
  - Ecosystems (such as mangroves, coral reefs, forests and coastal wetlands) were severely damaged.
  - Most vegetation and topsoil was removed up to 800 metres inland (see Figure 7).
  - Freshwater supplies and agricultural soil were contaminated by salt water.
- The overall economic cost came to over US$10 billion.

Figure 7 *Satellite photographs showing the settlement of Lhoknga, Sumatra, before and after the 2004 tsunami.*

Exam-style questions

AS 1 Explain the formation of tsunami. *(4 marks)*

A 2 Assess the severity of the various impacts of tsunami. *(12 marks)*

Over to you

1 In pairs, discuss and list the reasons why it's so difficult to predict which earthquakes are likely to produce tsunami and which aren't.

2 Using the radius of the waves in Figure 6, estimate at what times the tsunami was likely to reach places 6 (India) to 12 (South Africa) inclusive.

3 a Draw a table with four columns and four rows. In the left-hand column, write 'Economic impacts', 'Social impacts', and 'Environmental impacts'. Across the top, write 'Short-term (a few weeks)', 'Medium-term (up to a few months)', and 'Longer-term (up to a few years)'. Complete the table for the 2004 Asian tsunami.

   b Which seem to have been the most serious impacts? Explain why.

4 In what ways was the 2004 Indian Ocean tsunami an event of global importance?

On your own

5 Distinguish between a tsunami and a water column.

6 Create a flow chart to explain how the DART system works to give early warnings of tsunami.

7 How might the impacts of the 2004 tsunami have been reduced had there been a DART system in place?

In this section, you'll look at factors that determine vulnerability and resilience to natural hazards and disasters.

As we saw in Section 1.1, natural hazards are natural processes (such as earthquakes or volcanic eruptions) with the potential to affect people. Natural hazards can then become disasters if they strike a vulnerable population that can't cope using its own resources.

As Figure 1 shows, the level of risk that a country faces from natural hazards depends on a combination of factors. Some of these are directly related to the Hazard itself (H), such as – in the case of an earthquake – its magnitude, how long it lasts, and the time of day when it occurs. But human factors also play as much – if not more – of a role in determining Vulnerability (V) and the Capacity to cope (C).

$$Risk\ (R) = \frac{Hazard\ (H) \times Vulnerability\ (V)}{Capacity\ to\ cope\ (C)}$$

◀ **Figure 1** The hazard-risk formula measures an area's risk from natural hazards

Human factors, vulnerability and resilience

As the quote at the top of the page states, disasters start long before an earthquake, volcanic eruption or tsunami occurs. Every day, individuals and governments make decisions that affect how well they are prepared for, and can cope with, a hazard event. For example, governments decide how much money to invest in infrastructure to protect against hazards (such as flood defences or hurricane shelters) – a decision that can have a huge influence on the impact and scale of a hazard event.

A country's level of development also affects its vulnerability and resilience. For example, Haiti remains one of the world's poorest countries (see Section 1.7), and many Haitians live in informal or badly built homes in crowded and unsafe conditions (see Figure 2). Meanwhile, the soil near volcanoes is rich and fertile, making it good for agriculture. People live in these dangerous areas, despite the risk (see Figure 3).

> Disasters do not just happen – they result from failures of development, which increase vulnerability to hazard events.

From *Disaster Risk Reduction: A Development Concern* (a 2004 report by the UK Department for International Development)

Key word

Vulnerability – The ability to anticipate, cope with, resist and recover from a natural hazard.

Resilience – The ability to protect lives, livelihoods and infrastructure from destruction, and to restore areas after a natural hazard has occurred.

Hazard event – A natural hazard (such as an earthquake, volcanic eruption or tsunami).

⬆ **Figure 2** Densely packed slum housing in Port-au-Prince, Haiti. These buildings have been poorly constructed and are unlikely to withstand an earthquake.

◀ **Figure 3** Indonesian farmers continue to work in their fields in Tanah Karo, north Sumatra, as Mount Sinabung (in the background of the photo) begins to erupt in August 2010. Eventually, 30 000 people were evacuated from the danger zone.

Vulnerability and less-developed countries

Figure 4 shows how human factors and decisions affect vulnerability and resilience. Less-developed countries are generally more vulnerable to hazard events than wealthier countries – because they tend to have other, more-pressing problems (such as poverty and disease), they're able to spend less money on preparing for hazard events. This is one reason why natural hazards can quickly turn into disasters in less-developed countries.

▼ **Figure 4** *Human factors that affect vulnerability and resilience*

Governance (local and national) and political conditions
- The existence and – especially – the enforcement of **building codes and regulations** determine the quality and safety of buildings and other structures.
- The quality of the existing **infrastructure** (such as transport and power supplies) affects a country's recovery speed.
- The existence of disaster **preparedness plans** influences how quickly and effectively a country responds to and recovers from a hazard event.
- The efficiency of **emergency services** and response teams affects the speed and effectiveness of rescue efforts.
- The quality of **communication systems** affects the ability to inform people of a hazard in advance, and to coordinate rescue efforts.
- The existence of **public education and practised hazard responses** (such as earthquake drills, see page 32), influences a population's preparedness for and responses to a hazard event.
- The level of **corruption** of government officials and businesses influences how resources are used.

Economic and social conditions
- Their level of **wealth** influences people's ability to protect themselves and then recover from a natural hazard. For example, by affecting where they live and the quality of the buildings they live in (see Figure 2).
- People without **access to education** may be less aware of the risks of a hazard event and how to protect themselves.
- Poor-quality **housing** is less able to withstand the impact of natural hazards.
- Communities with poor **health care** suffer more disease and are less able to cope with and recover from a hazard such as a flood.
- A lack of **income opportunities** means that people cannot buy the resources they need to prepare for or cope with a hazard, and this can affect their health care and living conditions.

Physical and environmental conditions
- Areas with a **high population density** tend to have more low-quality housing (see Figure 2).
- Rapid **urbanisation** (large numbers of people moving to cities in a short space of time) creates a need for more housing – most of which is built quickly and of poor quality.
- The accessibility of an area affects how quickly rescuers and aid can arrive (see Section 1.1 on the Nepal earthquake).

Over to you

1 Create a mind map or table to (**a**) summarize the main factors that determine vulnerability to a hazard, and (**b**) give examples of how these factors make a community or country more or less vulnerable.

2 **a** Explain what is meant by the term 'failures of development' (from the quote on page 24).
  b Discuss and list the ways in which the community illustrated in Figure 2 represents a 'failure of development'.

On your own

3 Distinguish between the following pairs of terms: (**a**) hazard-risk formula and vulnerability, (**b**) vulnerability and resilience, (**c**) resilience and capacity to cope.

4 Explain, using examples, why it is difficult to define the term 'disaster'.

Exam-style questions

 1 Assess the reasons why, even within a country, some people are more vulnerable to hazards than others. *(12 marks)*

 2 Assess the relative importance of the concept of vulnerability in understanding hazard impacts. *(12 marks)*

In this section, you'll learn about the causes, effects and government responses to earthquakes in Haiti (a developing country), China (an emerging country) and Japan (a developed country).

In the field of natural disasters, governments are key players.

- Community infrastructures and housing depend upon **good planning**. Buildings *can* withstand earthquakes, but not all governments enforce building or planning regulations that allow them to do so.

- Even where buildings are of poor quality, a good deal can be done through **disaster preparedness** and planning.

- Finally, good financial management and decision-making enable aid and donations to reach the right people when they need it.

Contrasting impacts

Japan, China and Haiti are all located along tectonic plate margins, which makes them particularly vulnerable to earthquakes. Figure 1 shows that each country has experienced a powerful earthquake within the last decade. Yet, the death toll and the level of destruction experienced in those earthquakes varied hugely (see Figure 2).

As Figure 2 shows, despite having the lowest-magnitude earthquake, Haiti had the highest death toll by far, compared with the earthquakes in China and Japan. In fact, the Japan earthquake was 1000 times stronger than the one that struck Haiti. So why was the loss of life in Haiti so much greater? Basically, because it was more vulnerable.

Ranking (by the number of deaths caused)	Country (region)	Year	Magnitude	Number of deaths
1	China (Hebei)	1976	7.8	242 800
2	Indian Ocean tsunami	2004	9.0	226 408
3	Haiti	2010	7.0	222 576
4	Pakistan (Kashmir)	2005	7.6	73 328
5	China (Sichuan)	2008	7.9	69 195
6	Peru (Chimbote / Huaras)	1970	7.8	66 794
7	Iran (Manjil)	1990	7.7	35 000
8	Iran (Kerman / Bam)	2003	6.7	31 830
9	Armenia (Spitak)	1988	6.8	25 000
10	Guatemala	1976	7.5	22 870
11	India (Gujarat)	2001	8.0	20 023
12	Iran (Tabas)	1978	7.4	18 220
13	Turkey (Kocaeli)	1999	7.8	17 118
14	Japan (Eastern)	2011	9.0	15 840
15	China (Yunnan)	1970	7.8	15 621

Figure 1 *Earthquakes and tsunami with over 5000 deaths, 1970 – 2011*

	Haiti, 2010	China (Sichuan), 2008	Eastern Japan, 2011
Event	earthquake	earthquake	earthquake/tsunami
Magnitude	7.0	7.9	9.0
The number of dead and missing	316 000 (Haitian govt. estimate) 220 000 – 250 000 (UN estimate)	87 150	19 848
Injuries	300 000	375 000	6 065
Made homeless	1.3 million	5 million	130 927
Economic cost	US$14 billion	US$125.6 billion	US$240 billion
GDP per capita	US$1300 (2009 est.)	US$6600 (2009 est.)	US$39 473 (2009 est.)
Type of economy	Developing	Emerging	Developed

Figure 2 *A comparison of the earthquakes in Haiti, China and Japan*

Haiti – a developing country

As Figure 3 shows, Haiti is located on a fault between the North American and Caribbean Plates. On 12 January 2010, an earthquake with a magnitude of 7.0 struck near the capital. The resulting high death and injury toll made it one of the deadliest earthquakes on record.

Physical factors played a big part in making this particular earthquake so devastating:

♦ It had a shallow focus (13 km), which increased the amount of ground shaking.

♦ Liquefaction on looser soil caused many building foundations to sink.

♦ The epicentre (see Figure 4) was only 24 km from Port-au-Prince – the country's capital and its most densely populated city (home to 2 million people).

However, as well as physical factors, additional political, social and economic factors also helped to turn the 2010 earthquake into a disaster (see below and Figure 7):

♦ Haiti is a developing country – it's poor, and its limited resources were being spent on more immediate issues, such as disease, rather than earthquake preparations.

♦ A high level of corruption at both national and local government level had led to a lack of resources and commitment to improve the country's infrastructure and living standards.

♦ A lack of building controls and regulations meant that many of the buildings in Port-au-Prince were poorly built slum housing (see Figure 2 on page 24). These buildings could not sustain the ground shaking and simply collapsed. The dense urban environment also made it a difficult place for rescue teams to work.

♦ A lack of disaster preparation meant that government officials, police and emergency services (and ordinary Haitians) just didn't know what to do when the earthquake struck.

♦ Many Haitians were (and still are) living in poverty, so they didn't have the resources to prepare for or cope with the effects of the earthquake.

Key
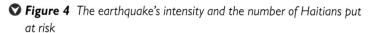
→ direction of plate movement
— plate boundary
⋆ epicentre
⌇ complex strike-slip fault

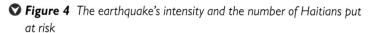

Figure 3 *The complex plate boundaries affecting Haiti*

Figure 4 *The earthquake's intensity and the number of Haitians put at risk*

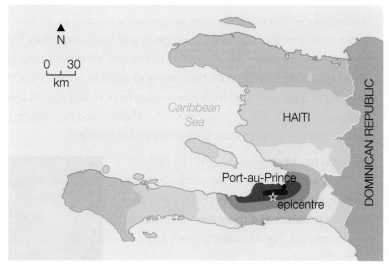

Key

estimated MMI intensity			population exposed to shaking
IV		moderate	5 887 000
V		slightly strong	7 261 000
VI		strong	1 049 000
VII		very strong	571 000
VIII		destructive	314 000
IX		ruinous	2 246 000
X		disastrous to catastrophic	332 000

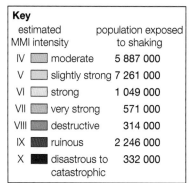

Impacts on Haiti

Much of Haiti's infrastructure, which was already poor, was severely damaged during the earthquake.

◆ Haiti had only one airport, several ports and a few main roads. When these became damaged (see Figure 5), crucial aid supplies were prevented from arriving or being distributed effectively – slowing down rescue efforts and ultimately leading to more deaths.

◆ Over a quarter of government officials were killed, and key government buildings were destroyed, making the government even less able to organise recovery and relief efforts.

◆ In October 2010, an outbreak of cholera (a potentially fatal infection caused by poor sanitation) occurred – and, as of 2016, was still ongoing. A lack of medical supplies and trained health care workers caused the disease to spread, so that by 2015 over 9000 Haitians had died and 720 000 had been affected.

Figure 5 *Haiti's already fragile infrastructure (like its main port, pictured here) suffered badly in the earthquake, which slowed down international relief efforts in the crucial first few days after the earthquake struck*

Haiti's recovery

By 2015, five years after the earthquake, Haiti was still recovering. Internationally, US$13 billion of aid had been donated, but most of it remained in the hands of international organisations and governments (with the Haitian government and Haitian organisations controlling less than 10%).

Initially – because so many Haitian government officials had been killed in the earthquake – international organisations were needed to provide emergency services. Later, concerns about political corruption and mismanagement meant that many of these organisations were unwilling to channel aid money through the Haitian government directly. Instead, they chose to manage projects themselves – often bringing in their own staff from overseas at huge cost. Experts argue that this has hampered Haiti's ability to become self-sufficient, and has meant that much-needed money has not gone to local businesses and industries.

Figure 6 *Five years after the earthquake, about 1000 Haitians were still living in tents along one of Haiti's busiest roads*

Progress has been slow. By 2015, 80 000 Haitians were still living in temporary housing or camps (see Figure 6), and cholera was an ongoing problem. But there have been some improvements. New buildings, roads and schools have been built, and health statistics have improved. And there are signs that the Haitian government is getting stronger and more able to cope with natural threats. In 2012, as another natural hazard (Hurricane Sandy) was heading towards Haiti, the government responded – warning citizens that the storm was coming and telling them to go to higher ground (an act that saved lives). After the storm, the government also took a leading role in organising international aid.

Pressure and release (PAR) model

In order to protect people from a hazard event, governments and other organisations must first understand how vulnerable a country is – and why. One tool commonly used to work this out is the disaster **pressure and release (PAR) model** (see Figure 7).

The PAR model looks at the underlying causes of a disaster. It's based on the idea that a disaster happens when two opposing forces interact: on one side are the processes that create vulnerability (shown in Figure 7 as root causes, dynamic pressures and unsafe conditions), and on the other side is the hazard event itself.

According to the PAR model, vulnerability is a process that starts with **root causes**. These are political and economic systems that control who has power in a society and who has access to resources (such as money). Through a series of processes called **dynamic pressures**, these root causes can lead to **unsafe conditions**. For example, a country that is poor (a root cause) will probably not spend time or money enforcing building codes (a dynamic pressure), which means that buildings may be poorly built (leading to unsafe conditions). This process – from root causes to unsafe conditions – is called the **progression of vulnerability**.

The progression of vulnerability

Root causes
- Haiti was heavily in debt to US, German and French banks. The Haitian government had to use much of its little available money for debt repayments, rather than improving the country's infrastructure.
- There was extensive corruption within Haiti's government.
- 80% of the population lived below the poverty line, on less than US$2 a day.
- 30-40% of the government budget came from foreign aid.

Dynamic pressures
- There was a lack of:
  - urban planning to control where and how buildings were constructed, and where people lived
  - disaster preparedness and management systems
  - effective education systems
  - disaster management systems.
- Macro forces. There was:
  - rapid urbanisation, which resulted in vulnerable, slum-like housing
  - a high population density (in the capital, Port-au-Prince, it was 306 people per square kilometre)
  - significant deforestation and soil degradation, as a result of creating large sugar plantations, which increased the risk of earthquake-related landslides.

Unsafe conditions
- The soft soil, on which many of Haiti's buildings were constructed, amplified the seismic waves – increasing ground shaking and damage.
- A lot of illegal housing was built in unsafe areas, such as hillsides.
- A low GDP per capita of US$1300 meant that buildings were constructed cheaply and quickly, which often resulted in poor-quality and vulnerable structures.
- Poor infrastructure limited access to people in need and made reaching them more expensive.
- Before the earthquake, only 39% of Haitians had access to safe water and 24% to sanitation.

Disaster

Hazards
Earthquakes
Landslides

$$\text{Risk (R)} = \frac{\text{Hazard (H)} \times \text{Vulnerability (V)}}{\text{Capacity to cope (C)}}$$

⬤ **Figure 7** *A disaster PAR model to help explain Haiti's vulnerability*

China – an emerging country

On 12 May 2008, an earthquake with a magnitude of 7.9 struck Sichuan, a mountainous region in south-west China (see Figure 8). Over 45.5 million people in ten provinces and regions were affected (5 million of whom were made homeless; the highest recorded homeless count from a disaster in history). The earthquake also triggered landslides that led to a quarter of the earthquake-related deaths.

Haiti and Sichuan both experienced devastating earthquakes that resulted in huge numbers of deaths and injuries, and a lot of structural damage. Before the earthquakes – in both countries – corrupt government officials often ignored building codes and accepted bribes to allow builders to take shortcuts. The resulting poorly constructed buildings could not withstand the ground shaking and collapsed (see Figure 9). The effects of the corruption were particularly evident in Sichuan, where thousands of schools fell down (killing 5335 children), while properly built government buildings nearby remained standing.

However, despite many similarities between the Sichuan and Haiti earthquakes, the overall outcome of the Sichuan earthquake was different in many respects:

◆ The earthquake's location meant that the damage was concentrated in rural areas and small towns – not a densely populated city like Port-au-Prince. This undoubtedly made a difference when comparing the death tolls in the two earthquakes (see Figure 1).

◆ China is also wealthier than Haiti. It's a large country with a growing economy (particularly in 2010), so it had the money available to pay for rescue and aid efforts.

◆ Unlike in Haiti (where the national government was effectively destroyed by the earthquake and had to wait for foreign help to arrive), China's strong central government was able to respond quickly and effectively to the disaster:

  ◆ Within hours, over 130 000 soldiers and relief workers were being sent to the affected areas. Troops parachuted or hiked into isolated mountainous areas to reach survivors (see Figure 10).

  ◆ Medical services were quickly restored, which helped to avoid the outbreaks of disease seen in Haiti.

  ◆ People in danger from landslides were safely relocated.

  ◆ The government pledged $US10 billion for rebuilding works, and Chinese banks wrote off the debts of any survivors who did not have insurance.

  ◆ Within two weeks, temporary homes, roads and bridges were being built.

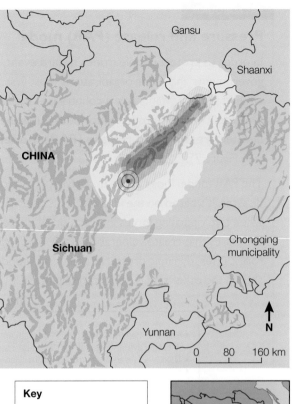

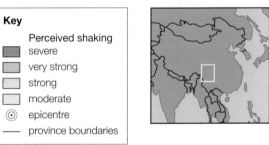

Key

Perceived shaking
- severe
- very strong
- strong
- moderate
- ⊙ epicentre
- — province boundaries

▲ **Figure 8** *The location and intensity of the 2008 Sichuan earthquake*

▼ **Figure 9** *Damage caused by the Sichuan earthquake*

At a national level, China has tough building codes; is investing in safer buildings and better infrastructure; and has the resources to respond quickly to a hazard event. But at the root-cause level, the corruption of local government officials and law enforcement means that unsafe building practices still continue. For example, after the 2008 earthquake, the government moved over 40 000 people to a newly built city called Yongchang. Almost immediately, cracks appeared in the brand-new homes – which led to the arrest of a local official for taking bribes.

In the longer term, the Chinese government saw the 2008 earthquake as an opportunity to rebuild the area from scratch. Two years after the earthquake:

◆ 97% of the planned 29 704 reconstruction projects in the region had started.

◆ 99% of the 196 000 farmhouses destroyed in the earthquake had been rebuilt.

◆ A total of 216 transport projects (including highways, main roads, railways and airports) were under construction or had been completed (see Figure 11).

These efforts have not only improved people's lives and the region's economy, but they have also made the area more resilient to future hazard events.

Japan – a developed country

On 11 March 2011, a magnitude 9.0 earthquake struck under the Pacific Ocean, 100 km east of Sendai on the eastern coast of the Japanese island of Honshu (see Figure 12). The resulting seawater displacement caused a tsunami to spread in all directions – at hundreds of kilometres per hour. The waves reached a staggering 10 metres high in places, and surged up to 10 km inland. The Fukushima Daiichi Nuclear Power Plant was severely damaged and released dangerous levels of radiation into the air (forcing 47 000 people to be evacuated). Today, an exclusion zone of 20 km still exists around Fukushima.

Yet, despite the magnitude of the earthquake and the force of the tsunami, Japan had fewer deaths and injuries than occurred in either the Haiti or Sichuan earthquakes – both of which were of a smaller magnitude. So why did Japan cope better than either Haiti or China? See page 32 for the answer.

⬆ **Figure 10** *Chinese soldiers during search and rescue operations in Sichuan*

⬇ **Figure 11** *A new high-speed train line from Chengdu to Dujiangyan in Sichuan Province was opened two years after the 2008 earthquake*

▶ **Figure 12** *The epicentre of the earthquake, and the areas hit by tsunami waves*

Key

	earthquake shaking intensity strong to severe
—	severe flooding more than 500 metres inland
☢	nuclear power plant
◎	epicentre

Japan's preparation

As a highly developed country, Japan had the financial resources and commitment to prepare for a hazard event:

◆ Good building construction:
  ◆ Strict building regulations meant that Japanese buildings were better able to withstand an earthquake (75% of buildings in Japan are constructed with earthquakes in mind, compared to none in Haiti).
  ◆ A low level of corruption meant that building regulations were strictly enforced.

◆ Well-developed disaster plans:
  ◆ Areas vulnerable to tsunami already had ten-metre-high walls, evacuation shelters and marked evacuation routes, which helped to reduce the loss of life.
  ◆ Many offices and homes were equipped with earthquake emergency kits (containing drinking water and basic medical supplies).
  ◆ An early warning system detected the earthquake one minute in advance – giving people some warning.

◆ Education and preparedness for earthquakes and tsunami. Emergency drills are regularly practiced in both schools and businesses (see Figure 14).

Yet, despite investing more money into disaster planning and management than most other countries, Japan failed to take into account the impact of a tsunami on a nuclear power plant. An investigation found that the plant had not been built to withstand such a large tsunami, and that a lack of basic safety procedures, planning, preparation and oversight by the government all contributed to the events at Fukushima.

Response

◆ The Japanese government responded immediately. Within 24 hours, 110 000 defence troops had been mobilised.

◆ Immediately after the earthquake, all radio and TV stations switched to official earthquake coverage, which told people what was happening and what they should do.

◆ The Bank of Japan offered US$183 billion to Japanese banks, so that they could keep operating (protecting the country's economy).

◆ Japan quickly accepted help from rescue and recovery teams from over 20 countries. By comparison, in Haiti foreign rescue teams were delayed by poor and damaged infrastructure and the lack of government coordination. In China, whilst the government was able to mobilise rescuers and equipment from across the country, it wasn't used to accepting help from other countries. As a result, it didn't have any procedures in place, so it took several days before international rescue teams were allowed in.

▲ **Figure 13** *A tsunami wave engulfs homes in Natori, Japan*

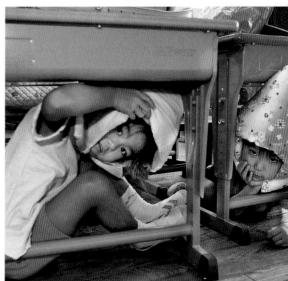

▲ **Figure 14** *Primary school students wearing padded hoods to protect their heads and necks during an earthquake drill at a school in Tokyo*

Japan's energy policy

As Figure 15a shows, before the earthquake and tsunami, 27% of Japan's electricity came from nuclear energy. Immediately after the earthquake, the nuclear power stations closest to the epicentre were shut down. Eventually, all 44 of Japan's nuclear reactors were closed. By 2013, the amount of electricity generated by nuclear power in Japan had dropped to just 1% (see Figure 15b).

Without its nuclear reactors to generate electricity, Japan had to start importing and using more fossil fuels. As a result:

◆ the price of electricity went up by about 20%

◆ the government's debt level rose, because it had to buy in more fossil fuels

◆ greenhouse gas emissions increased as a result of the increased use of fossil fuels.

Immediately after the accident at Fukushima, the government promised to reduce the country's use of nuclear energy. But high electricity prices, a slowing economy, and a reliance on imported fossil fuels, led the country's new government to reintroduce nuclear power as part of its energy policy. A 2014 energy announcement said that by 2030 nuclear energy would generate 20-22% of Japan's electricity.

The shutdown of Japan's nuclear power stations had an effect around the world. On the one hand, Japan's increased demand for natural gas pushed the price up; on the other, the events at Fukushima led Germany to shut down permanently all of its nuclear plants.

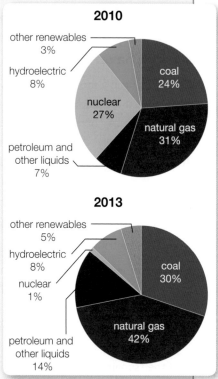

2010
- other renewables 3%
- hydroelectric 8%
- nuclear 27%
- petroleum and other liquids 7%
- coal 24%
- natural gas 31%

2013
- other renewables 5%
- hydroelectric 8%
- nuclear 1%
- petroleum and other liquids 14%
- coal 30%
- natural gas 42%

▲ **Figures 15a and 15b** *Japan's sources of electricity generation, before and after the 2011 earthquake and tsunami*

Over to you

1 Using Figure 1, assess the global importance of the three disasters in this section.

2 a In groups, make a copy of the PAR model (Figure 7) for (**i**) Sichuan, and (**ii**) Japan. Present these to the rest of your group.

   b Assess which of the three countries was best- and least-prepared to meet the pressures placed upon it by the disaster that occurred there.

3 a Draw and complete a table to show the economic, social, and environmental impacts of **each** of the hazard events in Haiti, Sichuan and Japan.

   b Go through each impact to classify it as short-, medium- or long-term.

   c Which one seems to have been the most serious disaster? Explain why.

4 How well would you rate the level of preparedness, vulnerability and resilience of Haiti, China and Japan to hazard events such as earthquakes?

On your own

5 Define and explain the pressure and release model in 100 words.

6 Assess the role of governments as key players in each of Haiti, China and Japan. For each country, identify examples of (**a**) poor governance, (**b**) good governance, and (**c**) what actions should be taken in future.

Exam-style questions

AS 1 Explain the impacts of one major tectonic disaster. *(6 marks)*

A 2 Assess the extent to which a country (or countries) has been able to meet the pressures placed upon it by a major disaster. *(12 marks)*

In this section, you'll look at geophysical disaster trends and hazard profiles.

A more hazardous world?

Are we living in a more hazardous world? In some ways, it seems as if we are. Figure 1 shows that, since 1960, the total number of reported natural disasters has risen quite dramatically – although the number of **geophysical disasters** has remained fairly steady. Several relevant factors are worth noting:

◆ Earth scientists believe that improvements in monitoring and recording events (e.g. through better seismographs) may be contributing to the rising trend in reported events.

◆ Improvements in communications technology – that would not even have been thought of in 1960 – now allow more news events (disasters included) to be reported. In 1960, even transatlantic satellite communications were still two years away! By contrast, the world watched live coverage of the tsunami that followed the 2011 Japan earthquake.

◆ The global population in 1960 was less than 3 billion, whereas by 2016 it had reached 7.3 billion and was still rising rapidly. Therefore, more people now occupy more hazardous space (particularly near rivers and coasts), and are aware of more storm or flood events taking place – because they affect those people directly.

◆ The increase in occupied living space also means that more of the world is now covered with concrete and other impermeable building materials (often on or close to flood plains). Therefore, the flood risk – in particular – has increased (as part of the effects of climate change).

However, Figures 2 – 4 present a slightly more complex picture, because they show that each disaster (such as an earthquake) is a very individual event – the significance of which can be determined using a number of different criteria. For example, Figure 2 identifies three very significant geophysical disasters, using the criteria of the number of people killed by each one. However, if you use different criteria, such as the number of people affected (Figure 3), or the amount of economic damage caused (Figure 4), other geophysical disasters become more significant.

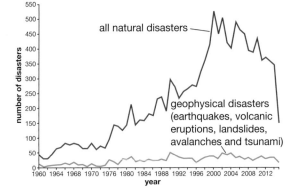

▲ *Figure 1* *The total number of reported natural disasters since 1960, together with the number of reported geophysical disasters*

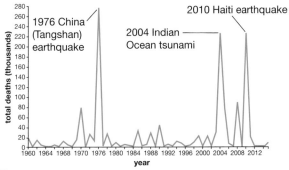

▲ *Figure 2* *The number of deaths due to geophysical disasters since 1960*

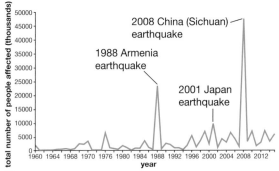

▲ *Figure 3* *The number of people affected by different geophysical disasters since 1960*

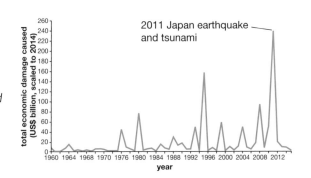

▶ *Figure 4* *The total economic damage caused by different geophysical disasters since 1960*

Current trends

High Income Countries are better able to cope with hazardous events. A country's level of economic development can influence the number of people killed by a disaster – but also its financial cost. For example, see Figure 4 for the economic cost of the magnitude 9.0 Japan earthquake and tsunami in 2011 – about US$240 billion – whereas the official death toll was 15 893. Compare that with the lower magnitude 7.0 Haiti earthquake in 2010 (Figure 2), which killed about 230 000 Haitians and had an economic cost of 'just' US$14 billion. Haiti is one of the poorest and least developed countries in the world, and Japan is one of the richest and most developed.

◆ Overall, the number of people being killed by disasters globally is falling. Better early warning systems, improved building codes and disaster preparedness have helped to reduce the death toll.

◆ However, between 1994 and 2013, the average number of people dying per disaster was over three times higher in low-income countries (322 deaths) than in high-income countries (105).

◆ Despite the falling overall death toll, the financial cost of disasters is rising. In the 1990s, the economic cost of natural disasters averaged US$20 billion per year. This increased to about US$100 billion per year between 2000 and 2010.

BACKGROUND

Data accuracy and reliability

Receiving accurate information about the frequency and impact of disasters is an important tool for governments, international organisations and aid agencies. However, the collection of disaster data is often incomplete or inaccurate:

◆ When a disaster strikes, the immediate focus is on organising the rescue and aid efforts – not on collecting data!

◆ No single organisation is responsible for collecting data. As a result, methods vary in the collection of data.

◆ There are even differences in the definitions of some of the key terms and categories used, such as 'disaster' and 'damage'.

◆ It's also difficult to gather data from remote areas, so it's likely that under-reporting of deaths and damage occurs in these areas, the very places in need of help and disaster planning.

Hazard profiles

A **hazard profile** is a diagram that shows the main characteristics of different types of tectonic hazard. It could be developed for a single hazard, or it could show multiple hazards – allowing comparisons to be made (see Figure 5).

Hazard profiles help governments and other organisations to develop disaster plans. For example, the profile for the 2004 Indian Ocean tsunami shows a high magnitude, rapid onset and widespread extent. Organisations using this profile to plan for future disasters might focus on making sure that early warning systems reach all countries that might be affected.

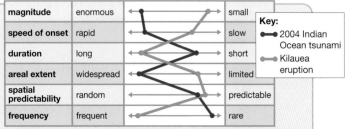

▲ **Figure 5** A hazard profile for the 2004 Indian Ocean tsunami and the ongoing eruption of Kilauea in Hawaii

However, the disaster plan for the ongoing eruption of Kilauea would probably be very different – the small magnitude and limited extent might lead the Hawaiian government to focus its efforts on developing evacuation plans for people living close to the volcano.

Over to you

1 Explain the pattern shown by Figure 1.

2 Explain how and why the trends shown in Figures 2, 3 and 4 are (a) similar, (b) different.

On your own

3 Draw a blank version of Figure 5, and (a) plot profiles for any two disasters in this chapter, (b) assess the two disaster profiles and consider reasons for any differences.

Exam-style questions

 1 Explain why some disasters are economically costly, while others are more costly in terms of human lives. *(6 marks)*

 2 Assess the statement that 'we are living in a more hazardous world'. *(12 marks)*

In this section, you'll use a case study of the Philippines to examine what makes a multiple-hazard zone.

Multiple natural hazards

Most countries face some kind of hazard, whether it's tectonic in nature (such as those covered by this chapter), or **hydrometeorological**. But some countries are exposed to multiple natural hazards – making them risky places to live.

If an area is at risk from multiple natural hazards, and is vulnerable, it is called a **multiple-hazard zone** (or disaster hotspot). Identifying multiple-hazard zones is important, because it helps decision makers: to understand a region's hazards, to set priorities for action, and to decide how to assign resources. A country or region identified as a multiple-hazard zone may also get more support from international aid agencies, as well as more resources to help with disaster-planning and prevention.

Key word

Hydrometeorological hazards – Natural hazards caused by climate processes (including droughts, floods, hurricanes and storms).

The Philippines: a multiple-hazard zone

The Philippines is considered one of the most disaster-prone countries in the world (see Figure 1). A report in 2015 found that – of the ten most at-risk cities from natural hazards in the whole world – eight of them were in the Philippines!

◆ The Philippines sits across a major convergent plate boundary (see Section 1.2), so it faces significant risks from both volcanoes and earthquakes (see Figures 1 and 2).

◆ Its northern and eastern coasts face the Pacific Ocean (the world's most tsunami-prone ocean).

◆ The Philippines lies within South-East Asia's major typhoon belts. In most years, it's affected by 15 typhoons and actually struck by 6 to 9 (see Figure 1). These events not only bring strong winds and heavy rainfall, but they also increase the risk of flooding and landslides.

◆ It has a tropical monsoon climate, so is subject to heavy annual rains.

◆ The Philippines has 47 volcanoes – 22 of which are active. Over 30% of the country's population lives within 30 km of a volcano.

◆ Landslides are common, due to a combination of steep topography, high levels of deforestation and high rainfall.

BACKGROUND

Area: The Philippines consists of 7107 islands, and is 25% bigger than the UK

Population: 101 million in 2015

Wealth: GDP per capita (PPP) in 2014 was US$7000; so, according to the World Bank, it's considered a middle-income country

Landscape: It is mostly mountainous, with coastal lowlands; many people live and work on steeply sloping land

▼ **Figure 1** *A summary of hazard events affecting the Philippines, 1960–2015*

	Number of events	Total deaths	Number affected	Economic damage US$ (thousands)
Droughts	8	8	6 553 207	64 453
Earthquakes	23	9384	5 796 577	583 178
Floods	105	2096	20 971 079	3 441 886
Landslides	32	2752	317 546	33 281
Volcanic eruptions	22	1077	1 734 708	231 961
Tsunami	1	32	Data unavailable	Data unavailable
Storms and typhoons	345	45246	163 682 029	19 620 701

Vulnerability

But it isn't only the Philippines' geography that puts it at such high risk of natural disasters. The combination of a growing population, rapid urbanisation, and poverty, increases the Philippines' vulnerability to hazard events.

♦ The Philippines is a rapidly developing lower-middle-income country. Its development, and a fast-growing population, has led to rapid urbanisation and a high population density.

♦ Many of the country's poor live in coastal areas, where sea surges, flooding and tsunami are made worse by poorly constructed housing and infrastructure.

♦ 25% of the population live in poverty (see Figure 2).

The challenges of multiple hazards

One hazard event can cause or increase other hazards. For example, an earthquake in the Philippines in 2006:

♦ killed 15 people, injured 100 and damaged or destroyed 800 buildings

♦ generated a local tsunami that was 3 metres high

♦ triggered landslides, which breached the crater wall of a volcano and fell into a lake, creating a flood that washed away houses.

Equally, different hazard events happening in a short space of time can leave communities and the government having to deal with a new disaster – just as they are trying to recover from the last one. This drains resources and stretches the ability of emergency systems to respond. For example, in 2013, the Philippines was struck by three natural disasters within three months – an earthquake in October that killed 223 people; Typhoon Haiyan in November that killed 6201 people (see Figure 3); and floods from a tropical depression in January 2014 that killed 64 people. This string of disasters left the Philippine government and aid agencies operating in a near-constant state of emergency.

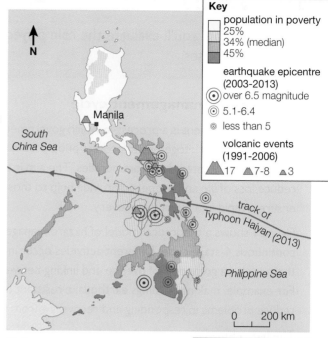

▲ **Figure 2** *The vulnerability of the Philippines to natural hazards*

▲ **Figure 3** *The devastation caused by Typhoon Haiyan at Tacloban city in November 2013*

Over to you

1 In pairs, (**a**) use the World Bank website to research a fact file of about ten socio-economic indicators (e.g. literacy rates, GNI) for the Philippines, (**b**) decide how developed you think the Philippines is.

2 Again in pairs, study Figure 1 and decide what you think is the most serious natural hazard that the Philippines faces, and why.

3 Discuss and decide (**a**) whether the damage caused by natural hazards in the Philippines is mainly social, economic, or environmental, (**b**) how vulnerable you consider the Philippines to be.

On your own

4 Define hydrometeorological hazards and consider their similarities and differences with tectonic hazards.

5 Use the hazard-risk formula (Section 1.6) to assess why the impacts of Typhoon Haiyan were always likely to be great.

Exam-style questions

AS 1 Assess the vulnerability of one named country to natural hazards. *(12 marks)*

A 2 Assess the extent to which hydrometeorological hazards can produce very similar impacts to hazards with tectonic causes. *(12 marks)*

In this section, you'll evaluate the role played by key players in hazard management and hazard responses

The hazard-management cycle

Hazard management is a process in which governments and other organisations work together to protect people from the natural hazards that threaten their communities. The aim is to: avoid or reduce loss of life and property; provide help to those affected; ensure a rapid and effective recovery.

Figure 1 shows a theoretical model of hazard management as a continuous 4-stage cycle. Different activities occur in each stage, but there is also a great deal of overlap and linking between the stages. For example, making buildings earthquake resistant (mitigation) will reduce problems in responding and recovering from earthquakes.

The hazard-management cycle involves key players. **Governments** at all levels – local, regional, and national – as well as **international organisations**, **businesses**, and **community groups** are involved in emergency planning.

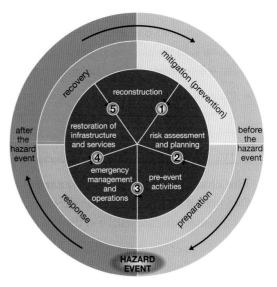

▲ *Figure 1* The hazard-management cycle

Stage	Focus	Actions	Takes place
Mitigation (Prevention) Preventing hazard events or minimising their effects	Identifying potential natural hazards and taking steps to reduce their impact. The main aim is to reduce the loss of life and property (largely by helping communities to become less vulnerable).	• Zoning and land-use planning • Developing and enforcing building codes • Building protective structures (such as tsunami sea defence walls)	Before and after hazard events
Preparedness Preparing to deal with a hazard event	Minimizing loss of life and property, and facilitating the response and recovery phases. Many activities are developed and implemented by emergency planners in both governments and aid organisations.	• Developing preparedness plans • Developing early warning systems • Creating evacuation routes • Stockpiling aid equipment and supplies • Raising public awareness, (e.g. by holding earthquake drills)	Before hazard events
Response Responding effectively to a hazard event	Coping with disaster. The main aims are to save lives, protect property, make the affected areas safe, and reduce economic losses.	• Search and rescue efforts • Evacuating people where needed • Restoring critical infrastructure (e.g. power and water supplies) • Ensuring that critical services continue (e.g. medical care and law enforcement)	During hazard events
Recovery Getting back to normal	*Short-term recovery* This focuses on people's immediate needs, so it overlaps with the response phase. Although called short-term, these activities may last for weeks. *Long-term recovery* This involves some of the same actions, but may continue for months or even years. It includes taking steps to reduce future vulnerability, which overlaps with the mitigation phase and the cycle continues.	*Short-term:* • Providing essential health and safety services • Restoring permanent power and water supplies • Re-establishing transportation routes • Providing food and temporary shelter • Organising financial assistance to help people rebuild their lives *Long-term:* • Rebuilding homes and other structures • Repairing and rebuilding infrastructure • Re-opening businesses and schools	After hazard events

▲ *Figure 2* Stages of the hazard-management cycle

The Park model (hazard-response curve)

The Park hazard-response curve (see Figure 3) is a model that shows how a country or region might respond after a hazard event. It can be used to directly compare how areas at different levels of development might recover from a hazard event.

◆ The impacts of a hazard event change over time – depending on factors such as the size of the hazard, the development level of the areas affected, and the amount of aid received.

◆ All hazard events have different impacts, so their curves are different. For example, hazard events that happen suddenly (such as the 2010 Haiti earthquake), and those that happen more slowly (such as the volcanic eruption in Montserrat that started in 1995 and lasted for years), have different responses.

◆ Wealthier countries have very different curves than developing countries, because they will be able to recover much faster.

◆ In hazard events that affect a number of countries (such as the 2004 Indian Ocean tsunami), each country has its own curve.

⬢ **Figure 4** *Relief and rehabilitation: international agencies help in the immediate aftermath of the 2010 Haitian earthquake*

⬢ **Figure 5** *Reconstruction: new housing built in Haiti for those made homeless by the 2010 earthquake*

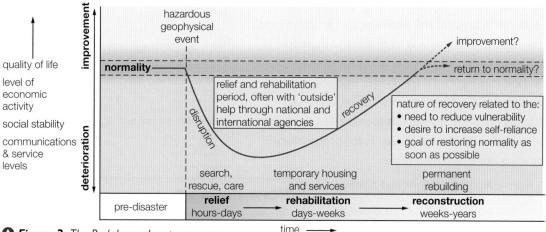

⬢ **Figure 3** *The Park hazard-response curve*

Over to you

1 a In pairs, select one disaster from this chapter each. Compare them in terms of the roles played by key players in: **(i)** mitigation, **(ii)** preparedness, **(iii)** response, and **(iv)** recovery.

b Explain how the hazard-management cycle helps us to understand disaster impacts better.

On your own

2 a Select any **two** major earthquakes in this chapter. Assess the seriousness of each in terms of the hazard-management cycle – mitigation, preparedness, response, and recovery.

b Re-draw Figure 3, the Park hazard-response curve, and explain how and why the response curve differs for these earthquakes.

Exam-style questions

AS 1 Explain the value of Park's hazard-response curve in understanding the management of the impacts of tectonic hazards. *(6 marks)*

A 2 Assess the usefulness of theoretical frameworks in understanding the prediction, impact and management of tectonic hazards. *(12 marks)*

Managing the impacts of tectonic hazards

In this section, you'll look at strategies used by key players to manage hazard events.

Managing hazard events

The sudden onset of many tectonic hazards, and the lack of reliable ways to predict them, highlights the importance of hazard management in protecting communities from the impacts of hazard events. Hazard management generally focuses on key players involved in:

◆ **hazard mitigation** – strategies meant to avoid, delay or prevent hazard events (e.g. land-use zoning, diverting lava flows, GIS mapping, and hazard-resistant design and engineering)

◆ **hazard adaptation** – strategies designed to reduce the impacts of hazard events (e.g. high-tech monitoring, crisis mapping, modelling hazard impacts, public education, and community preparedness).

Government hazard-mitigation strategies
1 Land-use zoning
Land-use zoning is a process by which local government planners regulate how land in a community may be used (e.g. as residential, industrial, or recreational). In areas at risk from volcanic eruptions and tsunami, land-use zoning is an effective way to protect both people and property. Those areas at risk, such as the land surrounding Mount Taranaki in New Zealand (see Figure 1), are divided into zones based on the likely type and level of damage from an eruption. Land-use planners, and others, use hazard maps like this to make decisions about the appropriate use of land in each zone, as well as for preparatory tasks such as determining safe evacuation routes.

In areas at high risk from volcanic eruptions and tsunami:

◆ any settlements tend to be limited, if they're allowed at all

◆ certain types of structures and facilities will be prohibited – such as those that pose a risk if damaged (e.g. nuclear power stations), or those that are critical for a community to function (e.g. hospitals)

◆ some communities may be resettled (e.g. people living along a coast deemed to be at risk of tsunami may be moved inland to higher ground)

◆ development in areas which provide natural protection will be limited (e.g. coastal mangrove forests, which act as buffers and reduce the impact of tsunami waves).

Land-use zoning is common in wealthy countries, but less so in some developing countries (see the Haiti information in Section 1.7). This is one reason why hazard events often cause more deaths and destruction in developing countries.

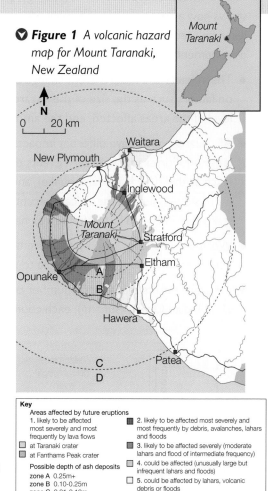

⊘ **Figure 1** A volcanic hazard map for Mount Taranaki, New Zealand

Key

Areas affected by future eruptions
1. likely to be affected most severely and most frequently by lava flows
☐ at Taranaki crater
☐ at Fanthams Peak crater

Possible depth of ash deposits
zone A 0.25m+
zone B 0.10-0.25m
zone C 0.01-0.10m
zone D < 0.01m

☐ 2. likely to be affected most severely and most frequently by debris, avalanches, lahars and floods
☐ 3. likely to be affected severely (moderate lahars and flood of intermediate frequency)
☐ 4. could be affected (unusually large but infrequent lahars and floods)
☐ 5. could be affected by lahars, volcanic debris or floods
☐ 6. unlikely to be affected by landslides, lahars or floods

2 Diverting lava flows
Historically, different methods have been attempted to divert lava flows away from people and communities. These methods have included building barriers and digging channels to try to divert the flows into safer directions. While these methods have led to some successes (e.g. barriers and channels successfully diverted a lava flow from the 1983 eruption of Mount Etna in Italy), in general, they've been fairly ineffective.

◆ The path taken by lava is hard to predict – making it difficult to know where to build the walls or dig the channels.

◆ The terrain has to be suitable (e.g. with a downward slope, so the diverted lava can easily flow away.

◆ Stopping the lava from flowing towards one community may push it towards another.

3 GIS mapping

GIS can be used in all stages of the disaster management cycle (see Section 1.10). For example, to identify where evacuation routes should be placed (the preparedness stage), or to help with rescue and recovery options (the response stage).

The GIS map shown in Figure 2 combines information about the 2015 Nepal earthquake (see Section 1.1), including:

◆ the locations and very rough population sizes of major towns and cities in Nepal

◆ the areas affected by the earthquake

◆ the locations of airports and airstrips.

Together, this information helped aid agencies to identify the areas most affected by the earthquake, and then to find the nearest location where aircraft or helicopters carrying emergency supplies and relief workers could land.

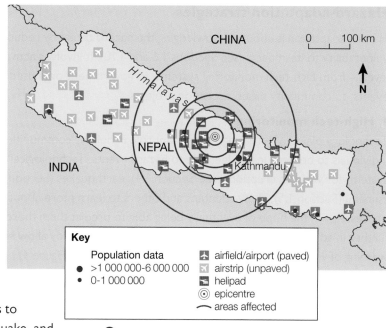

Key

Population data
- ● >1 000 000-6 000 000
- · 0-1 000 000

- ✈ airfield/airport (paved)
- ✕ airstrip (unpaved)
- ▣ helipad
- ◎ epicentre
- ⌒ areas affected

⬤ **Figure 2** *A GIS map showing information for relief operations after the 25 April 2015 Nepal earthquake and its aftershocks*

4 Hazard-resistant design and engineering defences

Collapsing buildings are one of the main causes of death and damage from tectonic hazards (see Section 1.7). Designing and constructing buildings that can withstand hazard events more effectively is key to protecting lives and property.

◆ New buildings and structures (such as bridges) can be designed to resist ground shaking during earthquakes (see Figure 3).

◆ The roofs of houses built near volcanoes can be sloped to reduce the amount of ash that builds up on them – and thus reduce the risk of them collapsing under the weight.

◆ Buildings at risk from tsunami can be elevated and also anchored to their foundations to stop them floating away.

◆ Existing buildings can be modified – called retrofitting – to make them safer, e.g. by strengthening the foundations.

◆ Protective structures, such as seawalls or retaining walls, can be built to stop or slow the impact of tsunami waves and landslides.

Much of the research into hazard-resistant design is undertaken by **engineers**. They study the impacts of tectonic events on structures and then develop ways to make them safer.

However, not all hazard-resistant design needs to be expensive and high-tech. In Pakistan, some houses have been built from bales of straw held together by strong plastic netting (sandwiched between layers of plaster). During an earthquake, the walls crack but they don't collapse.

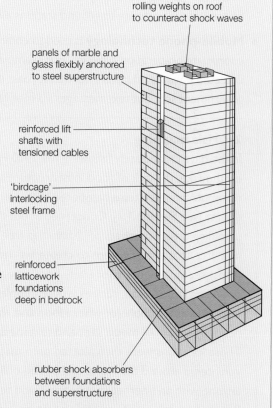

rolling weights on roof to counteract shock waves

panels of marble and glass flexibly anchored to steel superstructure

reinforced lift shafts with tensioned cables

'birdcage' interlocking steel frame

reinforced latticework foundations deep in bedrock

rubber shock absorbers between foundations and superstructure

⬤ **Figure 3** *A modern earthquake-proof building*

Hazard-adaptation strategies

Hazard adaptation is about acknowledging the hazard threat by reducing people's vulnerability to its impacts (see Sections 1.6 and 1.10). It involves activities at all levels – from high-tech monitoring systems, to government-organised preparedness events, to community based education projects.

1 High-tech monitoring

Technology plays an increasingly important role in helping communities and individuals to become less vulnerable to hazard events. Technological monitoring systems for volcanic eruptions (see Section 1.4), earthquakes (Section 1.3) and tsunami (Section 1.5), allow scientists and others to learn more about these natural processes – in the hope of eventually being able to predict them more accurately further in advance. In some cases, monitoring systems already allow some advanced warning of volcanic eruptions and tsunami to be given (see Figure 4).

◆ **GIS** helps to create hazard maps and manage hazards more effectively.

◆ **Early warning systems** use scientific instruments to detect signs that a volcanic eruption or tsunami is about to occur (see Sections 1.4 and 1.5). The relevant authorities can then be informed and rapid alerts issued to communities at risk.

◆ **Satellite-communication technology** helps to transmit the data from monitoring equipment, so that early warnings can be issued (e.g. the Indian Ocean Tsunami Warning System, by which scientific data collected from the seafloor is transmitted via satellite to ground stations every 15 seconds).

◆ **Mobile-phone technology** is used to communicate rapid warnings and coordinate preparation activities. For example, in 2011 – when seismographs detected P waves off Japan's northeast coast (see Section 1.3) – the Japanese government sent out text messages via mobile phone warning of the earthquake.

⬤ **Figure 4** *An early warning siren in Phuket, Thailand (set up as part of the Indian Ocean tsunami warning system after the 2004 tsunami), which can be used to alert residents and tourists about an approaching tsunami*

2 Crisis mapping

In the days immediately following the 2010 Haitian earthquake, the lack of good infrastructure and communication systems hampered rescue and aid efforts (see Section 1.7). Concerned about the unfolding crisis, members of Ushahidi (a free online resource that allows groups to create live interactive maps) set up a map site for Haiti. Local people began providing information, such as where people were trapped under rubble (or where food and water was needed), via social media sites and text messages. These locations were then plotted onto maps by volunteers around the world and placed online so that anyone with an Internet connection could see them. Rescue and aid workers quickly began to use these maps – which were constantly being updated – to decide how, when and where to direct resources.

This process, known as **crisis mapping**, uses **crowd-sourced** information (collected by volunteers in various locations) – as well as satellite imagery, other maps and statistical models – to accurately map areas struck by disaster. Now aid agencies are beginning to use crisis mapping *before* a disaster happens (by pre-mapping vulnerable areas around the world).

In Nepal, in the aftermath of the 2015 earthquake (see Section 1.1), volunteers riding bicycles with GPS trackers went to remote villages to collect data about the location of roads and tracks, the number of residents, and the quality of the buildings. This information is now being used to help build up a picture of the vulnerability of these communities, and will eventually provide the basis for projects to improve their infrastructure and disaster preparedness – as well as provide targeted help when the next earthquake occurs.

3 Modelling hazard impact

Computer models allow scientists to predict the impacts of hazard events on communities. Information is fed into computer systems, which then model the effects of a disaster (see Figure 5). They also allow scientists to compare the effects of different scenarios (e.g. the likely effects of a tsunami on a community if a seawall is built or not). These computer models can then be used by decision makers to help them develop plans and strategies to reduce the impact of hazard events and target resources more effectively.

4 Public education

Good education and better public awareness can help to reduce vulnerability and prevent hazards from becoming disasters. Public education helps people to understand what they can do to protect themselves before, during and after a hazard event.

It includes:

◆ regularly practising emergency procedures (e.g. in Japan children practise earthquake drills four times a year (see Section 1.7), and the Japanese government also holds an annual Disaster Prevention Day, in which over two million people regularly participate.

◆ encouraging households and workplaces to create emergency preparedness kits

◆ providing effective educational materials, such as information on constructing buildings to withstand earthquakes (see Figure 6).

Community preparedness and adaptation

Community based preparedness is becoming an increasingly important part of hazard management. People actually living in a community at risk from a natural hazard are often best placed to develop suitable preparedness plans and educate local residents. This is especially true in lower-income countries, where governments may not have the resources to invest heavily in disaster planning – or to reach all communities.

Local knowledge is an important part of community disaster preparedness. During the 2004 Indian Ocean tsunami, the elders of Thailand's Moken tribe (a small community of fishermen) noticed unusual movements in the Bay of Bengal. They immediately ordered villagers to run to the hilltop. Moments later the tsunami struck. As a result of the elders' actions, only one out of the 200 villagers died.

Community preparedness tends to be most effective when it's formalised, so that efforts can be ongoing and coordinated. For example, in some communities a committee is formed to develop plans, organise people and co-ordinate their efforts. The committees often liaise with local government, especially emergency planners, as well as schools.

Community preparedness activities usually include:

◆ creating a list of vulnerable people who may need special assistance (e.g. the elderly)

◆ organising practice evacuation drills

◆ providing first-aid courses.

In some developed countries, governments provide help to communities to develop preparedness plans and activities.

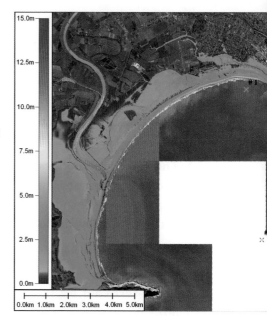

🔺 *Figure 5* *A computer model forecasting the depth of flooding from a tsunami generated by an undersea earthquake with a magnitude of 9.0 off the coast of New Zealand*

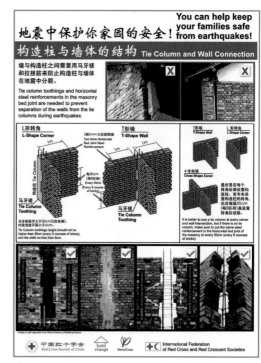

🔺 *Figure 6* *In China, posters like this show 'good practice' construction techniques to help local people to construct buildings that are better able to withstand earthquakes*

Key players in managing loss

During the recovery and response stages of the hazard-management cycle (see Section 1.10), efforts focus on helping communities to cope with personal, social and economic loss. This typically involves the delivery of aid, the efforts of communities to help themselves, and – in some cases – financial help from the insurance industry.

1 The role of aid donors

Most countries struck by disasters need some type of aid to help them recover and rebuild. Aid is generally divided into stages that follow the hazard-management cycle:

◆ **Emergency aid**, e.g. providing food, clean water, shelter.

◆ **Short-term aid**, e.g. restoring water supplies, providing temporary shelter.

◆ **Longer-term aid**, e.g. reconstructing buildings and infrastructure, redeveloping the economy and managing programmes to reduce the impact of future disasters.

Aid can be provided as cash, personnel, services or equipment. It can be distributed straight to the government of the affected country – which then uses it to manage the disaster recovery operation – or be controlled directly by aid agencies or foreign governments (as was the case in Haiti in 2010; see page 28). Many organisations provide aid, including governments (both that of the affected country and foreign governments), intergovernmental organisations (such as the United Nations), and non-governmental organisations (NGOs).

2 The role of non-governmental organisations

NGOs play a crucial role in disaster management. They are especially important in disasters where the local government is struggling to respond, or doesn't have the resources to do so (e.g. Haiti in 2010). They can provide funds, co-ordinate search-and-rescue efforts, and help to develop reconstruction plans.

Many NGOs are involved in all stages of the hazard-management cycle, and often remain in affected areas for years – helping communities to build up their resilience.

NGOs and the 2005 Pakistan earthquake

On 8 October 2005, a devastating 7.6 magnitude earthquake struck South Asia (see Figure 7). Pakistan and Pakistan-administered Kashmir were particularly badly hit:

◆ 73 000 people died, including 17 000 children.

◆ 128 309 people were injured, and 3.5 million were left homeless.

◆ Roads, water/sanitation facilities and communication systems were destroyed.

◆ The areas affected were largely mountainous, which made both search and rescue and longer-term reconstruction difficult.

NGOs responded immediately, by providing:

◆ over 500 000 tents and six million blankets (see Figure 8)

◆ safe water for over 700 000 people

◆ food and clothing

◆ emergency medical care.

After the immediate needs were met, short-term aid activities took over. For example:

◆ more permanent shelters were built

◆ water supplies were re-established

◆ roads closed by landslides were rebuilt or re-routed.

Figure 7 The location of the 2005 Pakistan earthquake

Over the next year, NGOs (such as the Red Cross and Red Crescent) continued to supply blankets, cooking kits and building supplies. Seeds and tools were also provided, so that families could start growing their own food again. Other NGOs (such as Oxfam) helped nearly 60 000 people rebuild their livelihoods by providing livestock and animal feed.

In 2007, most NGOs moved from relief operations into the recovery phase. Over the next five years:

◆ new schools, medical centres and homes were built

◆ community based disaster risk reduction programmes were developed.

3 The role of insurance in hazard management

Natural disasters are expensive – the economic costs can be staggering. And these costs are rising (see Section 1.8). In 2011, worldwide losses from earthquakes alone were US$54 billion.

Insurance coverage can help communities to recover from disasters. It provides individuals and businesses with the money they need to repair and rebuild. Yet, in many countries, few people have insurance for tectonic hazards. More pressing economic needs take priority over a hazard event that may not happen.

In some developed counties, such as Japan, governments and insurance companies work together to provide insurance for economic losses from disasters. However, these partnerships are either not available or are unaffordable in many developing countries.

4 The role of communities in managing loss

When a disaster strikes, it's local people who are the first to respond and who often play an important role in the community's recovery.

◆ They are crucial in the immediate search-and-rescue efforts (see Figure 9).

◆ In remote or isolated communities, it can take days or weeks for aid to arrive, so local people have to undertake the recovery steps themselves (e.g. creating temporary shelters or clearing debris from access roads). After an earthquake in Afghanistan in October 2015, villagers in mountain communities set up small groups to travel to the more remote areas to help with search and rescue.

◆ Community groups are also often involved in long-term strategies for rebuilding and improving resilience.

▲ **Figure 8** *Aid workers from an NGO speaking to a displaced family in a camp after the 2005 Pakistan earthquake*

▲ **Figure 9** *Local people clear rubble as they search for survivors after an earthquake in Afghanistan*

Over to you

1 a In pairs, draw a table of the advantages and disadvantages of the four hazard-mitigation strategies. Use examples of hazards throughout this chapter to illustrate your ideas.

b Go through each advantage and disadvantage, and score it on a scale from 1 (ineffective) to 5 (highly effective). Which strategies seem most effective, based on your scores?

2 Repeat questions 1a and 1b for hazard-adaptation strategies.

3 In class, prepare material for a debate motion that 'This house believes that hazard-mitigation strategies are always more effective than hazard-adaptation strategies'.

On your own

4 Distinguish between the following pairs of terms: (a) hazard mitigation and hazard adaptation, (b) GIS and crisis mapping, (c) public-education strategies and community preparedness.

5 Assess the roles of the four key players in managing loss in terms of where they are (a) most effective, (b) least effective.

Exam-style questions

 1 Assess the value of hazard-mitigation strategies. *(9 marks)*

 2 Evaluate the relative effectiveness of hazard-mitigation strategies. *(18 marks)*

Having studied Tectonic Processes and Hazards, you can now consider the three synoptic themes embedded in this chapter. 'Players' and 'Attitudes and Actions' were introduced on page 5. This page focuses on 'Futures and Uncertainties', as well as revisiting the three Enquiry Questions around which this topic has been framed (see page 5).

3 Futures and Uncertainties

People approach questions about the future in different ways. They include those who favour:

- **business as usual**, i.e. letting things stand. This might involve doing nothing, or only doing what's absolutely necessary when it's unavoidable. Governments often favour this approach, because it's cheaper.

- **more sustainable strategies**. In terms of tectonic hazards, this might involve questions about investment in technologies that either adapt or mitigate hazard threats (e.g. investment in a tsunami warning system). Scientists often favour these approaches.

> Working in groups, select two or three tectonic disasters from this chapter and then discuss the following questions in relation to each one:
>
> **1** What evidence is there that past decisions made by governments or planners have **(a)** worked, **(b)** not worked?
>
> **2** Is there evidence to suggest that more-sustainable strategies would help people either adapt to or mitigate hazard threats?

Revisiting the Enquiry Questions

These are the key questions that drive the whole topic:

1 Why are some locations more at risk from tectonic hazards?

2 Why do some tectonic hazards develop into disasters?

3 How successful is the management of tectonic hazards and disasters?

Having studied this topic, you can now consider answers to these questions.

Discuss the following questions in a group:

3 Consider Sections 1.1-1.5. What makes some locations at greater risk from tectonic hazards than others?

4 Consider Sections 1.6-1.8. How far are the factors that result in some tectonic hazards becoming disasters **(a)** physical or **(b)** human in nature? Explain your views.

5 Consider Sections 1.9-1.11. How good are people at managing tectonic hazards and disasters? Justify your views.

Books, music, and films on this topic

Books to read

1. *Richter 10* by Arthur C. Clarke and Mike McQuay (1996)

   A novel about how a young boy's life is crushed by the impacts of an earthquake, and how this inspires him to work as a seismologist to prevent a future earthquake from having such an impact.

2. *The day the island exploded* by Alexandra Pratt (2009)

   A novel based on a boy's quest for survival after experiencing a volcanic eruption and the destruction it caused.

3. *Volcano* by James Hamilton (2012)

   A factual book that assesses the impact of volcanoes on different places and cultures around the world.

Music to listen to

1. 'Earthquake' – Labrinth ft. Tinie Tempah (2011)

   This song describes the devastating effect that an earthquake can cause.

2. 'Volcano' – Jimmy Buffet (1979)

   Written about the uncertainty that an eruption of the Soufrière Hills volcano on the Caribbean island of Montserrat would cause to people's lives.

Films to see

1. *Pompeii* (2014)

   Inspired by the eruption of Mount Vesuvius in AD 79 that buried the Roman city of Pompeii, it looks at the terror caused by a volcanic eruption.

2. *San Andreas* (2015)

   A fictional film that portrays the destructive impact that a strong earthquake along California's notorious San Andreas fault line could have on nearby cities, such as Los Angeles and San Francisco.

3. *The impossible* (2012)

   Based on a survivor's experience of the 2004 Boxing Day tsunami in the Indian Ocean, which was caused by an undersea earthquake measuring over 9 on the Richter scale.

Chapter overview – introducing the topic

This chapter studies glaciated landscapes and the physical processes that form them. However, human activities are now changing these distinctive landscapes, which is threatening their future.

In the Specification, this topic has been framed around four Enquiry Questions:

1 How has climate change influenced the formation of glaciated landscapes over time?

2 What processes operate within glacier systems?

3 How do glacial processes contribute to the formation of glacial landforms and landscapes?

4 How are glaciated landscapes used and managed today?

The sections in this chapter provide the content and concepts needed to help you answer these questions.

Synoptic themes

Underlying the content of every topic are three synoptic themes that 'bind' or glue the whole Specification together:

1 Players

2 Attitudes and Actions

3 Futures and Uncertainties

Both 'Players' and 'Attitudes and Actions' are discussed below. You can find further information about 'Futures and Uncertainties' on the chapter summary page (page 94).

1 Players

Players are individuals, groups and organisations involved in making decisions that affect glaciated landscapes. They can be national or international individuals and organisations (e.g. inter-governmental organisations like the UN), national and local governments, businesses (from small operations to the largest TNCs), plus pressure groups and non-governmental organisations.

Players that you'll study in this topic include:

- Sections 2.11 and 2.12 – How **users** of glaciated landscapes (e.g. farmers and tourists) affect them, together with **traditional peoples** and **residents** of glaciated regions.

- Sections 2.11, 2.12 and 2.13 – How **environmentalists** and **climate scientists** are concerned about threats to biodiversity caused by climate change.

- Sections 2.11, 2.12 and 2.13 – How **national** and **international governments** try to manage those threats and the impacts of climate change through international agreements.

2 Attitudes and Actions

Actions are the means by which players try to achieve what they want. For example, climate scientists recognising the threats posed by climate change would seek international agreements to mitigate and adapt to its impacts.

Actions could pose the following questions:

1 **Economic versus environmental interests.** Faced with pressure from economic users, how can environmental damage to fragile landscapes be limited? Should such landscapes be exploited for their economic worth, or preserved? See Sections 2.1, 2.11 and 2.12.

2 **Resilience.** Faced with either human users, or more general threats from climate change, how far can the resilience of landscapes be developed to protect them? See Sections 2.11 and 2.12.

3 Faced with climate change and its impacts, how far should players impact upon natural systems either directly or indirectly? See Section 2.13 (plus the carbon and water cycles in Book 2).

In groups, consider and discuss two glaciated landscapes from this chapter where actions taken by governments (national and/or international), environmental or climate scientists have proved to be **(a)** effective, and **(b)** ineffective in the face of threats from different users. Explain your views.

Land of the polar bear

The Svalbard islands (which belong to Norway) lie in the Arctic Ocean, halfway between mainland Norway and the North Pole. As well as about 3500 polar bears, the islands are home to around 3000 people. Svalbard means 'cold coasts' – although, compared to other areas at the same latitude, the climate there is relatively mild. In Longyearbyen (the largest settlement on the islands) the average temperatures range from -14°C in winter to 6° in summer, although it can fall as low as -40°C.

About 60% of Svalbard is covered by ice (see Figure 1). There are over 2000 glaciers, and the island of Nordaustlandet has a large ice cap (the third largest in the world, after those covering Greenland and Antarctica). Much of the rest of the land is bare ground (rock, scree, moraines, fluvial deposits) – and only 10% is vegetated. **Permafrost** exists almost everywhere. The islands contain the largest wilderness area in Europe – untouched in places – which is both rugged and **fragile**.

⬆ **Figure 1** *Much of Svalbard is covered by ice*

Threats to Svalbard

Several human activities pose a threat to Svalbard's unique wilderness. They include: coal mining, scientific research and increasing tourism.

Coal mining

In the past, whaling and trapping were major economic activities in the islands – but Svalbard also has valuable mineral reserves. However – so far – little mineral extraction has taken place there (apart from coal mining). This is because mining is difficult in Svalbard, due to: the extreme cold; the long hours of winter darkness; challenging sea conditions affecting transportation to overseas markets; as well as the remoteness of the mines themselves. It wasn't until 1899 that the first Svalbard coal reached mainland Norway.

Most of the current coal mining takes place at Sveagruva (50 km south-east of Longyearbyen). The Norwegian state-owned mining company (Store Norske) employs about a third of all workers on Svalbard, and they extract high-quality coal. However, the company is now in economic and political difficulties – with job losses and calls from environmentalists to end mining on Svalbard. This would be disastrous for the local community, not least because the extracted coal supplies all of Svalbard's energy (see Figure 2).

Polar scientific research

Svalbard has a long history of polar scientific research, which involves studies of marine ecosystems, geology and meteorology. Norway, Russia and Poland all run permanent research stations on Svalbard.

A lot of current research is focused on analysing atmospheric changes that might be linked to climate change. The Arctic is expected to witness some of the most significant increases in temperature, and the impacts on ecosystems as well as physical systems (such as glaciers) are of great interest to scientists.

Key word

Permafrost – Where a layer of soil, sediment or rock below the ground surface remains almost permanently frozen.

Fragile environment – An environment susceptible to change and easily damaged.

⬇ **Figure 2** *The coal-fired power station at Longyearbyen supplies all of Svalbard's energy, but it's inefficient and needs upgrading or replacing*

Close to Longyearbyen is the SVALSAT receiving station, where huge antennae-studded 'golf balls' collect data from satellites orbiting the Earth (see Figure 3). Given Svalbard's high latitude, data can be collected quickly from passing satellites, which allows real-time modelling to take place.

Scientific research faces fewer regulations and restrictions than tourism or mining, but it does still result in environmental damage due to the associated infrastructure (such as the construction of research stations and access roads).

Tourism

Tourists have been visiting Svalbard for many years (usually by ship). However, since a new airport opened at Longyearbyen in 1975, tourist numbers have grown significantly. In 2013, 70 000 people visited Longyearbyen – 30 000 of whom were cruise ship passengers.

Figure 3 *SVALSAT – Svalbard satellite station, near Longyearbyen, Svalbard*

Most people visit Svalbard to explore the natural environment – the glaciers, fjords and the wildlife (polar bears, seals, walrus) – or to study the history of the islands. Adventure tourism is also becoming increasingly popular, with opportunities for hiking (see Figure 4), kayaking and snowmobile safaris. As a result, Longyearbyen has seen a significant growth in tourist facilities (such as hotels, shops, restaurants and tour operators).

However, the increasing numbers of tourists – as well as helping the local economy – also bring problems. These include: oil spills and waste discharges from shipping, air pollution from flights, and stress on wildlife and the fragile environment.

Protecting Svalbard

Figure 4 *Hiking in Svalbard*

Svalbard's economy does depend on the mining, research and tourism – but they all threaten the fragile environment. The Svalbard Environmental Protection Act came into effect in 2002 to protect the natural environment – its wilderness, flora, fauna – and the islands' cultural heritage. Two-thirds of Svalbard is now protected through national parks and nature reserves.

Over to you

1 Use Internet research to obtain six images of Svalbard focusing on its **(a)** landscape, **(b)** wildlife, **(c)** research stations, **(d)** traditional culture, **(e)** potential for mining, and **(f)** opportunities for tourism. Add brief descriptive notes to each.

2 Based on your annotated images, write a 400-word presentation about the economic, environmental, scientific and cultural value of Svalbard.

3 In pairs, design a mind map to explain why polar research, the protection of wilderness, mining, and tourism cannot easily coexist.

On your own

4 How far would you describe Svalbard's environment as fragile?

5 Write a 500-word summary about Svalbard, using the title 'To exploit or to preserve?'

6 Suggest how the relative importance of human activities on Svalbard may change in the future. Why might this happen?

In this section, you'll find out about the causes of longer- and shorter-term climate change, and how this has led to glacial periods.

The Pleistocene glaciation

Roughly every 200 – 250 million years in the Earth's history, there have been major periods of ice activity. The most recent and significant occurred during the **Pleistocene** epoch of the **Quaternary** period (see the text box on the right). Over the last 2 million years (since the start of the Quaternary and Pleistocene) – a blink of an eye in geological terms – temperatures on Earth have fluctuated considerably. This has led to cold periods (**glacials**) and warm periods (**interglacials**), as shown in Figure 1. In the last 1 million years, there may have been as many as ten glacial periods – separated by interglacials.

During the glacial periods, the climate cooled sufficiently for precipitation to fall as snow, rather than rain. This resulted in the formation and growth of huge ice masses, which – in the Northern Hemisphere – spread south over large parts of Europe, Asia and North America. During the warmer interglacial periods (some of which were considerably warmer than the conditions we experience today), much of the ice melted and the ice sheets and glaciers retreated. Figure 2 shows how temperature and ice volume has fluctuated in Antarctica over the last 450 000 years.

Look at Figure 3. It shows the approximate distribution of ice at its maximum extent about 20 000 years ago (towards the end of the Pleistocene). Vast ice sheets covered much of North America and Europe. There were also extensive glaciers and ice caps in South America and in mountainous regions. With so much water locked up as ice on the land, the global water cycle was significantly different and sea levels fell by 130 metres!

At this map's small scale, it looks as though the entire land area of the UK was covered by ice – but this was not the case. Although ice extended over much of the UK, the southern-most regions stayed ice-free (see Figure 7), despite being frozen and experiencing **periglacial** conditions (see page 57).

Geological time is divided into eras (e.g. Cenozoic), periods (e.g. **Quaternary**) and epochs (e.g. **Pleistocene** and **Holocene**). The Quaternary period began around 2 million years ago. The Pleistocene epoch lasted about two million years and ended around 11 700 years ago.

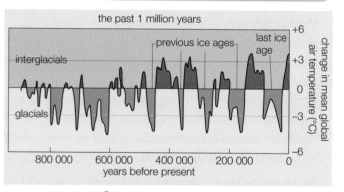

⬆ **Figure 1** General trends in mean global temperatures over the last 1 million years

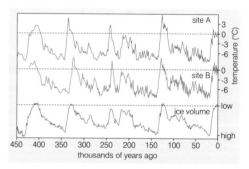

The two temperature lines (sites A & B) were constructed by measuring *deuterium* isotopes (heavy hydrogen) in ice cores in two different locations. The ice volume was reconstructed using measurements of *foraminifera* (microscopic plants) buried in ocean sediments.

⬆ **Figure 2** Temperature and ice volume fluctuations in Antarctica

◀ **Figure 3** The maximum extent of global ice coverage during the last glacial, about 20 000 years ago

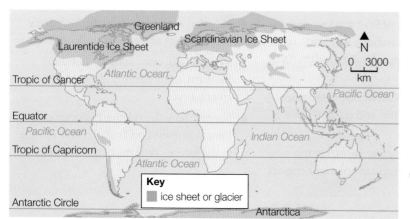

Pleistocene characteristics

The Pleistocene glaciation had three main characteristics:

1 It wasn't just a single ice age. Over the 2 million or so years during which it lasted, temperatures fluctuated enough to allow a number of ice advances and retreats.

2 The extent of the ice advance during each glacial was different.

3 There were also fluctuations within each major glacial. These relatively short-lived pulses of ice advance are known as **stadials**, and the warmer periods of retreat as **interstadials**. The most recent stadial in the British Isles was the Loch Lomond stadial (see page 52), which was completed in 1000 years.

It's possible to draw two main conclusions from the above characteristics:

◆ Medium- and large-scale glacial erosional landforms (such as those found in the Alps) are likely to be the result of several glacial advances.

◆ Depositional features tend to be the result of conditions and processes at work during the most recent glacial-interglacial cycle.

Long-term factors leading to climate change and glaciation

There are a number of theories about past climate changes, and – unlike present-day climate change – they are all natural.

Milankovitch cycles

Milutin Milankovitch was a mathematician and astronomer. Between 1912 and 1941, he carried out calculations which showed that the Earth's position in space, its tilt and its orbit around the sun, all change. These changes, he claimed, affect the amount of incoming solar radiation – and where it falls on the Earth's surface. They also produce three main cycles of 100 000, 41 000 and 21 000 years (see Figure 4). These cycles would be sufficient to start an ice age – or end one – and are called **Milankovitch cycles**.

▶ *Figure 4* *Milankovitch cycles*

a 100 000 year cycle
The Earth's orbit stretches from nearly circular to elliptical and back in a cycle of about 95 000 years. During the Quaternary, the major glacial-interglacial cycle was almost 100 000 years. Glacials occur when the orbit is almost circular, and interglacials when it is more elliptical.

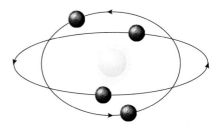

b 41 000 year cycle
It takes about 41 000 years for the Earth's axis to tilt, straighten and tilt again. When the tilt increases, summers become hotter and winters colder – leading to conditions favouring interglacials.

c 21 000 year cycle
As the Earth slowly wobbles in space, its axis describes a circle once every 21 000 years.

1 At present, the orbit places the Earth closest to the sun in the Northern Hemisphere's winter, and furthest away in summer. This tends to make winters mild and summers cool.

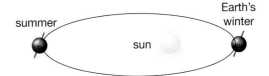

2 12 000 years ago, the position was in reverse.

Variations in solar output

Over 2000 years ago, Chinese astronomers began recording sunspots – 'flares' on the sun's surface which indicate that the sun's radiation is more active than usual. The sun emits variable amounts of radiation in the form of such flares; high levels of sunspots increase emissions, which increase the Earth's average temperature (see Figure 5). Changes caused by sunspots are small – generally between +0.5 and −0.5 °C globally. There are longer-term cycles, when sunspots seem to disappear almost completely, such as the period known as the Maunder Minimum between 1650–1700, which is believed to have caused the Little Ice Age. Similarly, warmer periods (e.g. the Medieval Warm Period) may have been caused by increases in sunspot activity.

Volcanic eruptions

Volcanic eruptions have the power to change the Earth's climate – but they need to be big and explosive. Eruptions produce ash and sulphur dioxide gas. If these rise high enough into the atmosphere, they will be spread around the stratosphere (10–50 km above the Earth's surface) by high-level winds. The blanket of ash and gas will then reflect solar energy back into space – and prevent it from reaching the Earth's surface. The temperature at the Earth's surface will then fall and the planet will cool.

In 1815, the volcano of Tambora in Indonesia erupted. It was the largest volcanic eruption in human history. As a result, temperatures around the world in the following year cooled so much that it was called 'the year without a summer' – and up to 200 000 people died in Europe alone as harvests failed. The effects of this eruption lasted for four or five years. Generally, though, volcanic eruptions only affect the climate for short periods.

Shorter-term climate events

The Loch Lomond stadial

A rapid drop in average temperature around 115 000 years ago triggered the Devensian period. As temperatures fluctuated, at least three pulses of ice advance and retreat occurred during this period. The most extensive occurred between 26 000 and 10 000 years ago (see Figure 6). Climate fluctuations caused two stadials of ice advance, separated by an interstadial.

The Loch Lomond stadial shows how ice accumulates and spreads in response to climatic conditions. It was sufficiently cold for an ice cap to develop over the uplands of Western Scotland (see Figure 7), and for small tongues of ice to flow from the deeper corries of Southern Scotland, the Lake District and Snowdonia.

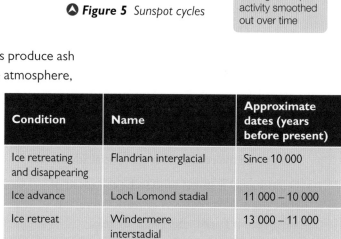

▲ **Figure 5** *Sunspot cycles*

Condition	Name	Approximate dates (years before present)
Ice retreating and disappearing	Flandrian interglacial	Since 10 000
Ice advance	Loch Lomond stadial	11 000 – 10 000
Ice retreat	Windermere interstadial	13 000 – 11 000
Ice advance	Dimlington stadial	26 000 – 13 000

▲ **Figure 6** *The most recent glaciation in the British Isles*

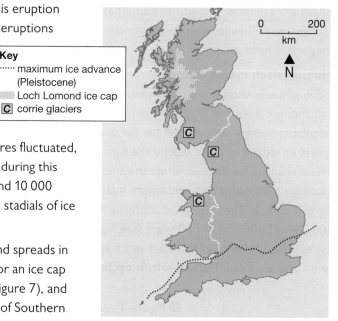

Key
...... maximum ice advance (Pleistocene)
▨ Loch Lomond ice cap
C corrie glaciers

▲ **Figure 7** *The Loch Lomond stadial*

The Little Ice Age

The Little Ice Age (during the **Holocene**) was a period of cooling which occurred after the Medieval Warm Period (roughly 950–1250). The Little Ice Age lasted from about 1550–1850, although some define it as starting as early as 1300. (Climatologists and historians don't agree about the start and end dates, because conditions varied in different places.) Different causes have been suggested for the Little Ice Age, including: volcanic activity (although climate change on a timescale of hundreds of years and 1–2°C can't be explained solely by volcanoes) and also low levels of solar radiation (caused by a lack of sunspot activity).

The Little Ice Age had a number of effects. Most were felt in Europe and North America.

◢ **Figure 8** *A frost fair held on the River Thames in London in 1683*

◆ It brought colder winters. For example, canals and rivers in the British Isles and the Netherlands were often frozen so hard and so deep that they could support ice skating and entire winter festivals (see Figure 8).

◆ Farms and villages in the Swiss Alps were destroyed as glaciers advanced. The Mer de Glace (sea of ice) in France extended to the floor of the Valle de Chamonix – and threatened to engulf the outer parts of the town (see Figure 9). Since then, the glacier has retreated by 2300 metres.

◆ Sea ice extended out from Iceland for miles in every direction – closing its harbours to shipping.

◆ Iceland's cereal crops also failed, and its people were forced to change from a grain-based diet.

◆ Greenland was largely cut off by ice from 1410 until the 1720s.

◆ Crop practices across Europe had to change to adapt to a shorter growing season – and there were many years of famine.

Key
- present
- 2001
- 1939
- 1842
- 1644
- built-up area

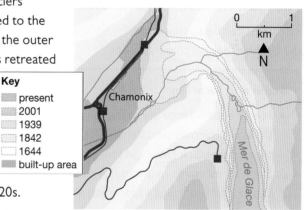

◢ **Figure 9** *Past positions of the Mer de Glace since 1644*

Over to you

1 Explain the evidence base by which we know that climate has changed in the past.

✦ 2 a Study Figures 1 and 2. Quoting evidence, write a brief summary comparing glacial and interglacial periods during the last 450 000 years.

    b Use Figure 2 to analyse the links between temperature changes and volumes of ice.

    c Based on the evidence shown in Figure 2, explain the link between the map in Figure 3 and the change to global temperatures shown in Figure 1.

    d How certain does the link seem to be between climate change and the landform evidence for past glacial/interglacial periods?

3 Based on the evidence of the Holocene (Little Ice Age), assess the likely impacts if a cooler period were to occur on the Earth now.

On your own

4 Distinguish between the following pairs of terms: **(a)** Pleistocene and Ice Age, **(b)** glacial and interglacial, **(c)** stadial and inter-stadial, **(d)** Pleistocene and Holocene, **(e)** Milankovitch cycles and sunspots.

5 Why is it important to understand past climates when trying to explain the formation of present-day landscapes?

6 Research a detailed study of one glacier and its fluctuations in flow since the end of the last Ice Age. Use www.grid.unep.ch/glaciers as a starting point.

Exam-style questions

 1 Explain two pieces of evidence to show that climate has changed in the last 450 000 years. *(4 marks)*

 2 Explain two long-term factors that can lead to climate change. *(6 marks)*

The cryosphere

Some places on Earth are so cold that water is permanently frozen there – stored as snow or ice. These places make up the **cryosphere**, which is the frozen part of the Earth's hydrological system.

- The cryosphere's largest element is made up of land surfaces – including the ice sheets found in Greenland and Antarctica, as well as ice caps, glaciers, and areas of snow and permafrost.

- The remainder is made up of the frozen areas of oceans, lakes and rivers, which occur mainly in polar or mountainous regions.

Together, these areas act as **stores** within the global hydrological cycle.

The components of the cryosphere play a vital role in the Earth's climate. Snow and ice reflect heat from the sun – called the **albedo effect** – which helps to regulate temperatures on Earth. Polar regions are among the most sensitive to temperature change, so the cryosphere is an important focus for climate scientists researching global climate change.

The classification of ice masses

Ice masses can be classified in different ways – by scale and location, and by their thermal characteristics.

Scale and location

- **Ice sheets** – These are vast expanses of ice, often over 1 km thick, which cover land surfaces. Antarctica is the largest of the two ice sheets. It covers 14 million km² and stores 90% of the Earth's freshwater. In places it is 4 km thick! Ice sheet margins can extend out to sea to form **ice shelves**.

- **Ice caps** – These are smaller masses of ice that are often associated with mountain ranges. Europe's largest ice cap is Vatnajokull in Iceland, which is over 800 km² in area and about 1 km thick.

- **Glaciers** – Most glacial landscapes are created by the movement of glaciers. There are two main types of glacier. **Cirque** or **corrie glaciers** are small and occupy armchair-shaped hollows on mountains. These may overspill to feed **valley glaciers**, which are larger masses moving from ice fields or corries and following river courses.

- **Ice fields** – These are areas of less than 50 000 km². They are extensive regions of interconnected valley glaciers. High peaks called nunataks rise above them (see Figure 1).

Thermal characteristics

- **Temperate glaciers** are **warm-based glaciers**. Water, which acts as a lubricant, is found throughout the ice mass – allowing the ice to move freely and erode the rock. The base of the glacier is at about the same temperature as the **pressure melting point**. Temperate glaciers move between 20 and 200 metres a year, but can move up to 1000 metres.

- The ice in **polar** or **cold-based glaciers** remains frozen at the base (because the base is much colder than the pressure melting point temperature), so there is little water or movement and very little erosion. Polar glaciers may advance only a few metres a year.

Key word

Pressure melting point – The temperature at which ice is on the verge of melting.

◀ **Figure 1** The Southern Patagonian ice field

The distribution of cold environments

Most of the world's cold environments are located in the far Northern Hemisphere. This reflects the latitudinal position of the landmasses – in the Southern Hemisphere, the equivalent latitudes coincide with ocean rather than land.

There are four main types of cold environment:

- **Polar (high latitude) regions** – areas of permanent ice (essentially the vast ice sheets of Antarctica and central Greenland), inside the 66.7° latitude of the Arctic and Antarctic Circles.

- **Periglacial (tundra) regions** – literally speaking, at the 'edge' of permanent ice. These are characterised by permanently frozen ground (**permafrost**), and include large tracts of northern Canada, Alaska, Scandinavia and Russia. These regions vary between areas that are permanently frozen and those that thaw in summer.

- **Alpine/mountain (high altitude) regions** – for example, the European Alps, Himalayas, northern Rockies and Andes – where high altitudes result in cold conditions. It's in these high altitudes that glaciers and glaciated landscapes are found.

- **Glacial environments** – found at the edges of the ice sheets and, in particular, in the highest mountainous regions, e.g. the Himalayas and southern Andes.

Evidence for the Pleistocene ice sheet

The UK's **relict glaciated landscapes** provide evidence (see Figure 2) that much of the country was covered by an ice sheet during the Pleistocene.

> **Depositional evidence** – drumlins (e.g. in the Vale of Eden, Cumbria), erratics (e.g. the Bowder Stone in the Lake District), moraine (in the Cairngorms). See Section 2.9 for further details about depositional landforms.

> **Erosional evidence** – found in the Cairngorms (Scotland), Snowdonia (Wales), and the Lake District (England). It includes: corries, arêtes and glacial troughs, along with roches moutonnées, crag and tail, and knock and lochan landscapes. See Section 2.8 for further details about erosional landforms.

> **Meltwater evidence** – meltwater channels (e.g. Newtondale, North Yorkshire), glacial till (e.g. the Holderness coast), eskers (e.g. Blakeney, Norfolk). See Section 2.10 for further details about landforms created by meltwater.

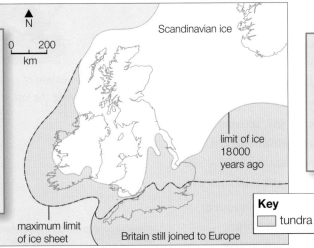

Scandinavian ice

limit of ice 18000 years ago

maximum limit of ice sheet

Britain still joined to Europe

Key
▢ tundra

◀ **Figure 2** *UK evidence for the Pleistocene ice sheet*

Over to you

1. a On a blank world map outline, use an atlas to identify, shade and name the four types of cold environment.
   b Use Internet research to find a located photograph of each environment.

2. a Compare your completed labelled map with Figure 3 in Section 2.2. Identify key similarities and differences.
   b Google the phrase 'extent of ice in last Ice Age' and compare the global coverage of ice then with that shown on your world map.

On your own

3. Distinguish between the following groups of terms: **(a)** ice sheet and glacier, **(b)** glaciers, valley glaciers and ice fields, **(c)** temperate and polar glaciers, **(d)** polar, periglacial and alpine regions.

4. Using a map of the UK, locate and shade the three types of landscape evidence still present for Pleistocene ice sheets.

Exam-style questions

AS 1 Compare the thermal characteristics of temperate and polar glaciers. *(3 marks)*

A 2 Assess the contribution of ice cover during the Pleistocene to landscapes in the UK. *(12 marks)*

What are periglacial landscapes?

Periglacial landscapes are found on the fringes of polar glacial environments. While they do not have a permanent covering of ice, they experience extreme cold for much of the year – with penetrating frosts and periodic snow cover – and they are underlain by **permafrost** (see below). Areas affected by permafrost today cover about 25% of the world's total land area, but in the past they were much more extensive – see Figure 1, which shows the extent of permafrost in the Northern Hemisphere during the last glacial maximum (26 500 – 19 000 years before present). It covered a significantly larger area than that covered by permafrost today. If you look at Figure 2 in Section 2.3, you can see the extent of tundra (and hence permafrost) across Britain 18 000 years ago.

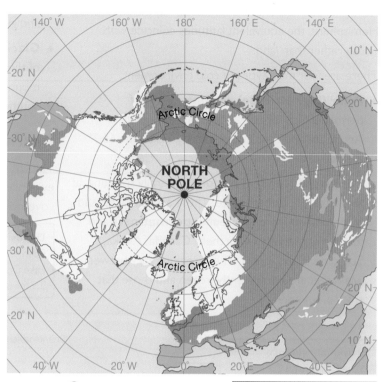

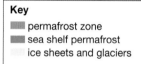

Figure 1 *The extent of permafrost during the last glacial maximum*

Key
- permafrost zone
- sea shelf permafrost
- ice sheets and glaciers

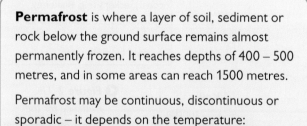

Permafrost is where a layer of soil, sediment or rock below the ground surface remains almost permanently frozen. It reaches depths of 400 – 500 metres, and in some areas can reach 1500 metres.

Permafrost may be continuous, discontinuous or sporadic – it depends on the temperature:

Type of permafrost	Temperature range
Sporadic	0°C to -1.5°C
Discontinuous	-1.5°C to -5°C
Continuous	-5°C to -50°C

Cycles of intense freezing and thawing mean that surface sediments thaw in summer – resulting in a saturated and possibly mobile seasonal **active layer**. Figure 2 shows deep permafrost in northern Alaska. Notice that there is a thin layer of brown unfrozen soil above the permafrost. This is the active layer. Drainage is prevented by the permafrost, so the active layer will often become saturated and boggy.

Figure 2 *Permafrost in periglacial soils in northern Alaska, USA*

What are periglacial landscapes like?

Periglacial landscapes – such as those found in Alaska and parts of Northern Europe, Russia and Canada – are typically wide expanses of largely featureless plains, either strewn with rocks (blockfields) or covered with low growing, marshy vegetation (see Figure 3). Lakes or streams are common in the summer, when the snow has melted and some of the permafrost has thawed.

The environment is characterised by tundra vegetation, which consists of low-growing plants such as mosses, lichens, grasses, sedges and dwarf shrubs. Their small, waxy leaves are well adapted to reduce water loss caused by exposure to strong winds. They flower and seed in just a few weeks. Tundra plants also have to cope with thin soils and waterlogging on flat, poorly drained land.

Thaw lakes are also common (see Figure 4). They form during the summer, when any lying snow melts (together with the thin active layer of thawed topsoil). Water retains heat, and its relatively dark surface also absorbs radiation from the sun. This warmth increases the depth of thawing of the underlying permafrost – forming unfrozen zones called **taliks**.

Geomorphological processes

Geomorphological processes are those that result in the modification of landforms on the Earth's surface – they 'shape the Earth' ('geo' means 'earth' and 'morph' means 'shape'). Although cold environments may appear inactive and barren, the geomorphological processes that act upon them can be surprisingly vigorous and effective.

Geomorphological processes are most active at the margins of cold environments – where precipitation amounts are high, liquid water is readily available, and temperatures hover above and below freezing.

Look at Figure 5. This photo shows proglacial lakes at the snout of Grimsvotn Glacier in Iceland. Notice the vast deposits of glacial debris in the foreground, which form hills and ridges. Rivers flowing from the lakes actively erode the landscape in the summer. The exposed mountains in the distance, with their covering of snow, will be gradually broken down and shaped by processes of **weathering** and mass movement. This periglacial landscape is clearly being actively shaped by geomorphological processes.

▲ **Figure 3** A periglacial landscape in the Northwest Territories of Canada

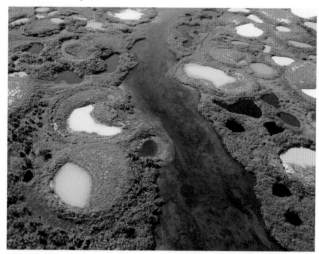

▲ **Figure 4** Thaw lakes north of Quebec, Canada

▼ **Figure 5** Active geomorphological processes at the snout of Grimsvotn Glacier, Iceland

Key word

Weathering – The breakdown or disintegration of rock *in situ*, at or just below the ground surface.

Periglacial processes and landforms

Geomorphological processes in periglacial landscapes are mostly associated with frost, ice and snow – together with meltwater. Figure 6 lists the processes and the landforms they create in periglacial areas. See also Figure 7.

	Processes	Landforms
Frost	Frost shattering / Freeze-thaw weathering	Blockfields, talus (scree)
Snow	Nivation	Nivation hollows
Ground ice	Ice crystals and lenses (frost-heave)	Patterned ground, sorted stone polygons
Ground ice	Ground contraction	Ice wedges
Ground ice	Groundwater freezing	Pingos
Meltwater	Solifluction	Solifluction lobes
Meltwater	Meltwater erosion	Braiding
Wind	Windblown	Loess

⊙ **Figure 6** *Periglacial processes and landforms*

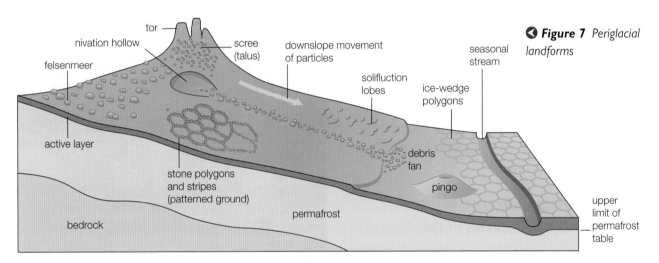

⊙ **Figure 7** *Periglacial landforms*

Frost action

Frost action is one of the most important processes in a periglacial environment. **Frost shattering** (also known as freeze-thaw) commonly affects bare rocky outcrops high on a mountainside. The process begins when water (rain or meltwater) seeps into cracks and holes (pores) within a rock. When the temperature falls to 0°C or below, the water turns to ice and expands by about 9%. This exerts stresses within the rock – enlarging cracks and pores. As the process of freezing and thawing is repeated, the cracks become larger until chunks of rock break away and pile up as scree at the foot of the slope.

Frost shattering can create extensive areas of broken up angular fragments of rock. These areas are called **blockfields** or **felsenmeer** (see Figure 8).

⊙ **Figure 8** *Blockfield (felsenmeer) developed on granite bedrock, Manitoba, Canada*

Snow

Nivation is an umbrella term used to cover a range of processes associated with patches of snow (see Figure 9). These processes are shown in Figure 10, and are most active around the edges of snow patches.

♦ Fluctuating temperatures, and the presence of meltwater, promote frost shattering.

♦ Summer meltwater will carry away any weathered rock debris, to reveal an ever-enlarging **nivation hollow**.

♦ Slumping may also take place during the summer, as saturated debris collapses due to the force of gravity.

As long as the freshly weathered material is removed by meltwater (or mass movement processes), the nivation hollow will continue to be enlarged. If climatic cooling takes place, the hollow will eventually be occupied by glacial ice – and it may then become enlarged to form a corrie (see page 71).

Ground ice

In extremely low temperatures, the ground **contracts** and cracks develop. During the summer, meltwater fills these cracks and then freezes in the winter to form **ice wedges**, which increase in size through repeated cycles of freezing and thawing (see Figure 11). These also affect the ground surface, by forming narrow surface ridges due to **frost heave** (see the text box on the right). In the summer, ponds of meltwater can then form in the hollows between these ridges (see Figure 12).

As ice wedges become more extensive, a polygonal pattern may be formed on the ground, with the ice wedges marking the sides of the polygons – **patterned ground** (see Figure 7).

Stone polygons are a feature of patterned ground, and are directly associated with ice wedges. Frost heave causes the ground to expand and lift soil particles upwards. Smaller particles may then be removed by wind or meltwater – leaving a concentration of larger stones lying on top of the ice wedges (marking out the polygonal pattern).

However, sloping ground can distort the polygons as the stones gradually slide or roll downslope – leading to the formation of **stone stripes** rather than polygons (see Figure 7).

○ *Figure 9* *Nivation snow patches*

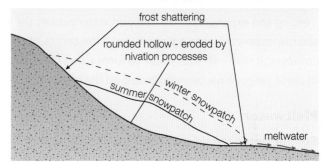

● *Figure 10* *Nivation processes that result in the formation of a nivation hollow*

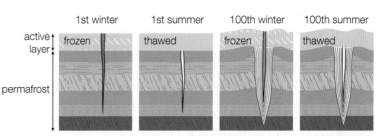

Ice can form within pores (**pore ice**) or as **ice needles** in soil and sediment. This can then force individual soil particles, or small stones, upwards to the surface. This process is known as **frost heave**.

● *Figure 11* *The formation of ice wedges*

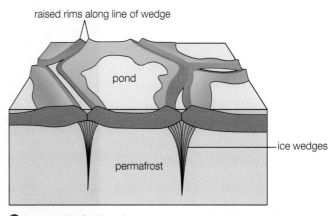

● *Figure 12* *Surface features associated with ice wedges*

Pingos

The freezing of water in the upper layers of the soil, where the permafrost is thin or discontinuous, leads to the expansion of ice within the soil. This causes the overlying sediments to heave upwards into a dome-shaped feature which is less than 50 metres high – known as a **pingo**. These dome-shaped hills may be 500 metres across, and are found mainly on sandier soils. This type of pingo is called an **open system** or East Greenland type.

Closed system, or Mackenzie type, pingos are typical of low-lying areas with continuous permafrost. On the site of small lakes, groundwater can be trapped by freezing from above and by the permafrost below as it moves inwards from the lakeside. Subsequent freezing and expansion of the trapped water pushes the overlying sediments upwards into a pingo form. If the centre of the pingo then collapses, it may infill with water to form a small lake. Over a thousand of these pingos have been recorded on the Mackenzie Delta in Canada.

Figure 13 A pingo in the Mackenzie Delta in Canada

Meltwater

Solifluction is the slow downhill flow of saturated soil. It is a common process in periglacial environments, where the active layer provides enough water to allow flow to occur. As the saturated soil slumps downhill during the summer it forms **solifluction lobes** (see Figure 14).

> **Key word**
>
> **Mass movement** – The downward movement of material under the influence of gravity. It includes a wide range of processes such as rockfalls, landslides, mudflows and also **solifluction** (soil flow).

Meltwater can also form **braided rivers** as it flows across glacial outwash plains (see Figure 15 and page 79)

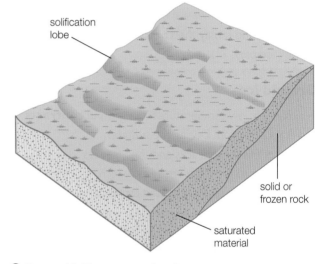

Figure 14 The process of solifluction

Figure 15 A braided river in Auyuittuq National Park, Baffin Island, Canada

Wind

A lack of vegetation and a plentiful supply of fine, loose material (i.e. silt) in glacial and periglacial environments, enables strong, cold winds to pick up large amounts of material and redeposit it far away from its source as **loess**. It covers large areas in the Mississippi-Missouri valley in the USA, as well as in north-west China (where in places the loess is over 300 metres deep).

Canada – periglacial landscapes

The largest island in the Arctic Archipelago is Canada's Baffin Island. Baffin Bay and Davis Strait on the eastern side of the island are often open during the summer, but the western side is usually closed by ice all year round. It is typical of a periglacial landscape, and has continuous permafrost.

Baffin Island has a number of ice caps – the largest being the Penny and Barnes ice caps. Its topography ranges from rugged mountains to flat lowlands. The coastal strip between the two northern peninsulas and the Foxe peninsula is littered with lakes and ponds fed by run-off streams.

Like all periglacial areas, Baffin Island has a tundra ecosystem. Globally, there are three different types of tundra – Arctic, Antarctic and Alpine – and all have similar conditions. In Canada, the southern boundary of the tundra extends from the Mackenzie Delta to southern Hudson Bay, and then northeast to Labrador.

Winters in Baffin Island are long, dark and cold. However, during the short summer, the snow and soil layers above the permafrost (the active layer) melt. This creates and feeds a vast network of lakes, streams, rivers and wetlands (see Figure 17). The waterlogged soil and 24-hour sunshine during the summer period boosts rapid plant growth (with densely packed low-lying plants growing in the tundra's lower latitudes).

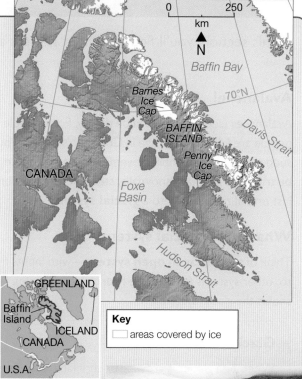

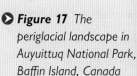

▲ **Figure 16** The Baffin Island area

Key
☐ areas covered by ice

▶ **Figure 17** The periglacial landscape in Auyuittuq National Park, Baffin Island, Canada

Over to you

1 a In pairs, list in a table the distinctive **(i)** weathering and **(ii)** mass movement processes in periglacial regions.

b Select any two of each and develop a PowerPoint presentation to explain them.

c Find photos of four landforms or landscapes that result from these processes and add them to your presentation.

2 a Make your own copy of Figure 11, showing the development of ice wedges. Write detailed labels to describe the processes involved.

b Explain why ice wedges change over time, as well as the effect they have on the surface landforms.

3 In pairs, devise simple labelled sketches to explain how patterned ground forms.

4 Use Figure 14 to create a labelled diagram to show how solifluction operates on a slope.

On your own

5 Distinguish between the following pairs of terms: **(a)** periglacial and glacial, **(b)** permafrost and active layer, **(c)** weathering and mass movement, **(d)** active layer and meltwater, **(e)** felsenmeer and scree, **(f)** pingos and ice polygons.

6 Describe the typical characteristics of a periglacial landscape. Consider over what timescale and in what ways this landscape is likely to change.

7 Study Figure 13. Make a careful sketch of the pingo in the photograph. Write annotations to describe its characteristics and likely mode of formation.

Exam-style questions

AS 1 Explain two ways in which periglacial processes have contributed to upland landscapes. *(6 marks)*

A 2 Assess the contribution of periglacial processes to upland landscapes. *(12 marks)*

In this section, you'll find out about glaciers as systems, as well as glacial mass balance.

Avalanche!

Avalanches (see Figure 1) can be deadly. For example, every year 27 people are killed on average in Colorado's Rocky Mountains by the sudden, massive downhill movement of snow and ice. Avalanches can be triggered by heavy snowfalls, steep slopes and vibrations – and are just one of the inputs into a **glacial system**.

What is the glacial system?

Think of a glacier as an **open system** – with inputs from, and outputs to, other systems (such as the hydrological system). Figure 3 (opposite) shows the glacial system.

△ *Figure 1* An avalanche at Val d'Isere in the French Alps

Glacial mass balance

Whether a glacier is growing or shrinking, depends on the balance between accumulation and ablation. The year-to-year change in this ice budget is known as the **mass balance** of the ice. It's calculated by dividing the glacier into two zones:

◆ The **accumulation zone**, where there is a net *gain* of ice over the course of a year. Here inputs exceed outputs.

◆ The **ablation zone**, where there is a net *loss* of ice during a year. Here outputs exceed inputs.

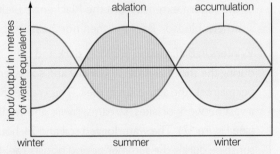

△ *Figure 2* Seasonal variations in the mass balance

The mass balance varies over the course of a year (see Figure 2). Ablation is at its highest during the summer (due to rapid melting of the ice). However, during the winter, higher amounts of snowfall and limited melting result in accumulation exceeding ablation.

Why do glaciers advance and retreat?

Glaciers advance and retreat in response to long-term trends in the mass balance:

◆ If *accumulation* exceeds *ablation* (leading to a positive mass balance), the glacier's mass increases and it advances.

◆ If *ablation* exceeds *accumulation* (leading to a negative mass balance), the glacier's mass decreases and it retreats.

The advance or retreat of a glacier is usually extremely slow. It is recorded annually – or even over decades – by the position of the glacier's snout. The snout is often accompanied by a terminal moraine (see page 74), which is pushed ahead of an advancing glacier. If the glacier then retreats, the terminal moraine will be left behind on the valley floor.

Changes in a glacier's mass balance affects how it works and creates landforms:

◆ If the climate cools, the ice thickens, moves faster and advances. It then erodes and transports more vigorously.

◆ However, during warmer periods, a glacier shrinks – the ice becomes thinner and retreats. Its movement slows and it erodes and transports less debris – and deposits more.

Changes in mass balance can occur year-on-year (with variations in ablation and accumulation), as well as over decades and much longer periods. For example, the Rhone Glacier in France has progressively retreated over the last 50–80 years; and ice sheets and glaciers in the last glacial retreated over much longer periods.

Inputs

- The main *input* is direct **precipitation** in the form of snowfall. As this snow is increasingly compacted over many years, it turns from low-density white ice crystals (snowflakes) to high-density clear glacial ice.
- Avalanches from mountainsides.
- Wind deposition – strong winds at high altitudes blow snow onto the glacier.

Outputs

- The main *output* is water, which results from melting close to the glacier's snout (where temperatures are higher).
- Where the ice front extends over water (e.g. ice shelves in Antarctica), huge chunks of ice may break off to form icebergs. This process is called **calving**.
- The processes of evaporation and **sublimation** also act as outputs.

Key word

Sublimation – The change from the solid state (ice) to gas (water vapour) with no intermediate liquid stage (water).

Energy

A glacier's mass combines with the force of gravity to generate potential energy. As the glacier moves, this movement is converted into kinetic energy, which enables the glacier to carry out the processes of *erosion, transportation* – and ultimately *deposition*. Meltwater facilitates the conversion of the potential energy into kinetic energy (work).

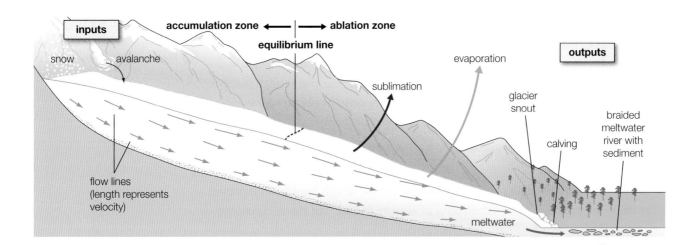

Stores/components

The main *stores* are snow and ice. There may be seasonal variations in the size of these stores (particularly in more temperate regions, where there can be significant winter snowfall and summer melting). Over the last 30 years or so, the stores of many of the world's glaciers have shown a decline in mass, which has been attributed to climate change and global warming.

Flows/transfers

There are many flows and transfers of energy and material. These include processes such as evaporation, sublimation, meltwater flow and the processes of glacial movement (**internal deformation** and **basal slip**) see page 66. Flows and transfers are more pronounced and active in warmer environments (where there are significant seasonal variations in temperature). In the world's coldest environments, such as Greenland and Antarctica, glacial systems are less active.

Feedback loops

- **Positive** and **negative feedback loops** are significant aspects of all geomorphic systems:
- Negative feedbacks regulate systems to establish balance and equilibrium.
- Positive feedbacks enhance and speed up processes, promoting rapid change.
Both positive and negative feedbacks occur in glacial systems.

Dynamic equilibrium

Many physical systems move towards a state of **dynamic equilibrium**, where landforms and processes are in a state of balance. In a glacial system, the **equilibrium line** marks the boundary between the **accumulation zone** (glacial inputs) and the **ablation zone** (glacial outputs). If the glacier is in a state of balance – where inputs equal outputs – the equilibrium line will remain in the same place. As this balance shifts, the equilibrium line will move up or down the glacier (hence the term 'dynamic' equilibrium).

🔺 *Figure 3* *The glacial system*

Glacier health

A glacier's health can be assessed by determining its mass balance. Collecting data to show whether a glacier is gaining or losing mass is essential when determining its health.

The Gulkana Glacier, Alaska

The Gulkana Glacier (see Figure 4) is one of two 'benchmark' Alaskan glaciers that the United States Geological Survey (USGS) has studied since the 1960s. The purpose of this study has been to understand glacier dynamics and hydrology, as well as to assess the glaciers' responses to climate change.

The USGS uses several fixed index sites on the glacier (see Figure 5) and, together with meteorological and runoff data, scientists have been able to plot trends and calculate the glacier's mass balance.

The Gulkana Glacier varies from 1160 metres to 2470 metres in height. In 2011, it occupied a total area of 16.7 km^2. The Alaskan climate is continental – with a large temperature range and irregular, light precipitation (compared to more coastal areas).

◆ Field visits to measure and maintain stakes at the three index sites are made each spring (at the start of the melt season), and again in early autumn (near its completion). By collecting data near the balance maxima (spring) and minima (autumn), direct measurements closely reflect maximum winter accumulation and the annual balances at each location.

⚫ Figure 4 *The Gulkana Glacier, Alaska*

◆ The density of the material gained or lost is measured by digging a snow-pit or extracting an ice core.

◆ Since 1975, both the stakes and the glacier surface elevations have been surveyed at the index sites – to allow calculations of velocity and surface elevation change.

Figures 6 and 7 show the results of the study.

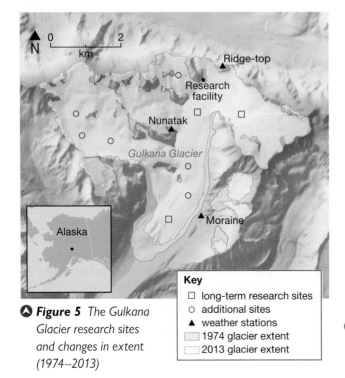

⚫ Figure 5 *The Gulkana Glacier research sites and changes in extent (1974–2013)*

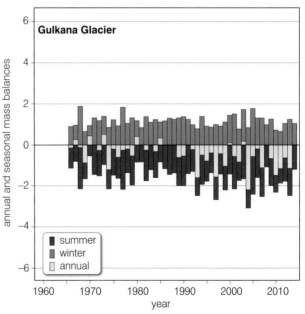

⚫ Figure 6 *The Gulkana Glacier mass balance (1966–2014) in metres of water equivalent*

Year	Winter balance	Summer balance	Net balance	Cumulative balance	Equilibrium line altitude
1990	1.36	−2.04			1794
1991	1.31	−1.37			1704
1992	0.98	−1.22			1758
1993	0.82	−2.49			1880
1994	1.37	−1.96			1777
1995	0.94	−1.65			1806
1996	0.87	−1.39			1768
1997	0.99	−2.68			1865
1998	0.79	−1.43			1793
1999	1.04	−2.15			1842
2000	1.44	−1.49			1704
2001	1.40	−2.08			1790
2002	0.76	−1.83			1833
2003	1.79	−1.80			1718
2004	0.93	−3.22			1851
2005	1.73	−1.99			1758

▲ **Figure 7** *Mass balance data for the Gulkana Glacier, Alaska (1990–2005). The values in the table are in metres of water equivalent.*

Over to you

1 a Use a photo of a glacier and Figure 3 to identify and label these parts of a glacier system: inputs, outputs, stores, flows, feedback.
  b Explain how and why the glacier is in a state of dynamic equilibrium
  c Explain why a glacier is considered as an *open* system.

2 Design a flow diagram to show reasons why glaciers move, advance and retreat.

3 Study Figure 5.
  a Describe how the extent of the glacier changed, 1974–2013.
  b How could evidence from Figure 5 **alone** lead to inaccurate conclusions about the glacier's changing mass balance?

4 Study Figure 7.
  a Explain positive mass balance values in winter and negative mass balance values in summer.
  b Copy the table and complete the calculations in the blank columns.
  c Present the cumulative balance data as a line graph. Describe the trends.
  d Draw a similar graph to show changes in the altitude of the equilibrium line. Describe any trends.

On your own

5 Distinguish between the following: **(a)** inputs and outputs, **(b)** positive and negative feedback loops, **(c)** accumulation and ablation, **(d)** mass balance and glacier health.

6 Study Figure 6. Describe the changes in mass balance between 1966 and 2014.

7 a Using all of the evidence, assess the health of the Gulkana Glacier.
  b Use GIS resources on the USGS website (http://www.usgs.gov) to update the evidence available on the Gulkana Glacier's improving or declining health.

Exam-style questions

AS 1 Explain the characteristics of a glacier as a system.
(6 marks)

A 2 Assess the health of one named glacier.
(9 marks)

In this section, you'll learn about the different processes that affect glacial movement, as well as variations in the rate of that movement.

How does ice move?

The Jakobshavn Glacier in Greenland is one of the world's fastest-moving glaciers – flowing at around 20 metres a day at its snout! Ice moves in two main ways: internal deformation and basal slip.

Basal slip

Basal slip (or sliding), see Figure 1, involves movement that usually occurs in a series of short jerks. This happens in temperate glaciers (where meltwater helps to lubricate the base of the ice). Such movement can be up to 2–3 metres a day.

When a glacier encounters an obstacle (e.g. an outcrop of hard rock), the resistance to movement on the upslope side causes an increase in stress and pressure, which may result in **pressure melting**. This allows the glacier to move over the obstacle. The meltwater often refreezes on the downslope side, where pressure is reduced. Melting and freezing that depends on pressure is called **regelation**, and the associated movement is called **regelation creep**.

Downhill movement can raise the temperature of the base ice, due to increased pressure and friction. This **positive feedback** may lead to further melting of the basal ice, which then allows the glacier to slip more easily over its bed.

Internal deformation

Internal deformation (see Figure 1) occurs through:

◆ *inter*-granular movement, where individual ice crystals slip and slide over each other.

◆ *intra*-granular movement, where individual ice crystals become deformed or fractured due to the intense stresses within the ice (as exerted by the glacier's mass under the influence of gravity). Gradually, the mass of ice deforms and moves downhill in response to gravity.

Internal deformation occurs in both polar and temperate glaciers, where ice moves up to 1–2 cm a day.

◗ **Figure 1** *Internal deformation and basal slip*

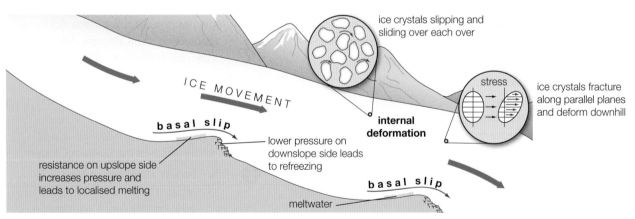

ice crystals slipping and sliding over each over

ICE MOVEMENT

basal slip

resistance on upslope side increases pressure and leads to localised melting

lower pressure on downslope side leads to refreezing

internal deformation

stress

ice crystals fracture along parallel planes and deform downhill

basal slip

meltwater

Variations in the rate of movement

The rate of ice movement is affected by many factors.

◆ Increases in gradient cause the ice to flow faster, so it becomes 'stretched' and thinner. This is called **extensional flow**, which creates crevasses on the surface.

◆ However, reductions in gradient force the ice to slow, 'pile up' and thicken. This is called **compressional flow**. Any crevasses previously opened then close.

◆ Between the zones of extensional and compressional flow, the ice moves in a **rotational** manner.

Compressional flow increases the mass and erosional power of the glacier, leading to a steeper gradient, faster extensional flow, a thinning of the ice and a reduction in potential erosion – a theoretical **negative feedback** loop.

Factors affecting the rate of movement

◆ **Altitude** affects precipitation and temperature; greater precipitation and lower temperatures increase the supply of snow and ice, and so its **mass balance**.

◆ **Gravity** and **gradient** (**slope**). Gravity causes ice to move; the steeper the gradient, the faster it flows (see page 66).

◆ **Friction** exerted by the ground has to be overcome for ice to move. Glaciers flow faster towards their centre (away from its effects).

◆ The heavier the ice (i.e. the greater its **mass**), the more force is needed to overcome increased friction caused by extra weight.

◆ In temperate zones, movement is faster over **impermeable** surfaces, because basal meltwater is retained – aiding slippage.

◆ **Meltwater** lubricates the base of the ice, enabling it to slip downhill.

◆ Ice **temperature**. In Antarctica, ice is so cold that it's frozen to the bedrock. So polar glaciers move more slowly than temperate glaciers.

Over to you

1 Explain the different rates of movement in polar and temperate glaciers.

✦ 2 Use Section 7.5 on central tendency to help you with the following:
   a Using Figure 2, create a dispersion diagram for daily rate of flow for each of Glaciers A and B, and then calculate the range and inter-quartile ranges for each set of data.
   b Calculate the mean, median and mode for each glacier.
   c Compare the data for each glacier.

On your own

3 Distinguish between the following: **(a)** pressure melting and regelation creep, **(b)** extensional, compressional and rotational flow.

4 Draw a spider diagram to show factors determining **(a)** how glaciers move and **(b)** how fast they move.

Exam-style questions

AS 1 Describe one way in which glaciers move. *(3 marks)*

A 2 Explain why glaciers move at different rates in different parts of the world. *(6 marks)*

How fast do glaciers move?

Polar glaciers move almost exclusively by internal deformation, so their movement is slow (as are their rates of erosion and sediment transfer). Greenland's Jakobshavn Glacier (see Figure 2A) moves unusually quickly; polar glaciers generally move only a few metres a year. Canada's Saskatchewan Glacier (see Figure 2B) is more typical of glacier speeds.

With temperate glaciers, internal deformation combined with basal slip results in greater movement and higher rates of erosion and sediment transfer. Temperate glaciers move at between 2 and 200 metres a year (with, for example, the Mer de Glace in the French Alps moving at about 70 metres a year – like Canada's Saskatchewan Glacier).

Glacier A Day No.	Metres per day	Glacier B Day No.	Metres per day
1	19.6	1	0.24
2	19.3	2	0.23
3	18.6	3	0.24
4	18.8	4	0.26
5	18.9	5	0.25
6	18.4	6	0.25
7	18.1	7	0.27
8	18.7	8	0.27
9	18.5	9	0.25
10	18.8	10	0.25
11	18.8	11	0.22
12	19.2	12	0.28
13	19.7	13	0.26
14	19.6	14	0.26
15	19.7	15	0.25
16	19.8	16	0.24
17	20.0	17	0.24
18	20.2	18	0.24
19	20.2	19	0.24
20	20.5	20	0.25
21	20.9	21	0.24
22	20.5	22	0.25
23	20.3	23	0.26
24	20.2	24	0.26
25	20.2	25	0.25

▲ **Figure 2** *Daily rates of flow in two glaciers in metres. Glacier A is Jakobshavn Glacier in Greenland; Glacier B is Saskatchewan Glacier in Canada.*

In this section, you'll find out about glacial processes and the types of landforms and landscapes they create.

Glacial processes

Melting glacial ice can reveal many strange things, such as clothes, personal belongings, weapons – even the frozen preserved bodies of climbers, soldiers and travellers lost in the mountains many years before. The aircraft fragments in Figure 1 were transported by, and melted out of, the Gigjokull Glacier in Iceland. However, transportation is just one of the glacial processes operating within the glacial system.

Glacial erosion and entrainment

There are two main types of glacial erosion: **abrasion** and **plucking** (also see page 70):

◆ Abrasion occurs because of **entrainment** – ice includes angular frost-shattered material, which scours the landscape. Large rocks carried below the ice often scratch the bedrock to form **striations** or scratches.

◆ Plucking or **quarrying** occurs when meltwater freezes part of the underlying bedrock to the base of a glacier. Any loosened rock fragments are then 'plucked' away as the glacier subsequently slips forward.

Glacial transportation

Glaciers act like giant conveyor belts, which transport material from mountainous areas onto adjoining lowlands. Material can be carried in three ways:

◆ **Supraglacial** – mainly weathered material carried on top of the ice (see Figure 2).

◆ **Englacial** – formally supraglacial material, but now buried by fresh snowfall and carried within the ice.

◆ **Subglacial** – material carried below the ice, which is dragged and pulverised by the overlying glacier.

Water, as well as ice, has an important role in the transportation of material. In temperate environments, water flows on top of glaciers – leading to fluvial transport. This water may flow down crevasses or holes in the ice (moulins), and thus transport material into and beneath the glacier. Temperate glaciers also have meltwater streams flowing under the ice, which carry material to the glacier's snout and beyond.

Glacial landforms

Glacial landforms develop in different environments where different processes are operating:

◆ **Subglacial** – below a glacier or ice-sheet. Erosion (abrasion and plucking) is active here, forming striations and roche moutonée (see pages 70 and 73). Meltwater below the ice can create eskers as a result of subglacial deposition (see page 78). These features occur at very localised or **micro-scales**, or those extending over parts of glaciated landscapes, at **meso-scales**.

▲ *Figure 1* *Fragments of a crashed Second World War aircraft transported by the Gigjokull Glacier in Iceland*

Key word

Entrainment – The process by which surface sediment is incorporated into a fluid flow (e.g. air, water or ice) as part of the process of erosion.

Glacial deposition

The deposition of sediment transported by the ice occurs when it melts – mainly in the ablation zone close to the glacier's snout. Here the sediment on – and in – the ice simply melts out. Water may then carry the sediment further away from the ice, sometimes over distances of many kilometres. See pages 74–77 for more information about the landforms created by glacial deposition.

▼ *Figure 2* *Supraglacial transport on the Mer de Glace in France*

- **Marginal** – at the sides or end of a glacier or ice sheet. Weathering and deposition operate here to create landforms such as moraines (page 74).

- **Proglacial** – in front of, at, or immediately beyond the margin of a glacier or ice sheet. Fluvioglacial processes operate here (with meltwater eroding, transporting and depositing sediment) to create outwash plains, meltwater channels and proglacial lakes (see pages 79–80).

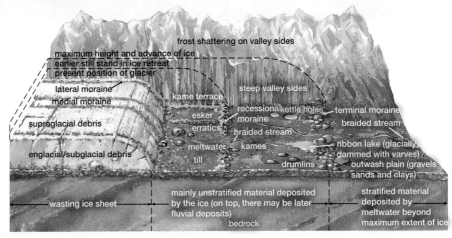

▲ *Figure 3 Glacial landforms*

- **Periglacial** – exist at the edges of ice sheets and glaciers and can be extensive (see pages 56–61). The processes operating here are mostly associated with frost, ice and snow, as well as meltwater, and they create a range of landforms such as blockfields, ice wedges and pingos (see pages 58–60). Because periglacial margins are often extensive, landforms associated with periglacial processes occur at a **macro-scale**.

Figure 3 shows a range of different glacial landforms found in subglacial, marginal and proglacial environments.

Landform evidence

Landforms created by glacial processes differ in upland and lowland areas. Upland areas are characterised by erosional landforms, e.g. corries, glacial troughs and hanging valleys. Lowland areas tend to have depositional landforms, e.g. outwash plains, glacial till deposits. The boulder clays of East Anglia and the Holderness coast show former outwash plains formed here. Corries, tarns and glacial troughs in the Lake District show that glaciers were active there.

Figure 4 shows glacial and meltwater deposits across the British Isles. These deposits, along with other landforms, tell us where glaciers operated and how far they extended in the past.

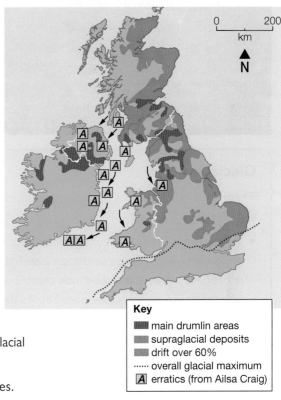

Key
- ▨ main drumlin areas
- ▨ supraglacial deposits
- ▨ drift over 60%
- ⋯⋯ overall glacial maximum
- Ⓐ erratics (from Ailsa Craig)

▲ *Figure 4 Glacial and meltwater deposits across the British Isles*

Over to you

1 a With the help of diagrams, describe the different forms of glacial transportation.
   b Explain the sequence of events by which the aircraft on the Gigjokull Glacier in Figure 1 was hidden, then exposed.

On your own

2 Distinguish between the following: **(a)** abrasion and plucking; **(b)** supraglacial, englacial, and subglacial; **(c)** glacial and fluvial erosion; **(d)** proglacial and periglacial.

3 Sketch a copy of Figure 3. Categorise all the features and landforms shown into subglacial, marginal and proglacial.

Exam-style questions

 1 Describe how glaciers transport material. *(3 marks)*

 2 Explain how landforms help to determine the extent of ice cover in upland and lowland areas. *(6 marks)*

In this section, you'll find out how glacial erosion creates distinctive landforms and landscapes.

What is a glacially eroded landscape?

Figure 1 was taken close to the summit of Helvellyn, in the English Lake District. The photo is looking roughly east/north-east over one of the most iconic glacial landscapes in the UK. The steep ridges, bare rocky outcrops, lakes and wide valleys are typical of a glacially eroded landscape.

Red Tarn – a corrie lake

Corrie – a scooped out hollow in the landscape

Striding Edge – a classic example of an arête

Glacial troughs or U-shaped valleys.

◀ **Figure 1** *The Lake District – a glacially eroded landscape*

Glacial erosion

There are two main types of glacial erosion – abrasion and plucking:

◆ **Abrasion** is the sandpapering effect of ice as it grinds over and scours a landscape (see Figure 2). It happens because **freeze-thaw** creates sharp, angular rock fragments (see page 58). As the rock fragments become trapped under the ice, they become extremely effective abrasive tools. Large rocks carried beneath the ice often scratch the bedrock to form **striations**. These scratches provide useful clues for scientists studying the direction of ice flow in a post-glacial environment. Over time, these rocks become pulverised by the weight of the ice to become fine **rock flour**. This finer material then tends to smooth and polish the underlying bedrock.

▶ **Figure 2** *The processes of abrasion and plucking in the formation of a roche moutonnée*

◆ **Plucking** or **quarrying** occurs when basal meltwater freezes around part of the underlying bedrock at the base of a glacier. Any loosened rock fragments are 'plucked' away as the glacier subsequently slips forward. It's a bit like pulling out a loose tooth! This process is particularly common where a localised reduction in pressure under the ice has led to regelation (refreezing of meltwater).

In addition, the weight and pressure exerted by thick ice can be sufficient to **crush** the surface of even poorly jointed bedrock.

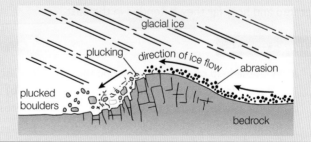

How glacially eroded landforms develop

A number of different erosional landforms are associated with cirque and valley glaciers (as the following pages explain).

◆ **Cirque (or corrie) glaciers** are small masses of ice that occupy armchair-shaped hollows in mountains. They often overspill from the hollows to feed valley glaciers.

◆ **Valley glaciers** are larger masses of ice that move down from either an ice field or a cirque. They usually follow former river courses and are bounded by steep sides.

Corries

A **corrie** (also known as a cirque in France and a cwm in Wales) is an enlarged, often deep, hollow on a mountainside. Its characteristic features include a steep, cliff-like back wall – often with a large pile of scree at its base. There is also usually a raised rock lip at the front of the hollow, which acts as a dam to trap water and form a small lake – or **tarn** (see Figure 1).

To understand how a corrie develops, we need to look at the processes at work before, during and after glaciation (see Figure 3).

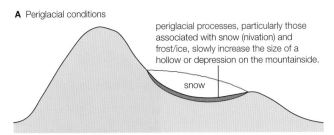

A Periglacial conditions

periglacial processes, particularly those associated with snow (nivation) and frost/ice, slowly increase the size of a hollow or depression on the mountainside.

snow

B Glacial conditions: As the climate cools, snow turns to ice in the depression and a corrie glacier develops. Accumulation at the top of the glacier increases its mass and rotational sliding results in a 'scooping out' of the hollow by abrasion.

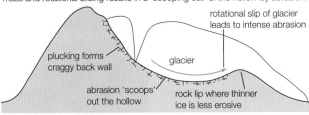

rotational slip of glacier leads to intense abrasion

plucking forms craggy back wall

glacier

abrasion 'scoops' out the hollow

rock lip where thinner ice is less erosive

C Post-glacial conditions: As the climate warms, first periglacial and then temperate processes (including frost and water action) modify the shape of the corrie to create what we see today.

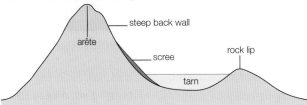

steep back wall

arête

scree

rock lip

tarn

🔺 **Figure 3** *How a corrie forms*

Arêtes and pyramidal peaks

When two neighbouring glaciers cut back into a mountainside, the narrow, knife-edge ridge that forms between the two corries is known as an **arête**. Arêtes are common in both present-day glacial landscapes (such as the Alps) and also in post-glacial landscapes, such as the Lake District (see Figure 1).

Where three or more corries erode back-to-back, the ridge becomes an isolated peak called a **pyramidal peak** (see Figure 5).

Corrie orientation

In Britain, as elsewhere in the Northern Hemisphere, corries are nearly always orientated between the north-west (315°), through to the north-east (where the frequency peaks) to the south-east (135°). This is because – in the UK:

◆ northern slopes receive much less solar insolation, so glaciers on those slopes lasted much longer than those facing in more southerly directions (where more sun meant faster glacial melting)

◆ western slopes face the sea, and – although still cold – the relatively warmer winds which blow from that direction were better at melting the snow and ice (so more snow accumulated on east-facing slopes)

◆ the prevailing westerly winds cause snow to drift into east-facing hollows.

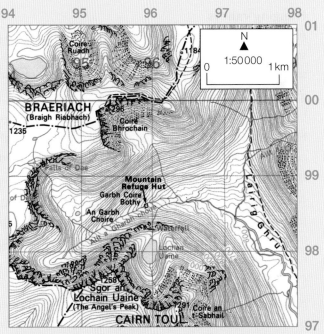

🔺 **Figure 4** *An OS 1:50 000 extract of the Cairngorms in Scotland. Corries are shown by semi-circles of black cliff symbols, marking their back edges.*

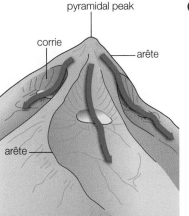

pyramidal peak

corrie

arête

arête

◀ **Figure 5** *How arêtes and pyramidal peaks form*

Glacial troughs

Glaciers are immensely powerful agents of erosion. They can be several hundred metres thick, and their vast mass enables them to create spectacular and dramatic landscapes. The Athabasca Glacier is shrinking and, as it does so, more and more of its impressive valley (**glacial trough**) is being revealed (see Figure 6).

Glacial troughs are characteristically steep-sided, mainly flat-bottomed and deep – often several hundred metres deep. They tend to be largely straight, because of the power and inflexibility of the glaciers that erode them. The main process of erosion is abrasion through basal slip – plucking does take place, but more selectively than abrasion.

Glacial troughs may contain deep and narrow lakes, called **ribbon lakes**. These result from localised overdeepening due to enhanced erosion caused by:

◆ weaker bedrock allowing increased vertical erosion

◆ the merging of a tributary glacier, which can lead to greater erosion of the valley floor due to the increased ice mass

◆ a narrowing of the valley and the resulting thicker ice, which leads to increased vertical erosion.

Hanging valleys

A small glacier in a tributary valley doesn't have the same mass as a much larger glacier, so it won't be able to erode down as far as the larger glacier. When the ice melts, the smaller valley is left 'hanging' above the main valley as a **hanging valley** (see Figure 7).

Truncated spurs

In a typical upland river valley, the river flows around interlocking spurs of rock. However, when a valley such as this is occupied by ice, the rigid, more-powerful glacier cuts off or 'truncates' the tips of the rocky spurs as it moves downhill – leaving behind steep cliffs called **truncated spurs**.

Although abrasion is the dominant process in the formation of truncated spurs, plucking also takes place on the bare rocky surfaces. Sub-aerial processes, such as frost shattering (freeze-thaw) and rockfalls (a type of mass movement) also act on the exposed rocky cliffs protruding above the ice.

Look at Figure 8, which shows part of Greenland's Watkins Range Mountains. Enormous glaciers are currently sculpting the landscape to form several of the landforms described previously. Notice how truncated spurs tend to form between hanging valleys.

Figure 6 *The Athabasca Glacier in Canada*

Figure 7 *This impressive hanging valley in New Zealand is marked by Stirling Falls plunging into Milford Sound (a fjord formed by the sea flooding a huge glacial trough). Notice that the hanging valley has the same steep U-shaped profile as the main glacial trough.*

A Arête
C Corrie
P Pyramid peak
↘ Ice movement

Figure 8 *Glaciers and glacial landforms in Greenland's Watkins Range Mountains*

Hanging valleys will form here when the glaciers melt

Truncated spurs are being formed here

The formation of landforms due to ice-sheet scouring

Crag and tail

The underlying geology plays a part in the formation of **crag and tail** landforms. A crag and tail is formed when a very large resistant object, or crag, obstructs the flow of a glacier (e.g. the basalt volcanic plug on which Edinburgh Castle sits). The ice is forced around the obstruction, eroding weaker rock. However, material immediately in the lee of the obstruction is protected by the crag, which leads to the formation of a gently sloping **tail** of deposited material.

Roche moutonnée

Roches moutonnées are bare outcrops of rock on the valley floor that were sculpted by moving ice. They demonstrate two processes of glacial erosion – abrasion and plucking. Look at Figure 9.

◆ Upstream side (left): Increased pressure due to the resistance of the outcrop to moving ice caused localised pressure melting. This led to basal slip and abrasion as the glacier slid over the outcrop. The abraded upstream side shows polishing and striations.

◆ Downstream side (right): reduced pressure caused meltwater to freeze – forming a bond between rocky outcrop and overlying glacier. As the glacier moved forward, it plucked away loose rock, leaving a jagged surface.

Knock and lochan landscapes

This is the name given to a glacially scoured lowland area, which has alternating roches moutonnées (known as cnoc or knock, which means a small rock hill in Gaelic) and eroded hollows, which often contain small lakes (lochans). These landscapes are closely linked to rock structure, along features such as joints and minor faults. They are most often found where alternate resistant and weakly jointed rocks allow differential erosion.

⬤ **Figure 9** *A roche moutonnée at Honister Pass in the Lake District*

Over to you

 1 a Using the OS map extract (Figure 4), identify the landforms of glacial erosion (corries, arêtes, and the glacial trough).

b Draw a sketch map of these landforms. Include a scale and north arrow.

c Calculate the height of each corrie, the width of each corrie basin, and the direction in degrees from north that each faces (known as its orientation).

 2 Use the data from question 1c and Section 7.6.

a Carry out Spearman's rank correlations for height of basin, size of basin and orientation.

b Comment on the significance of the correlation.

c Explain the results you have found out.

On your own

3 Distinguish between the following pairs of terms: **(a)** abrasion and scouring; **(b)** cirque glaciers and valley glaciers; **(c)** glacial troughs and ribbon lakes; **(d)** knocks and lochans.

4 Draw a sketch of Figure 9. Add annotations to show the characteristics and processes responsible for the formation of this roche moutonnée.

5 Distinguish between the processes that form a roche moutonée and a crag and tail.

Exam-style questions

 1 Assess the importance of any two processes of glacial erosion in the development of upland landscapes. *(12 marks)*

A 2 Assess the significance of the processes of glacial erosion in making upland landscapes distinctive. *(12 marks)*

In this section, you'll learn how glacial deposition creates distinctive landforms and landscapes.

What is a landscape of glacial deposition like?

Like many Alpine glaciers, the Steingletscher Glacier (shown in Figure 1) is shrinking and retreating. As it does, it's revealing some classic landforms of glacial deposition – a debris-strewn valley floor, ridges of sediment and proglacial lakes.

Glaciers act like giant conveyor belts – transporting rock debris from upland erosion and depositing it on valley floors or lowland plains. The debris beneath a glacier is pulverised by the sheer weight of ice above to form fine splintered **rock flour**. It is this fine-grained sediment that gives meltwater streams and proglacial lakes a milky blue colour.

Rock debris deposited in situ is called **till**. It is characteristically angular and very poorly sorted. Meltwater streams from the glacier snout can carry sediment for many kilometres – eventually depositing it as a vast, gravelly, well-sorted **outwash plain**. Figure 2 shows some of the main depositional landforms associated with lowland landscapes.

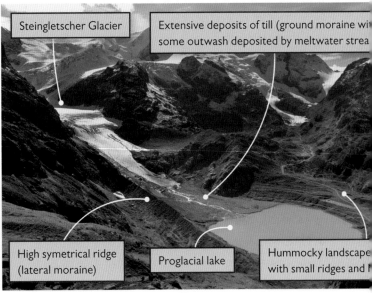

Steingletscher Glacier

Extensive deposits of till (ground moraine wi[th] some outwash deposited by meltwater strea[m]

High symetrical ridge (lateral moraine)

Proglacial lake

Hummocky landscape with small ridges and [...]

▲ **Figure 1** *The Steingletscher Glacier in the Swiss Alps*

▼ **Figure 2** *Landforms of glacial deposition*

Ice contact depositional features

Moraines

Moraine is a generic term for landforms associated with the deposition of till from within, on top of, and below, a glacier – so it consists of poorly sorted, mainly angular sediments. Figure 2 shows the distinctive characteristics of moraines.

- **Ground moraine.** Sediment transported beneath a glacier that is smeared over underlying bedrock. It can be several metres thick, often forming an irregular, hummocky surface topography.

- **Terminal moraine.** A ridge of sediment piled up at the furthest extent of an advancing glacier. It commonly appears as a line of hills (rather than a solid ridge) due to the erosive action of meltwater streams from the retreating glacier.

- **Recessional moraine.** Retreating glaciers may experience periods of stability, when a secondary ridge of sediment forms at the snout (called recessional moraine). This has the same characteristics as terminal moraine, but doesn't mark the furthest extent of the ice.

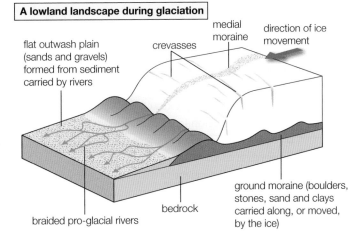

A lowland landscape during glaciation

flat outwash plain (sands and gravels) formed from sediment carried by rivers

crevasses

medial moraine

direction of ice movement

braided pro-glacial rivers

bedrock

ground moraine (boulders, stones, sand and clays carried along, or moved, by the ice)

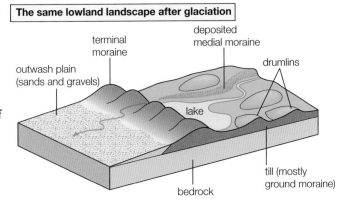

The same lowland landscape after glaciation

outwash plain (sands and gravels)

terminal moraine

deposited medial moraine

lake

drumlins

bedrock

till (mostly ground moraine)

◆ **Lateral moraine.** Figure 1 shows a superb example of lateral moraine – a high and almost symmetrical ridge, formed along the outer edge of a glacier (primarily from scree). It can be several metres high.

◆ **Medial moraine.** This is formed when lateral moraines from two merging glaciers join up – leaving a line of debris in the centre of the combined glacier's flow. As the combined glacier melts, the medial moraine is deposited to form a low ridge.

Drumlins

A typical **drumlin** is an oval or 'egg-shaped' hill – made up of glacial till and aligned in the direction of ice flow (see Figure 7 and page 319). Drumlins vary in size, but are commonly 30–50 metres high and 500–1000 metres long. They usually occur in clusters or 'swarms' on flat valley floors or lowland plains in parts of northern England, Scotland, Ireland, Sweden, Finland and Canada – all previously glaciated regions.

There is controversy about drumlin formation. Although some have a rocky core (with sediment moulded around it), most do not. Some consist, at least partly, of fluvial sediments as well as glacial till – suggesting that meltwater played a part in their formation. It may be that they owe their origin to several processes.

Lowland depositional features

A till plain

A **till plain** is an extensive plain created by the melting of a large ice sheet that detached from a glacier. The till effectively levels out the topography to create a mostly flat landscape. Angular, unsorted till is often divided into **lodgement till** (dropped by moving glaciers) and **ablation till** (dropped by stagnant or retreating ice).

> **BACKGROUND**
>
> ### Working out previous ice extent, movement and origins
>
> It's possible to use landforms created from glacial deposition to work out how far ice extended, and in which direction it flowed – as well as the origin of depositional features. For example, terminal moraine provides evidence of a glacier's furthest extent, while a crag and tail (see page 73) – created by erosion *and* deposition – indicates the direction of ice flow.

Till fabric analysis

Till fabric analysis involves the study of the orientation and 'plunge' of rock fragments within a till deposit (see Section 7.3) which can suggest the direction of ice flow at the time of deposition. Studies involving 50 or more rock fragments – conducted at several localities – provide glaciologists with evidence of ice-flow patterns, especially in association with other evidence, e.g. erratics (see next page) and drumlin orientation. The angle of 'plunge' can suggest thrust motions within the till as the glacier deforms it.

Data are recorded in categories (classes), say every 10° or 20°, and presented in a rose diagram (see Figure 3), allowing orientations to be identified and compared with other evidence. In Figure 3 there is a clear NW-SE orientation. Data in Figure 4 show pebble orientation for a sample of rock particles taken from a till deposit.

Class (degrees)	Midpoint (degrees)	Number of particles
350–009	000	2
010–029	020	7
030–049	040	6
050–069	060	4
070–089	080	3
090–109	100	–
110–129	120	–
130–149	140	2
150–169	160	3
170–189	180	10
190–209	200	9
210–229	220	2
230–249	240	–
250–269	260	–
270–289	280	–
290–309	300	–
310–329	320	1
330–349	340	1

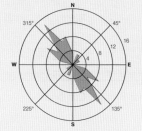

◁ **Figure 3** *Till fabric analysis showing the orientation of 50 pebbles from the Endon till in the Endon Valley in Leek, Staffordshire*

▷ **Figure 4** *Rock particle orientation from a till sample*

Erratics

An **erratic** is a boulder or rock fragment deposited far from its origin. The one in Figure 5 is clearly different from the surrounding landscape. Erratics give clues to establish the direction of previous ice movement, and also help to recreate **paleo-environments** (fossil or past environments). By working out the source of the erratic, the origin of the ice can be established, e.g. volcanic material from Ailsa Craig in the Firth of Clyde has been found 250 km south on the Lancashire plain.

▲ *Figure 5* *An erratic in the Yorkshire Dales consisting of Silurian rock, deposited on top of limestone*

> **BACKGROUND**
>
> ### Using GIS to work out past ice extents
>
> Geographers at the University of Sheffield have produced a glacial map of the UK, showing the extent of ice during the last Ice Age. It used over 1000 academic publications, and BGS mapping. Relevant data were extracted, digitised and entered into a GIS. The GIS contains over 20 000 features spilt into thematic layers such as moraines, eskers, drumlins, meltwater channels and ice-dammed lakes.

Drumlin orientation

One of the best-known drumlin fields covers much of the Vale of Eden in Cumbria:

- South of Appleby, debris-laden ice moved east and was squeezed through the Stainmore col (see Figure 6). The drumlin field it left behind (see Figure 7) indicates the direction of movement, and the volume of material eroded, transported and deposited.

- South and west of Brough (see Figures 7 and 8), 308 drumlins have been mapped. They vary in length from 150 to 1000 metres, and in width from 75 to 500 metres. Most are moulded entirely from till.

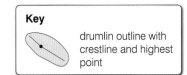

Key

drumlin outline with crestline and highest point

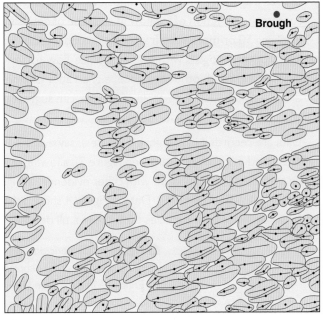

▲ *Figure 7* *The drumlin field near Brough, Cumbria*

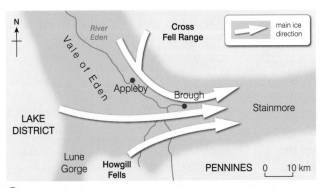

▲ *Figure 6* *The main sources and movements of ice in the southern Vale of Eden*

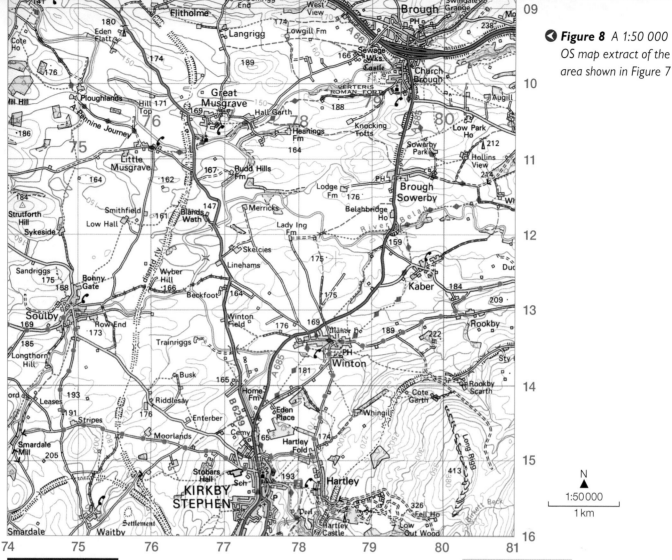

◀ **Figure 8** A 1:50 000 OS map extract of the area shown in Figure 7

Over to you

 1 Figure 4 shows pebble orientation for a sample of particles from a till deposit. Use Section 7.3 to help you with the following activities.

   a Construct a rose diagram for the data.

   b Suggest a preferred orientation of the particles.

   c Critically evaluate the level of confidence in your conclusion.

   d What additional evidence would you seek from a field locality to increase confidence in drawing conclusions about the direction of ice flow?

 2 Investigate one area of glacial drift in the UK (e.g. the southern Vale of Eden in Cumbria) to try to reconstruct past ice extent and ice flow direction. Use: **(a)** an OS map of the area, **(b)** geological drift maps from the British Geological Survey website (bgs.ac.uk), **(c)** Figures 6 and 7, **(d)** GIS sources from Sheffield University (sheffield.ac.uk).

3 Identify ways in which fieldwork results, such as Figure 4, can further aid this investigation.

On your own

4 Use the photos in this section, and others sourced from the Internet, to describe the typical landscape of glacial deposition. Annotation on the photos may help with your answer.

5 Draw a simple annotated diagram to describe the location and formation of the different types of moraine.

Exam-style questions

AS **1** Explain how different features can be used to reconstruct previous ice extent and movement. *(6 marks)*

A **2** Evaluate the statement that 'glacial deposition creates landforms that are distinctive'. *(20 marks)*

In this section, you'll learn about the role of glacial meltwater in creating distinctive landforms and landscapes.

What are fluvio-glacial landscapes?

Fluvio-glacial landscapes – it sounds more complex than it is. Fluvio-glacial means river (fluvio from Latin) and glacial. The landscapes created by fluvio-glacial processes are associated with flowing water – essentially meltwater – in glacial or periglacial environments. Meltwater is seasonally abundant in temperate glacial and periglacial environments, and is often seen flowing out from under the ice at a glacier's snout (see Figure 1). However, meltwater is much less common in the world's coldest environments, which are characterised by cold-based glaciers.

What are fluvio-glacial processes?

Just like normal rivers, meltwater erodes, transports and deposits sediment. It also forms many features typical of rivers, such as meandering channels, levees and deltas. Occasionally, a huge amount of meltwater becomes trapped, either beneath the ice or as surface lakes. When these eventually burst (called a **glacial outburst**), the surging meltwater has the power to carve deep channels or gorges.

Ice can produce huge quantities of water when it melts. This can flow on top of a glacier (in **supraglacial** channels), within the ice (**englacial** channels), or under it (**sub-glacial** channels).

Meltwater discharged from supraglacial and sub-glacial streams is higher during warmer summer months. It often flows at high velocity, and is usually very turbulent, so it picks up and transports a larger volume of material than a normal river of a similar size. Such material then erodes vertically, creating sub-glacial valleys and large potholes. Deposition occurs when there is a decrease in discharge.

There are two distinct types of fluvio-glacial landforms:

◆ Ice-contact features   ◆ Pro-glacial features

Ice-contact features

Eskers

Eskers are long, winding ridges of sand and gravel, up to 30 metres high and several kilometres long. They usually take the form of meandering hills running roughly parallel to the valley sides, which suggests that they were formed by sub-glacial river deposition during the final stages of a glacial period, as the glacier retreated. Today, eskers form discontinuous hills, having been eroded by meltwater and post-glacial rivers.

One of the best examples of an esker in the USA is the Dahlen esker in North Dakota (see Figure 2). It's about 7 kilometres long, 120 metres wide and up to 25 metres high, comprising a mixture of fluvial sands and gravels. In places, it has a distinctive native prairie vegetation, due to its sandy soil, whereas the flat and fertile till plains on either side are intensively farmed.

⊙ **Figure 1** Meltwater flowing from beneath the Athabasca Glacier in the Canadian Rockies

⊙ **Figure 2** The Dahlen esker in North Dakota, USA

Kames

Kames form on the ice surface. They consist largely of sand and gravel deposited by streams in the final stages of a glacial period. As Figure 3 shows, there are three different types:

- **Kame terraces** result from the infilling of a marginal glacial lake. When the ice melts, the kame terrace is left as a ridge on the valley side.

- **Kame deltas** are smaller features that form when a stream deposits material on entering a marginal lake. It forms small, mound-like hills on the valley floor.

- **Crevasse kames** are small hummocks deposited on the valley floor as a result of sediment deposited in surface crevasses.

Pro-glacial features

An outwash plain (sandur)

An **outwash plain** (or **sandur**) is an extensive, gently sloping area of sands and gravels that forms in front of a glacier. It results from the 'outwash' of material carried by meltwater streams and rivers. During a glacial period, the creation of meltwater and the deposition of material is mainly restricted to summer. However, at the end of a glacial period, huge quantities of material are spread out over the outwash plain by great torrents of meltwater. Some of the most extensive plains exist in Iceland and Alaska, where large **braided rivers** choked with sediment meander across vast floodplains (see Figure 4).

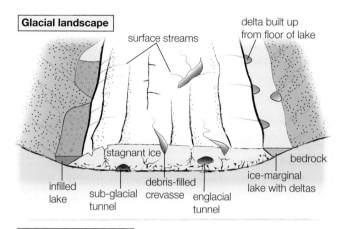

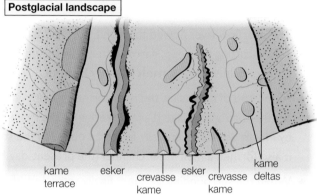

Figure 3 *How kames and eskers are formed*

Figure 4 *An outwash plain in Canada with a braided river system*

Meltwater channels and pro-glacial lakes

When ice sheets expand and dam rivers, they can create **pro-glacial lakes**. For example, during the last ice advance (70 000–10 000 years ago), the North Yorkshire Moors remained largely unglaciated. As Figure 5 shows, they formed an ice-free 'island' surrounded by a vast ice sheet:

◆ Lake Eskdale, a pro-glacial lake, was formed when the North Sea ice sheet blocked the mouth of the River Esk. The level of the lake rose until its water found a new route over a low point in its southern watershed on the North Yorkshire Moors.

◆ The overflow river flowed through Lake Glaisdale before cutting the deep, narrow, steep-sided and flat-floored Newtondale Valley – a **meltwater channel**.

◆ At the end of this valley, the river formed a delta where it flowed into another pro-glacial lake (Lake Pickering). Lake Pickering, also dammed by North Sea ice, eventually found an outlet to the south-west, where it also formed a meltwater channel (the present-day Kirkham Gorge).

◆ When the surrounding ice sheet finally retreated, these pro-glacial lakes emptied. Today, Newtondale forms a narrow, wooded gorge 80 metres deep and 5 kilometres long; the site of the former Lake Pickering is now the fertile, flat-floored Vale of Pickering.

Kettle holes

Kettle holes form when large blocks of ice (left behind when glaciers retreat) are covered by deposits from meltwater streams. When the ice melts, a depression is left behind which then fills with water to form a kettle hole.

Glacial and fluvio-glacial deposits

Figure 6 shows a typical fluvio-glacial landscape – with a wide, multi-channelled (braided) river flowing over a vast area of sediment. This environment consists of two types of sediment: glacial and fluvio-glacial deposits. Their characteristics are described in Figure 7.

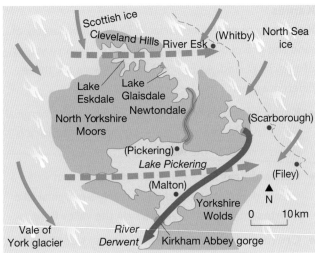

🔺 **Figure 5** *Pro-glacial lakes and meltwater channels in North Yorkshire*

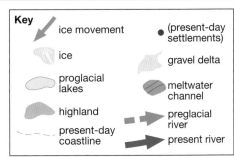

🔻 **Figure 6** *A meltwater river flowing from the Solheimajokull Glacier in Iceland*

Glacial deposits	Fluvio-glacial deposits (*These include deposits once deposited by ice and then re-deposited by meltwater*)
The deposits are unstratified; it's difficult to identify layers	The deposits are **stratified**; there are layers due to seasonal variations in sediment accumulation
The deposits are unsorted; when the ice melts, it deposits material regardless of size	The deposits are **sorted**; as the meltwater loses energy, larger rocks and boulders are deposited first
The material is angular (from physical weathering and erosion), as well as being of various shapes and sizes (from boulders to rock flour)	The material is smooth and rounded (due to attrition), as well as being sorted and **graded**

🔺 **Figure 7** *Characteristics of glacial and fluvio-glacial deposits*

Deposits are affected by **imbrication**, where glacial deposits are orientated (or aligned) – overlapping each other like toppled dominoes. They are aligned in two ways:

◆ Usually, the longest axis (see Section 7.3, where it's referred to as the 'a-axis') lies parallel to direction of flow. This is typical of sediment carried in suspension, where speed of flow decreases rapidly and sediments are deposited without any significant rolling. It occurs where deposits are churned up (known as **reworking**) by flash floods.

◆ In other cases, the a-axis lies perpendicular to direction of flow, where sediment has been carried along, rolling along the riverbed.

Figure 8 shows how sediment size and shape changes with distance from the glacier.

Fluvio-glacial landforms have different depositional characteristics. On **outwash plains**, the coarsest and largest material is deposited first, near the glacier). The finest material (clay) travels furthest across the plain. This is referred to as **sorting**.

◆ **Eskers** consist of sorted coarse material, usually sand and gravel, and are often stratified.

◆ **Kames** are mounds of coarse sand and gravel, which are sorted and stratified.

◆ **Meltwater streams** crossing outwash plains are **braided**, where variations in volume of meltwater lead to streams becoming choked with coarse material.

◆ A **varve** is a distinct layer of silt lying on top of a layer of sand, deposited annually in lakes near glacial margins. Coarser, lighter-coloured sand is deposited during late spring – when meltwater streams have peak discharge and carry maximum load. When temperatures fall towards autumn, and discharge decreases, finer, darker-coloured silt then settles. Each band of light and dark material represents one year's accumulation (see Figure 9). Because greater melting causes increased deposition, variations in the thickness of each varve indicates warmer and colder periods.

Site Number	Distance in metres from snout	Mean sediment size (cm)	Average sediment shape (Cailleux index). Greater R values equate to greater roundness.
1	2	17.4	160
2	18	16.3	147
3	48	11.2	211
4	76	5.9	239
5	107	3.5	218
6	145	4.7	498
7	256	2.4	652
8	356	2.1	755
9	487	1.8	766
10	702	2.4	812

⊙ **Figure 8** *How sediment changes with distance across an outwash plain*

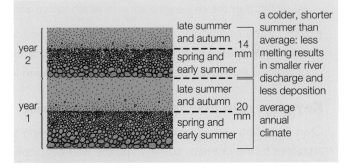

⊙ **Figure 9** *The formation of varves*

Over to you

1 In pairs, draw a spider diagram which shows the features of a fluvio-glacial landscape.

2 **a** Use Section 7.10 to help you carry out a Student t-test on sediment size and shape in outwash plains, using the data from Figure 8.

   b Use Section 7.5 to help you analyse the data in Figure 8 for central tendency in size and shape.

   c To what extent are the sediments in Figure 8 sorted by size and shape?

On your own

3 Distinguish between the following:
   (a) ice contact and pro-glacial features;
   (b) supraglacial, englacial and sub-glacial;
   (c) imbrication and sorting.

4 Draw a table to show the similarities and differences between eskers, kames and varves.

Exam-style questions

AS 1 Explain the formation of varves.
(6 marks)

A 2 Assess the significance of the contribution of glacial meltwater to the formation of glaciated landscapes (12 marks)

Glaciated landscapes in the UK

There are no glaciers remaining in the UK today. However, a number of distinctive **relict glaciated landscapes** – that is, relics of a previous climate – still exist in the UK. These landscapes developed when the climate was much colder and a large part of the UK was covered by ice (see Figure 1):

◆ 18 000 years ago, the upland areas of the Lake District, the Cambrian mountains, and Scotland were shaped by valley glaciers, which eroded the underlying rocks to create distinctive landscapes with features such as U-shaped valleys, ribbon lakes and corries (or cwms). These landscapes have enormous environmental, economic and cultural value, but they are challenging areas in which to live and work.

◆ Relict glaciated landscapes can also be found in low-lying areas that were covered by glacial deposits, such as the boulder clay of Holderness (which was deposited by Scandinavian ice).

Key players in managing the Lake District

Over 16 million visitors arrive in the Lake District every year – attracted by the dramatic scenery and the well-developed tourist facilities. The local economy benefits hugely from this influx (in 2014, visitors spent about £1.1 billion, which helped to support local shops, hotels, pubs and activity centres).

Many services that benefit tourists also benefit local people (e.g. better public transport and roads). Some of the money that tourists bring in is also used to protect the environment that visitors come to see. Tourism provides over 16 000 jobs (full-time equivalents) in the Lake District National Park, as well as boosting the local economy through **the multiplier effect**.

However, both the landscape and ecology of the Lake District are fragile and under threat from overuse. Activities such as walking, climbing and camping can lead to footpath erosion (see Figure 2), trampling and littering, which challenges the area's **resilience**. And the Lake District is also facing other problems:

◆ Additional tourist traffic causes congestion and pollution.

◆ Jobs in the tourism industry are often poorly paid and seasonal.

◆ An increased demand for housing (often for second or holiday homes) is driving up house prices, so local people can no longer afford to buy homes where they grew up. Up to a fifth of houses in the National Park are now second or holiday homes.

Footpath erosion

There are almost 2000 miles of footpaths and rights of way in the Lake District. However, the very people who come to enjoy these footpaths can inadvertently create major problems:

◆ Walkers destroy vegetation and compact the soil, which reduces infiltration rates.

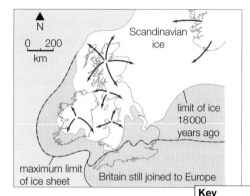

▲ **Figure 1** The ice cover over the UK 18 000 years ago created the relict glaciated landscapes of today

▲ **Figure 2** Coledale in the Lake District – before and after footpath repair. It costs about £100 for a metre of stone pitching or £20 000 for a helicopter to lift material onto the highest slopes.

- The exposed soil is also more easily washed away by heavy rain (which the Lake District receives a lot of).

- Gullies may then form along footpaths, which channel even more water and cause further erosion (see Figure 2).

- When walkers try to avoid badly eroded sections of footpath, they inevitably end up widening the path.

- Increased storms (as a result of climate change) are likely to worsen the problem.

The **Lake District National Park Authority (LDNPA)** has been working to repair footpaths for over 40 years. Teams of rangers and volunteers use various techniques for this, but it's a slow and expensive task.

Water storage, forestry and farming

Tourism is particularly important to the Lake District's economy, but the area also has other important uses, including farming, water storage and forestry.

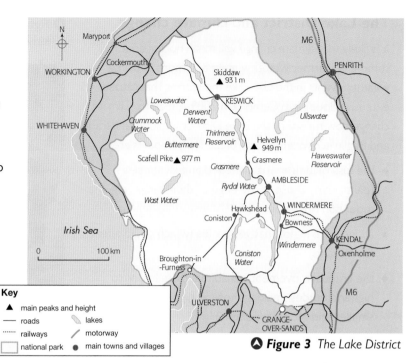

Key
- ▲ main peaks and height
- — roads
- ⋯⋯ railways
- ▢ national park
- ⬓ lakes
- ╱ motorway
- ● main towns and villages

⬣ **Figure 3** *The Lake District*

Water storage and forestry

Thirlmere lies between Keswick and Grasmere (see Figure 3). The dam at the northern end was constructed by Manchester Corporation in the late nineteenth century to provide water for the city's expanding population. Before the reservoir was built, this glacial valley contained two small tarns and a small hamlet with a pub; creating the reservoir submerged the tarns and the settlement. A 96-mile aqueduct, first connected in 1894, was built to carry water from Thirlmere to Manchester. This aqueduct is still in use today, but now it also carries water from Haweswater (which was turned into a reservoir in 1935).

The land surrounding Thirlmere is forested. This reduces soil erosion (which could cause siltation of the reservoir), and also generates income from selling the timber. The reservoir and its surrounding forest are both managed by United Utilities, who are responsible for looking after the wildlife and protecting the reservoir from pollution.

Farming

Langdale in the Lake District (see Figure 4) is typical of the landscape that tourists come to see. This landscape is dominated by farming. The trend is towards fewer and larger hill farms, but there has also been an increase in the number of the smallest holdings.

⬣ **Figure 4** *Farming in Langdale*

1 Fell tops over 600 metres (60% of all land). Used for sheep grazing in summer.

2 Lower slopes (30% of all land). Fields separated by dry-stone walls, used for raising sheep (for wool and lamb) between autumn and spring.

3 Flat valley floor (10% of all land). The most sheltered and fertile land, used for growing winter feed crops (e.g. hay), for jobs such as shearing, and for keeping a few cattle (for beef and milk).

The Lake District and climate change

It is likely that climate change will mean hotter drier summers, warmer wetter winters and more extreme weather events in the UK – like the storms that caused extensive flooding in Cumbria in 2009 and 2015. Climate change now threatens the Lake District's unique landscape and fragile ecosystem. Some of the likely impacts are:

◆ the loss of indigenous plant and animal species (particularly those at the edge of their range, e.g. the mountain ringlet butterfly)

◆ an increase in non-native species (which could affect food chains)

◆ the gradual movement of habitats from lower to higher altitudes (making those at higher altitudes smaller and more vulnerable)

◆ an increase in insect species, e.g. the midge which can infect cattle with Bluetongue disease (seen for the first time in Britain in 2007)

◆ that the heavy rain will wash more soil and farm chemicals into the lakes – causing siltation and eutrophication

◉ *Figure 5* *Storm Desmond caused extensive flooding in Carlisle, Keswick and other parts of the Lake District in December 2015*

◆ that the peat on the fells will dry out in the warmer summers (releasing stored carbon), and the dry moorland will also be more prone to fires

◆ that the forests will be at greater risk of damage from gales in winter and forest fires in summer

◆ that roads and properties will be damaged and cut off by winter floods (see Figure 5).

Key players in managing for the future

In 1951, the Lake District was designated as one of England's first National Parks. The LDNPA is responsible for managing the area, but only owns a tiny proportion of the land (3.9%). The rest of it is owned by a variety of organisations and private landowners.

The Lake District National Park Partnership was formed in 2006 to give organisations involved in the Park more say in its management. Twenty-five organisations (or **stakeholders**) are involved, which between them represent the public, private, community and voluntary sectors. They include: Action with Communities in Cumbria, Cumbria Association of Local Councils, Cumbria County Council, Environment Agency, Historic England, LDNP, National Trust and United Utilities.

Tackling climate change

In 2008, the 'Low-carbon Lake District' initiative was launched (see Figure 6). It's a programme to tackle climate change, and is working with local businesses, communities and other agencies to reduce greenhouse gases (**mitigation**) and prepare for the impacts of climate change (**adaptation**). Work to create a Low-carbon Lake District includes the following initiatives:

◆ Using a low-carbon budget, which measures carbon emissions from the local area and works to meet reduction targets.

◆ The GoLakes travel programme, with the aim of transforming how visitors get to – and travel around – the Central and Southern Lake District.

◆ Planning policies to meet the highest energy efficiency standards and integrate low-carbon energy generation where possible.

◉ *Figure 6* *Low-carbon Lake District official logo*

Low-carbon
Lake District

The vision for 2030

The Lake District National Park Partnership's vision for 2030 is spelled out in the text box on the right. Figure 7 shows how this will be achieved. The Lake District National Park is focussed on the future. Strategies cover everything from biodiversity, water quality, farming, skills and training – to employment, housing, transport and tourism.

The vision for the Lake District National Park in 2030

The Lake District National Park will be an inspirational example of sustainable development in action. It will be a place where a prosperous economy, world-class visitor experiences and vibrant communities all come together to sustain the spectacular landscape, its wildlife and cultural heritage. Local people, visitors, and the many organisations working in the Lake District – or who have a contribution to make to it – must be united in achieving this.

Elements of the vision	Examples of how will this be achieved
Creating a prosperous economy	Businesses will locate in the National Park, because they value the opportunity, environment and lifestyle that it offers. Traditional industries will be maintained to ensure a diverse economy.
Achieving a world-class visitor experience	Developing Lakes Culture (a cultural tourism plan). Negotiating a return of the Cumbrian stage in the Tour of Britain cycle race. Installing new steamer jetties on Lake Windermere.
Helping vibrant communities	Limiting holiday home ownership. Improving access and travel in the Central Lakes. Developing a car-parking strategy across the National Park.
Maintaining a spectacular landscape, as well as its wildlife and cultural heritage	Improving and protecting water bodies in the Lake District. Improving biodiversity. Assessing the value and benefits of the land and agreeing priorities for its management. Bidding for World Heritage status for the Lake District National Park in 2016.

Figure 7 *Achieving the vision for 2030*

Over to you

1 Create a table to summarise the opportunities and the threats that large numbers of visitors bring to the Lake District.

2 a In pairs, read this section and identify the main stakeholders (or players) in deciding the future of the Lake District.

  b Identify the values of each player – are these economic, social or environmental in nature? Explain.

  c Decide which players are likely to agree about **(i)** increasing the economic resilience of the Lake District and how this might be done, **(ii)** seeing the landscape as something of special quality which needs to be preserved.

3 Extreme weather events have become more frequent, such as the major flooding caused by Storm Desmond in December 2015, and drought in June 2010.

  Choose one of these events and use the Internet to research its causes, its impacts, and the resilience of the Lake District in coping with it.

On your own

4 Define these terms: **(a)** resilience, **(b)** mitigation and **(c)** adaptation.

5 To what extent does tourism reduce the resilience of the Lake District landscape?

6 Discuss in 500 words the view that climate change could bring even greater challenges to the Lake District in future.

Exam-style questions

 1 Assess the value of one glaciated landscape *(12 marks)*

 2 Assess a range of different threats to one glaciated landscape *(12 marks)*

In this section, you'll learn about the value of the Sagarmatha (Everest) National Park, the threats facing it and how these are managed.

The Sagarmatha National Park

The Sagarmatha National Park (see Figure 1) includes the highest mountain on Earth: Everest (at 8848 metres). It's an area of dramatic mountains, glaciers and deep valleys – and its unique and fragile ecosystem is home to rare species such as the snow leopard. The National Park is a well-known destination for mountain tourism (over 37 000 people visited the area in 2014), and is also home to over 6000 Sherpa people.

Sagarmatha was established as a National Park in 1976, and as a UNESCO World Heritage Site in 1979. Like many **active glaciated areas**, the National Park has enormous environmental and cultural value. However, it also faces threats from both natural events and human activities.

Tourism – opportunities and threats

The tourism industry in the Sagarmatha National Park is largely in the hands of the Sherpa people. However, outsiders from other parts of Nepal – as well as foreigners – are increasingly establishing businesses in the area, and the number of migrant workers is also rising. Increasing tourist numbers have boosted the local economy, which has led to improved standards of living, better health care, education and infrastructure.

However, despite attempts to protect the National Park, tourism does cause environmental damage and socio-economic change. The direct impacts include:

◆ footpath erosion
◆ the construction of illegal trails
◆ water pollution
◆ problems with waste disposal (see Figure 2)
◆ a greater demand for forest products, such as firewood
◆ an increased demand for new hotels and lodges.

There are also indirect impacts – particularly on the Sherpa community, whose way of life is changing as they adapt to visitor demands.

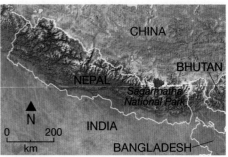

▲ **Figure 1** The Sagarmatha National Park in Nepal

▲ **Figure 2** Climbers and trekkers often leave behind piles of rubbish, which blight the area and cause environmental problems.

BACKGROUND

The Sherpa people

The Sherpa people are a tribe of Tibetan origin. Most live in north-eastern Nepal, around Mount Everest. In the Tibetan language, their name means eastern people, from 'shar' (east) and 'pa' (people).

93% of Sherpas are Buddhists. In addition to Buddha, and other Buddhist divinities, the Sherpa believe in numerous deities who are thought to inhabit mountains, caves and forests. Many of the Himalayan mountains are worshipped, and the Sherpas call Mount Everest Chomolungma and worship it as the 'Mother of the World'.

Their culture and religion includes the restriction of animal hunting and respect for all living things. These beliefs and conservation practices, have contributed to the successful management of the Sagarmatha National Park.

Traditionally, Sherpa economic activities are based on farming and trade. The Sherpas also have a reputation as expert mountaineers and guides, which has brought some of them a new source of income.

Climbing Everest

As tourists with more money and leisure time seek out unusual holidays, some of them turn to adventure tourism in wilderness areas such as the Himalayas. Commercial companies guide these people, and the chances of a successful 'summit' have now increased due to improvements in equipment (and the number, and standard, of mountain guides). Improved weather forecasting also

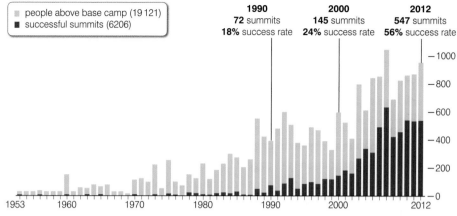

1990 — 72 summits, 18% success rate; 2000 — 145 summits, 24% success rate; 2012 — 547 summits, 56% success rate

people above base camp (19 121)
successful summits (6206)

Figure 3 *More people are climbing Everest than ever before, but it remains dangerous and can lead to serious injury or death*

enables group leaders to select 'weather windows' which give them the best opportunity to successfully, and safely, take their clients to the summit. Over 6000 people have now reached the summit of Everest (see Figure 3). About 90% of these were guided clients (many with limited climbing skills, but willing to pay thousands of pounds).

The two main routes up Everest are dangerously crowded during the peak climbing season, and badly polluted with piles of rubbish, abandoned equipment and human waste (see Figure 2). New rules mean that all waste is now removed from Everest Base Camp, but camps at higher altitudes remain seriously polluted.

The increasing numbers of climbers on Everest (see Figure 4) are causing problems, and some mountaineers are now calling for tighter controls, including:

♦ limiting the number of permits given to climbers each year

♦ restricting group sizes to reduce dangerous traffic jams, particularly at the Hillary Step on the southern ascent from Nepal

♦ insisting that all guides are properly qualified and experienced.

Figure 4 *A long line of climbers making their way up the Lhotse Face towards the summit of Everest, looming above them*

Climate change and shrinking glaciers

Although Nepal generates a negligible amount of greenhouse gas emissions itself, it's particularly vulnerable to the impacts of climate change, and there is growing evidence that glaciers in the Himalayas are retreating as a result. In 2013, researchers at the University of Milan found that some glaciers around Mount Everest had shrunk by 13% in the last 50 years – with the snowline 180 metres higher than it was 50 years ago. Using satellite and topographical images, they suggested that glaciers are disappearing faster each year (with some smaller glaciers now only half the size they were in the 1960s).

A reduction in the size of Himalayan glaciers will have dire consequences for farming and hydropower generation downstream. Glacier retreat can also lead to the creation and growth of lakes dammed by glacial debris. Avalanches and earthquakes can then breach these natural dams, causing catastrophic floods (see page 88).

Glacial outburst floods

When glaciers melt, glacial mass balances (see page 62) are disrupted, which in turn risks disrupting the hydrological cycle (see page 93), and in the long term will lead to a reduced water supply for downstream communities.

The formation and growth of glacial lakes is closely related to deglaciation (see page 87). Lakes blocked by moraine are often found close to the snouts of valley glaciers. These lakes increase in size if glaciers melt, and the moraine wall may collapse – causing a glacial lake outburst flood. In August 1985, one such flood caused a 10-15 metre high surge of water and debris to flood down the Bhote Koshi and Dudh Koshi rivers in Nepal – destroying, amongst many things, the Namche Small Hydro Project. Continued increases in summer warming could lead to more floods and the possible destruction of settlements, irrigation systems and domestic water supplies in communities downstream.

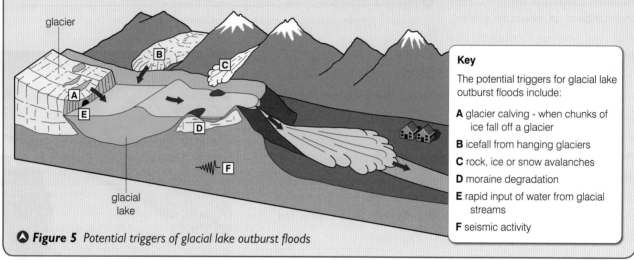

Key

The potential triggers for glacial lake outburst floods include:

A glacier calving - when chunks of ice fall off a glacier

B icefall from hanging glaciers

C rock, ice or snow avalanches

D moraine degradation

E rapid input of water from glacial streams

F seismic activity

🔺 *Figure 5* *Potential triggers of glacial lake outburst floods*

Avalanches

On 25 April 2015, a 7.8 magnitude earthquake struck Nepal. Mount Everest was approximately 220 km east of the epicentre, but the earthquake triggered several large avalanches on and around the mountain. One avalanche swept into South Base Camp and then through part of the Khumbu Icefall. It was the deadliest avalanche ever experienced on Everest (killing at least 22 people, including ten Sherpas). In April 2014, an earlier avalanche killed 16 Nepalese guides (mostly Sherpas).

Deforestation and landslides

Glaciated landscapes face threats not only from natural hazards (such as avalanches and glacial outburst floods) but also from human activities, which can reduce the resilience of these fragile environments.

Nepal was once heavily forested, but less than 30% of the country's natural forest now remains. This deforestation has been caused by:

- farming (pressure on land means that steep hillsides are cleared of natural vegetation and replaced with crops, and cows, water buffalo and goats are also allowed to graze uncontrolled)
- the use of firewood as the main source of fuel
- the clearance of forested areas to build roads, reservoirs, hydroelectricity projects, etc.

The dire consequences of this deforestation include:

- the loss of wildlife habitats (and hence a loss of biodiversity)
- the exposure of the soil, which causes nutrients to be washed away – leaving the soil infertile and lowering crop yields
- the erosion of the exposed soil, particularly on steep hillsides
- a significantly increased risk of landslides
- the disruption of the water cycle.

Managing Sagarmatha for the future

The stunning scenery and environment are Sagarmatha's greatest assets. Sustainable management is vital in order to balance its use and preservation. Some **stakeholders** involved in managing the National Park are shown below. Strategies to protect the Park include:

◆ establishing plant nurseries (Figure 6) to provide seedlings to re-establish forests on hill slopes and reduce erosion

◆ setting up projects sponsored through the Sir Edmund Hillary Foundation, including building schools, hospital and bridges

◆ banning goats (to protect the mountain vegetation)

◆ using kerosene (not firewood) for cooking and heating

◆ building micro HEP stations to generate electricity for local use

◆ limiting some development projects (including the extension of the Sanboche airport).

▶ *Figure 6* *Local children being taught about re-afforestation at a plant nursery in Sagarmatha National Park.*

Managing the Sagarmatha National Park – key stakeholders and players

1 Global organisations
◆ A UNESCO World Heritage Site since 1979.

2 Government agencies
◆ The National Park was established in 1976 and is managed by the National Park and Wildlife Conservation Office. A buffer zone was added in 2002, to enhance its protection.

3 Local residents and stakeholders
◆ 6000 people live in the area – mostly Sherpa.

◆ Other residents are incomers who have set up businesses, e.g. guest houses, adventure holidays

◆ A Park Advisory Committee, with local leaders, village elders, head lamas (Buddhist spiritual leaders) and park authority representatives has been established.

4 NGOs
◆ Sagarmatha Pollution Control Committee is a community based NGO involved in pollution control and rubbish disposal.

Over to you

1 Study Figure 3. Explain the changes shown in the numbers of people climbing above Everest base camp.

2 **a** Consider the players in the Sagarmatha National Park. Identify those players whose indirect actions are altering natural systems.

  b Draw a spectrum ranging from exploitation to preservation of landscapes and resources. Identify where each player belongs, as suggested by their actions and attitudes.

On your own

3 **a** How could earlier snow melt and climate change affect river discharge patterns?

  b What impacts might this have on people living downstream?

4 The Himalayas are vulnerable to the impacts of climate change. Explain why management strategies have focussed on adaptation, rather than on reducing greenhouse gas emissions.

Exam-style questions

 1 Assess the nature of the threats from tourism facing one named fragile glaciated upland landscape. *(12 marks)*

 2 Evaluate the varying ways in which fragile glaciated upland landscapes can be managed. *(20 marks)*

A fragile ecosystem

It's said that if you tread on tundra vegetation, your footprint might be visible for a decade. Tundra (found in glacial and periglacial landscapes; see Figure 1) is unique and fragile – one of the Earth's coldest and harshest ecosystems.

Tundra ecosystems are surprisingly diverse, but they also develop very slowly. Tundra plants and animals have adapted to the harsh climate and short growing season, but they are very sensitive to change – once damaged, they may never recover. They are highly vulnerable to environmental stresses, such as reduced snow cover and warmer temperatures caused by global warming.

Arctic tundra

The Arctic is a harsh environment, with long, dark winters and short summers. The growing season only lasts for about three months – when average temperatures rise to 12°C. At this time the surface layer of permafrost melts to form bogs and shallow lakes. These attract insects, which in turn attract migrating birds. The Arctic tundra ecosystem also supports a variety of large animals, including Arctic foxes, grey wolves, snow geese and musk oxen (see Figure 2). In winter, temperatures fall well below freezing. Plants – which many animals rely on for food – must survive under the snow to re-emerge and flower quickly once temperatures rise again in the spring.

How global warming is changing the Arctic tundra

Rapid warming in the Arctic has contributed to more extensive melting of the sea ice in the summer months, as well as greatly reduced snow cover and a reduction in the permafrost. Shrubs and trees that previously could not survive in the tundra environment have now started to grow there. The same is true of animals (e.g. in Alaska, the red fox has started to spread northwards, and is now competing with the Arctic fox for food and territory). The Arctic region is now warming twice as fast as the global average. This phenomenon is known as **Arctic amplification**.

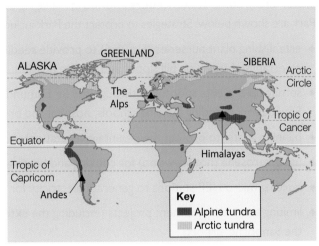

▲ *Figure 1* *Tundra ecosystems are found in the Arctic and on mountains close to the snowline – where the climate is cold, rainfall is low and the land is snow-covered for much of the year*

▲ *Figure 2* *Musk oxen in Greenland's Arctic tundra – a periglacial landscape*

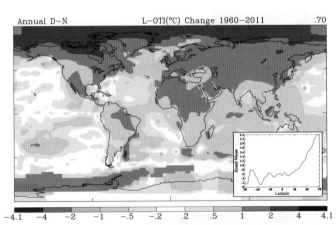

▲ *Figure 3* *Trends in mean surface air temperature (1960 to 2011) show an increase over the Arctic region of more than 2°C. This is supported by the inset graph of temperature increase relating to latitude.*

What is 'Arctic amplification'?

The Arctic acts like a fridge by giving off more heat than it absorbs. Scientists are concerned that this process may be affected by three positive feedback loops explained below, which are expected to worsen as climate changes.

1 As the extent of the white polar sea ice reduces, a smaller fraction of the sun's radiation is reflected back into space. More solar energy is then absorbed by the darker ice-free water – raising temperatures that further reduce global ice cover.

2 When permafrost melts, it releases trapped carbon into the atmosphere as carbon dioxide (CO_2) and methane (CH_4) – thus increasing concentrations of greenhouse gases in the atmosphere. This in turn leads to increased global temperatures and further melting (see Figure 4).

> ❱ *Figure 4* *Melting permafrost positive feedback loop*

3 Lower levels of snow cover increase areas of bare rock exposed to solar energy. Heat absorption from the sun then increases, leading to increased temperatures and snowmelt.

Together, these three feedback loops exacerbate global warming – hence the term Arctic amplification.

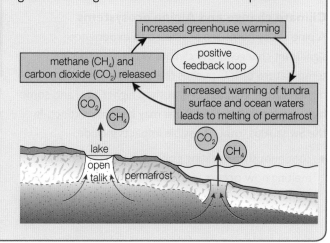

Changes to the carbon cycle

As much as 14% of the Earth's carbon is stored in permafrost, accumulating because the cold climate limits decomposition of organic matter.

However, not all agree that melting permafrost will release stored carbon as CO_2 or CH_4, (see above). Some studies show that – as permafrost thaws – stored carbon actually remains in the soil to be used by new vegetation.

Although warmer temperatures accelerate decomposition – releasing carbon – they also increase the release of nutrients. Nutrients encourage plant growth and the removal of carbon from the atmosphere through photosynthesis, decreasing the level of greenhouse gases in the atmosphere. Increased plant growth is therefore a **negative feedback** opposing Arctic amplification.

Scientists at Alaska's Arctic Long Term Ecological Research Site set up artificial warming plots over 20 years, and found increased growth of shrubs at the expense of mosses, sedges and grasses. Measurements show no change in the amount of carbon below ground, which means that carbon released due to melting permafrost has been balanced by carbon inputs into the soil as a result of more plant growth and the decomposition of plant litter.

However, scientists are cautious about what this might mean for the future, for two reasons:

1 The effects observed may be transitional, and further warming could lead to a major release of carbon.

2 Although carbon is still being stored on land, the opposite is true of lakes, streams and bogs – where more carbon is released than stored (see Figure 5).

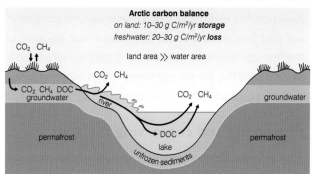

CO_2 is taken in by plants through photosynthesis. When plants die and decompose, their carbon either goes back into the air, or moves down into groundwater and unfrozen sediments in the form of CO_2, CH_4 or dissolved organic carbon (DOC). The CO_2, CH_4 and DOC are then carried into rivers and lakes, where sunlight and microbes break down DOC and release CO_2 and CH_4 back into the air.

▲ *Figure 5* *The Arctic carbon balance*

Alpine tundra

The vegetation in mountain or Alpine tundra consists of grasses and low-lying plants. Hardy cushion plants survive by growing in rock depressions where it's warmer and they are sheltered from the wind. In the warmer summer months, wildflowers attract insects and birds. This biome is also home to mountain goats and sheep, as well as marmots (large burrowing rodents).

Climate change and Alpine ecosystems

Alpine ecosystems are predicted to experience significant change as the climate warms – a rise in temperature of 1°C pushes the tree line 100 metres higher. Other vegetation is pushed higher too, and some high-mountain flora and fauna may become extinct. In the Swiss Alps, other possible impacts include:

◆ continuously retreating glaciers, because significant melting now occurs in the summer months. This in turn may threaten glacial lake outburst floods (see page 88) – causing dangerous flooding and mudslides.

◆ the possible closure of ski resorts at lower altitudes, as research suggests that by 2050 only those resorts at 1500 metres and above will be able to offer skiing throughout the winter.

◆ increasing melting of the permafrost, which threatens rock avalanches and mudslides. Permafrost covers about 5% of Switzerland. Several mountain settlements are under threat, and so are ski lifts (the supports of which are anchored in permafrost rather than solid rock).

Research has estimated that a rise in temperature of 2°C will cost Switzerland about £2 billion a year, because of the impacts on winter tourism and the measures needed to deal with the increase in flooding and other natural disasters caused by rising temperatures.

Protecting and conserving the Alps

The Alpine Convention is an international treaty and its members – the EU and Alpine countries (Austria, France, Germany, Italy, Liechtenstein, Monaco, Slovenia and Switzerland) – are working together to protect the natural environment, while promoting its economic development. It's an attempt to work on sustainable development within a **legislative framework**. The Convention covers a range of subjects, including water, soil conservation, landscape protection, transport, energy and climate. Two examples are included here.

▲ **Figure 6** *Alpine tundra in the Rocky Mountains, USA*

> **Key word**
>
> **Legislative framework** – A legally binding set of rules, such as the Alpine Convention.

An uncertain future

Successful management of tundra regions is challenging. And some difficult questions remain that we don't yet know the answers to, such as:

◆ How much carbon is locked up in permafrost?

◆ How much will be released into the atmosphere as the permafrost thaws?

◆ How much carbon could be released as CH_4, and how much as CO_2?

If governments around the world knew the answers to these questions, it could help them to decide what to do. There is a clear need for coordinated approaches at a global scale. The 2015 Paris climate agreement (COP21) made little mention of the Arctic, or the indigenous peoples living there, and while developed countries pledged funds to help developing nations with climate **adaptation** and **mitigation**, similar pledges were not made for Arctic communities.

> **The Alpine Convention – nature and landscape protection**
>
> More than 20% of the Alpine area consists of national parks or protected areas. These areas have a great range of biodiversity. The Alpine Convention includes measures to protect, care for and restore ecosystems as well as preserve the natural living environments of wild animal and plant species.

The Alpine Convention – Water

The Alps are known as the 'water towers of Europe'. Alpine rivers transport an average of 216 km³ of water a year to nearby regions. In the summer months, much of Europe's water comes from the melting of Alpine glaciers. But, according to climate scientists, in less than 100 years the Eastern Alps (as well as a large part of the Western Alps) will be completely ice-free.

The **hydrological cycle** shown in Figure 7 would be affected by this change. Precipitation in the form of snow would diminish and rainfall patterns change. River discharge patterns would also change with greater flooding in winter and droughts in summer. Not only that, but as the climate changes and the glaciers melt, water flows will increase initially, with equivalent increases in sediment yield. As the glacier retreats, however, river discharge would be reduced, together with sediment yield. Water would increasingly become cloudy and sediment-laden, decreasing water quality.

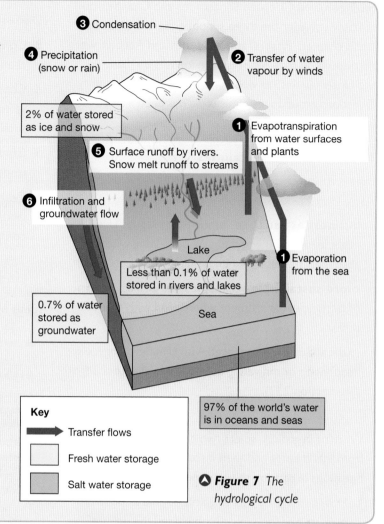

Figure 7 The hydrological cycle

Over to you

1 Draw a food web to show the unique biodiversity of the Arctic tundra, researching the Internet for examples of plants and animals.

2 a In pairs, identify the players whose indirect actions are likely to alter (or are already altering) natural systems in the tundra through climate change.

   b Explain how and why these actions threaten tundra ecosystems and the likely impacts on people and the environment.

3 Look at Figure 7.

   a Explain the role played by tundra ecosystems in maintaining the global water cycle.

   b Identify ways in which global warming may impact on or alter the water cycle in Arctic or Alpine tundra regions.

On your own

4 a Study Figure 3. Show how the Arctic region is at greatest risk from global warming.

   b Describe how positive feedback loops may exacerbate climate change and its consequences.

   c Copy Figure 4 and annotate it to show **(i)** positive feedback loops, **(ii)** the negative feedback loop suggested by some scientists.

5 a Research the 2015 Paris climate conference agreement (COP21). Identify those parts of the agreement which may impact on the Arctic.

   b Suggest reasons why many indigenous peoples of the Arctic were unhappy with the outcome of COP21.

Exam-style questions

 1 Explain how the impacts of climate change in the Arctic may affect the hydrological cycle. *(6 marks)*

A 2 Evaluate the indirect actions of players which threaten fragile active and relict glaciated upland landscapes *(20 marks)*

Having studied Glaciated Landscapes and Change, you can now consider the three synoptic themes embedded in this chapter. 'Players' and 'Attitudes and Actions' were introduced on page 47. This page focuses on 'Futures and Uncertainties', as well as revisiting the four Enquiry Questions around which this topic has been framed (see page 47).

3 Futures and Uncertainties

People approach questions about the future in different ways. They include those who favour:

- **business as usual**, i.e. letting things stand. This might involve doing nothing, or only doing what's absolutely necessary when it's unavoidable.

- **more sustainable strategies**, particularly in terms of landscape management when faced with tourist threats.

- **radical action** when faced with climate change, e.g. adapting to or mitigating threats from climate change (see Section 2.13).

Working in groups, select two glaciated landscapes from this chapter and then discuss the following questions in relation to each one:

1 How far have threats to glaciated landscapes from overuse and/or exploitation passed the point of no return?

2 Is there evidence to suggest that radical actions are needed to manage the threats to glaciated landscapes from climate change?

Revisiting the Enquiry Questions

These are the key questions that drive the whole topic:

1 How has climate change influenced the formation of glaciated landscapes over time?

2 What processes operate within glacier systems?

3 How do glacial processes contribute to the formation of glacial landforms and landscapes?

4 How are glaciated landscapes used and managed today?

Having studied this topic, you can now consider answers to these questions.

Discuss the following questions in a group:

3 Consider Sections 2.1-2.4. How far is climate change just a natural global process?

4 Consider Sections 2.5-2.7. How far are glacier systems different from other landscape systems, such as rivers or coasts? Explain your views.

5 Consider Sections 2.8-2.10. What makes glaciated landscapes unique? Justify your views.

6 Consider Sections 2.11-2.13. How far are glaciated landscapes able to cope with pressures from competing users?

Books, music, and films on this topic

Books to read

1. *Living Ice: Understanding Glaciers and Glaciation* by Robert P. Sharp (1991)

   This book explains how dynamic glaciers are – and how fragile they are.

2. *Environmental and Human Security in the Arctic* by Gunhild Hoogensen, Gjorv, Dawn Bazely, Goloviznina Marina, Andrew Tanentzap (2012)

   This book explores in depth how the security of the Arctic is essential to humans, as well as the environmental and political threats that it faces.

3. *After the Ice Age* by E. C. Pielou (2012)

   This book assesses the impacts of change by which glaciated landscapes transformed gradually into the forests, grasslands, and wetlands of today.

Music to listen to

1. 'Glacier' – James Vincent (2014)

   This song describes some of the properties that glaciers possess.

2. 'Glacier Song' – Front Country (2014)

   The lyrics describe what a glacier does, as well as some of the threats facing glaciers around the world.

Films to see

1. *Chasing Ice* (2012)

   A documentary film that captures breath-taking images of the impacts of climate change on glaciers in Alaska, Greenland and Iceland.

2. *Ice and the sky* (2015)

   A documentary about the work of French glaciologist Claude Lorius – a man credited with being involved in many discoveries about glaciers, as well as measuring the impact that climate change has had, is having, and could have in the future.

3. *Ice Age: Continental Drift* (2012)

   An animated film that looks at life on a glaciated landscape millions of years ago, and how a changing environment such as this affects wildlife.

Coastal landscapes and change

Chapter overview – introducing the topic

This chapter studies coastal landscapes and the physical processes that form them. These distinctive landscapes are being changed by physical and human processes, which are affecting their future.

In the Specification, this topic has been framed around four Enquiry Questions:

> 1 Why are coastal landscapes different and what processes cause these differences?
>
> 2 How do characteristic coastal landforms contribute to coastal landscapes?
>
> 3 How do coastal erosion and sea level change alter the physical characteristics of coastlines and increase risks?
>
> 4 How can coastlines be managed to meet the needs of all players?

The sections in this chapter provide the content and concepts needed to help you answer these questions.

Synoptic themes

Underlying the content of every topic are three synoptic themes that 'bind' or glue the whole Specification together:

> 1 Players
>
> 2 Attitudes and Actions
>
> 3 Futures and Uncertainties

Both 'Players' and 'Attitudes and Actions' are discussed below. You can find further information about 'Futures and Uncertainties' on the chapter summary page (page 146).

1 Players

Players are individuals, groups and organisations involved in making decisions that affect coastal landscapes. They can be national or international individuals and organisations (e.g. inter-governmental organisations like the UN), national and local governments, businesses (from small operations to the largest TNCs), plus pressure groups and non-governmental organisations.

Players that you'll study in this topic include:

- Sections 3.8 and 3.10 – How **economic users** of coastal landscapes (e.g. tourists and businesses) as well as **residents** affect coastlines.

- Sections 3.8, 3.9, 3.10 and 3.11 – How **local and national governments** attempt to manage coastlines through policy decisions.

- Sections 3.9 and 3.10 – How coastal management involves **planners** and **engineers** in modifying coastal processes through engineering defences.

- Sections 3.8, 3.10 and 3.11 – How **environmentalists** are concerned about the impacts of human activities on coasts, and their management.

2 Attitudes and Actions

Actions are the means by which players try to achieve what they want. For example, those attempting to manage coastlines seek policies that allow them to manage competing demands, while climate scientists seek to mitigate and adapt to coastal futures threatened by climate change.

Actions could pose the following questions:

1 **Economic versus environmental interests.** Faced with pressure from economic users, how can coastal environments be protected? Or should such landscapes be exploited for their economic worth? See Sections 3.10 and 3.11.

2 **Management.** How far have the actions of different players altered natural systems operating in coastal landscapes, and how might they have unforeseen consequences? See Sections 3.8 and 3.10.

> In groups, consider and discuss two coastal landscapes from this chapter where actions taken by governments (national and/or international), environmental or climate scientists have proved to be **(a)** effective, and **(b)** ineffective in the face of threats from different users. Explain your views.

In this section, you'll find out how different coastal landscapes result from the systems that produce them.

How coastal landscapes vary

The UK's coastline is an incredible 31 368 km long (if you include all the main islands). It varies enormously – from tropical-looking Luskentyre in the Outer Hebrides, to the jagged rocky coasts of Cornwall; from the low-lying muddy estuarine coast of The Wash, to the broad sandy beach and dunes at Bamburgh in Northumberland (all illustrated in Figure 1 opposite).

The role of geology

Resistant rock coastlines

As the UK's most southwesterly peninsula, Cornwall bears the brunt of the worst of the weather rolling in from the Atlantic Ocean. Due to its **geology** (rock type), Cornwall's **rocky coastline** can withstand the frequent winter storms without suffering from rapid erosion. Like the rest of western Britain (as shown in Figure 1), much of Cornwall consists of older resistant rocks, including:

◆ **igneous** rocks (such as basalt and granite)

◆ older compacted **sedimentary** rocks (such as old red sandstone)

◆ **metamorphic** rocks (such as slates and schists).

◆ These rocks are all resistant to the erosive power of the sea, wind and rain.

Coastal plain landscapes

In comparison with western and northern Britain, eastern and southern coasts consist of areas of weaker and younger sedimentary rocks – including chalks, clays, sand and sandstone. The Wash (shown in Photo C) is an area of **low, flat relief** – referred to as a **coastal plain**. At 20 km wide and 30 km long, The Wash is the largest **estuary** system in the UK (formed by the four rivers: Great Ouse, Witham, Welland and Nene), with a range of habitats from tidal creeks to mud flats, salt marshes and lagoons. However, much of the coast of eastern England consists of low-lying **sandy** beaches, such as Bamburgh beach in Northumberland (see Photo D) and the holiday resorts on the Lincolnshire and Norfolk coasts.

In reality, many coasts are a mixture of high- and low-energy environments. For example, while some coasts may be predominantly low energy (such as Holderness in the East Riding of Yorkshire), winter storms can still create a short-term high-energy erosional environment – and local geology can also create headlands (such as the chalky cliffs at Flamborough Head). Similarly, estuaries in western Britain, such as the Camel estuary in Cornwall, can still form if they are sheltered from the power of the worst Atlantic waves.

High-energy coastlines

Rocky coasts are generally found in **high-energy environments**. In the UK, these tend to be:

◆ stretches of the Atlantic-facing coast, where the waves are powerful for much of the year (such as Cornwall or north-western Scotland)

◆ where the rate of erosion exceeds the rate of deposition.

Erosional landforms, such as headlands, cliffs, and shoreline platforms (sometimes referred to as wave-cut platforms, even though they are not!), tend to be found in these environments.

Low-energy coastlines

Sandy and estuarine coasts are generally found in **low-energy environments**. In the UK, these tend to be:

◆ stretches of the coast where the waves are less powerful, or where the coast is sheltered from large waves (such as Lincolnshire and Northumberland).

◆ where the rate of deposition exceeds the rate of erosion.

Landforms such as beaches, spits and coastal plains tend to be found in these environments.

▶ **Figure 1** *The geology of the UK and how coastal landscapes vary*

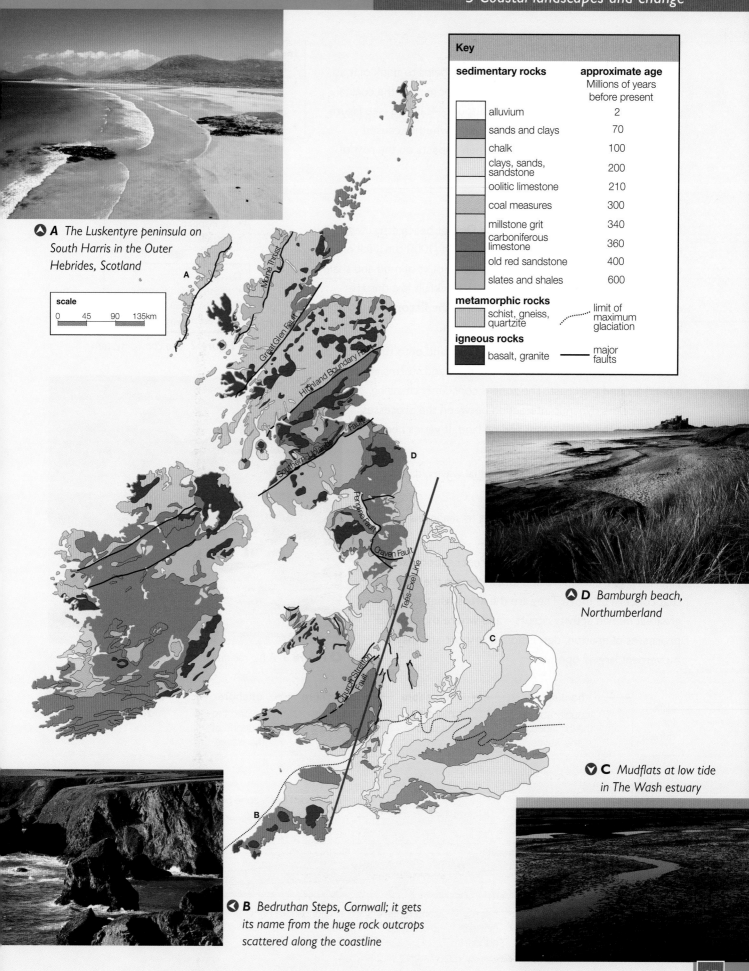

A The Luskentyre peninsula on South Harris in the Outer Hebrides, Scotland

scale

0   45   90   135km

Key

sedimentary rocks	approximate age
	Millions of years before present
alluvium	2
sands and clays	70
chalk	100
clays, sands, sandstone	200
oolitic limestone	210
coal measures	300
millstone grit	340
carboniferous limestone	360
old red sandstone	400
slates and shales	600

metamorphic rocks

schist, gneiss, quartzite   ········ limit of maximum glaciation

igneous rocks

basalt, granite   —— major faults

Moine Thrust

Great Glen Fault

Highland Boundary Fault

Southern Upland Fault

Pennine Fault

Craven Fault

Tees-Exe Line

Church Stretton Fault

D Bamburgh beach, Northumberland

C Mudflats at low tide in The Wash estuary

B Bedruthan Steps, Cornwall; it gets its name from the huge rock outcrops scattered along the coastline

The coast as a system

Because the coast is constantly changing, it helps to think of it as a system driven by wave energy. Each component of the coastal system – the inputs, processes and outputs – is linked (see Figure 2). Any change to a particular component (whether caused naturally or by human intervention) then impacts on the rest of the system.

The littoral zone

In December 2015, a big rockfall on a Dorset beach attracted many fossil hunters. Over 100 metres of cliff (1000 tons) fell onto the beach at Charmouth – exposing fresh blocks of mud and shale that were full of Dorset's famous fossils. Rockfalls like this are common on Dorset's coast. They occur in the **littoral**, or coastal, **zone** – the boundary between land and sea.

The littoral zone stretches out into the sea and onto the shore. It's a zone, rather than a line, because tides and storms affect a band around the coast. The zone is constantly changing, because of the dynamic interaction between the processes operating in the seas, oceans and on land. It varies because of:

◆ short-term factors (such as individual waves, daily tides and seasonal storms)

◆ long-term factors (such as changes to sea levels or climate change).

The littoral zone is divided into different sections (see Figure 4). The backshore and foreshore are the sections that concern us most. They are the areas where the greatest human activity occurs – and where the physical processes of erosion, deposition, transport and mass movement largely operate.

Inputs
- Marine – waves, tides, storm surges
- Atmospheric – weather/climate, climate change, solar energy
- Land – rock type and structure, tectonic activity
- People – human activity, coastal management

Processes
- Weathering
- Mass movement
- Erosion
- Transport
- Deposition

Outputs
- Erosional landforms
- Depositional landforms
- Different types of coasts

Figure 2 The coastal system

Figure 3 A big rockfall at Charmouth beach, Dorset

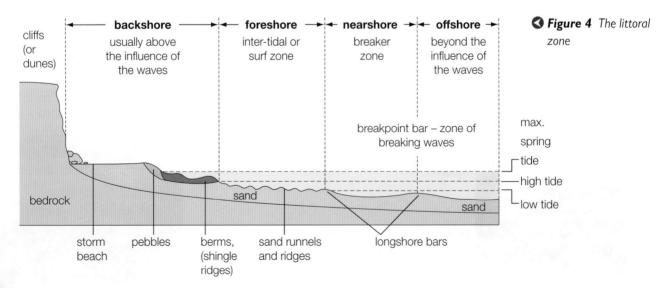

Figure 4 The littoral zone

Sediment supply

The processes of weathering and erosion (see Figure 2) produce output in the form of sediment, which is then transported and deposited to produce coastal landforms. The sources of sediment are complex. In The Wash (Figures 1 and 5), the sediment originates as follows:

◆ The main source is from cliffs eroding between West Runton and Weybourne, east of The Wash (along the north Norfolk coast). These cliffs have retreated at about 1 metre per year for thousands of years. Because they are sandstone, 60% of sediment consists of sand.

◆ Some sediment comes from **tidal currents**, which pick up glacial deposits from the shallow sea floor.

◆ The erosion of the Holderness cliffs further north also provides some sediment, which is carried southwards in suspension.

◆ Sand is carried southwards along the Lincolnshire coast.

◆ Four rivers also discharge into The Wash – bringing very fine sediment.

The fact that the sediment comes from two different directions – from the north and the east – illustrates what are known as **sediment cells** (see pages 116–117). The Wash and the Norfolk coastline form part of one of eleven sediment cells around the English and Welsh coasts.

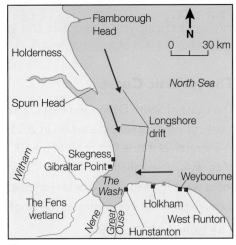

▲ *Figure 5* *The sources of the sediment for The Wash and north Norfolk beaches*

Classifying coasts

Coasts can be classified depending on:

◆ their **geology** – which can create rocky, sandy and estuarine coasts, as well as concordant and discordant coasts (see pages 102–04)

◆ the level of **energy** – creating high- or low-energy coasts (see page 96)

◆ the **balance** between erosion and deposition (which of them is the more dominant process) – creating either erosional or depositional coasts and their associated features (see pages 108–17).

◆ changes in **sea level** – creating either emergent or submergent coasts (see pages 123–25)

But no classification system is definitive – so Cornwall's high-energy coast is mainly rocky, but also has long stretches of sand and some estuaries. Similarly, the low-lying coasts of eastern and southern England still have high cliffs (e.g. at Beachy Head).

In this section, you'll find out the geological reasons why coastal landscapes differ.

The Jurassic Coast

The coast of South Devon and East Dorset is widely regarded as being amongst the most-stunning scenery in the UK. In 2001, UNESCO awarded 'World Heritage' status to a stretch of it (the first coast in the UK to be given this title). This particular stretch of coast is often called 'The Jurassic Coast', after the geological period during which its rocks were formed. The World Heritage status comes from its unique 'geological walk through time' – it demonstrates the whole Jurassic period, with abundant fossils.

The Lulworth Crumple

Stair Hole, on the Jurassic Coast, lies less than half a mile west of the more famous Lulworth Cove. At Stair Hole, the sea has eroded through limestone and clays to create a small cove. It's the best place to see the 'Lulworth Crumple' – one of the best-known examples of limestone folding (see Figure 1). Here, thin beds of Purbeck limestone and shale are clearly visible in the side of the cliff. These layers of rock were folded (or crumpled) in response to tectonic movements about thirty million years ago.

▲ *Figure 1* *The Lulworth Crumple at Stair Hole*

Coasts and geological structure

Coastal morphology is related not only to the underlying geology, or rock type, but also to its geological structure (known as its **lithology**). Lithology means any of the following characteristics:

- **Strata** – Layers of rock.

- **Bedding planes** (horizontal cracks) – These are natural breaks in the strata, caused by gaps in time during periods of rock formation.

- **Joints** (vertical cracks) – These are fractures, caused either by contraction as sediments dry out, or by earth movements during uplift.

- **Folds** – Formed by pressure during tectonic activity, which makes rocks buckle and crumple (e.g. the Lulworth Crumple).

- **Faults** – Formed when the stress or pressure to which a rock is subjected, exceeds its internal strength (causing it to fracture). The faults then slip or move along **fault planes**.

- **Dip** – This refers to the angle at which rock strata lie (horizontally, vertically, dipping towards the sea, or dipping inland).

The **relief** – or height and slope of land – is also affected by geology and geological structure. There is a direct relationship between rock type, lithology, and **cliff profiles**. The five diagrams that make up Figure 2 help to illustrate this.

Key words

Coastal morphology – The shape and form of coastal landscapes and their features.

Coastal recession – Another term for coastal erosion.

Lithology – The physical characteristics of particular rocks.

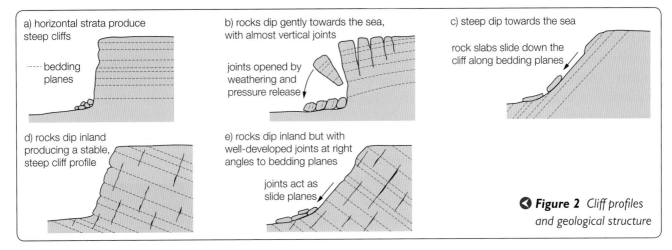

a) horizontal strata produce steep cliffs

----- bedding planes

b) rocks dip gently towards the sea, with almost vertical joints

joints opened by weathering and pressure release

c) steep dip towards the sea

rock slabs slide down the cliff along bedding planes

d) rocks dip inland producing a stable, steep cliff profile

e) rocks dip inland but with well-developed joints at right angles to bedding planes

joints act as slide planes

◀ **Figure 2** *Cliff profiles and geological structure*

Geology and rates of coastal recession

The geology and lithology at the coast affects the speed at which it erodes or **recedes**.

- Igneous rocks (e.g. granite) are crystalline, resistant and **impermeable**.

- Sedimentary rocks (e.g. limestone, chalk, sandstone and shale) are formed in strata. Jointed sedimentary rocks (e.g. sandstone and limestone) are **permeable**. Other sedimentary rocks (e.g. chalk) have air spaces between the particles – making them **porous**. Shale is fine grained and compacted – making it impermeable.

- Metamorphic rocks (e.g. marble and schist) are very hard, impermeable and resistant.

- Unconsolidated materials are loose, such as the boulder clay of the Holderness coast (see pages 126–27). They are not cemented together in any way and are easily eroded.

Coasts differ considerably. Figure 3 in Section 3.1 shows alternating strata in the cliff. Some strata project into the sea as headlands – indicating their resistance to erosion. Some strata are also more permeable than others. Geology and lithology play key roles in the type of cliff profile produced, and the rate at which coasts erode or recede. Other processes, such as weathering and mass movement (see pages 118–21), also affect the rate of erosion or recession (see Figure 3).

▼ **Figure 3** *Factors affecting the rate of coastal recession*

rock type	example location	annual rate of recession
glacial till	Holderness coast, Yorkshire	1–10 metres
sandstone	Devon	1 cm–1 metre
limestone	Dorset	1 mm–1 cm
granite	Cornwall	1 mm

weathering and mass movement rocks prone to weathering and mass movement will be subject to severe erosion

wave energy high-energy waves, driven by strong prevailing winds and a long fetch, will increase the rate of recession

other factors leading to rapid rates of cliff recession include the absence of a beach, rising sea levels, and human activities such as coastal defences elsewhere leading to increased erosion

Concordant coasts

Located in East Dorset, near Poole, the Isle of Purbeck is the eastern gateway to the Jurassic Coast. Because it's only surrounded on three sides by water, it isn't really an island but a peninsula. It has distinctive coastal features that are clearly linked to its geology and lithology.

The combination of different rock types on the Isle of Purbeck has led to coastal landscapes ranging from Lulworth Cove (Figure 4) to Kimmeridge Bay (Figure 5). Along its southern coast, the different rock types run in bands parallel to the coast – forming what is known as a **concordant coast**.

Key word

Concordant coast – This is where bands of more-resistant and less-resistant rock run parallel to the coast.

Over time, the sea gradually eroded the resistant Purbeck limestone at the entrance to Lulworth Cove. Then, rapid erosion of the less-resistant clays behind the limestone led to the formation of a cove or bay.

🔺 **Figure 4** *Lulworth Cove*

The cliffs at Kimmeridge Bay consist of less-resistant clays, where fossils can easily be found.

🔺 **Figure 5** *Kimmeridge Bay*

The rock type on Dorset's coast varies between resistant Purbeck limestone, which forms steep cliffs, to less-resistant clays and sands. These rock types alternate along the coast (see Figure 6), so that where a resistant rock is eventually eroded (e.g. at the entrance to Lulworth Cove) – allowing the sea to break through to the less-resistant rocks behind – erosion follows more quickly. At Lulworth, this has led to the formation of a small bay or cove (helped by a local stream, the valley of which makes erosion inland easier).

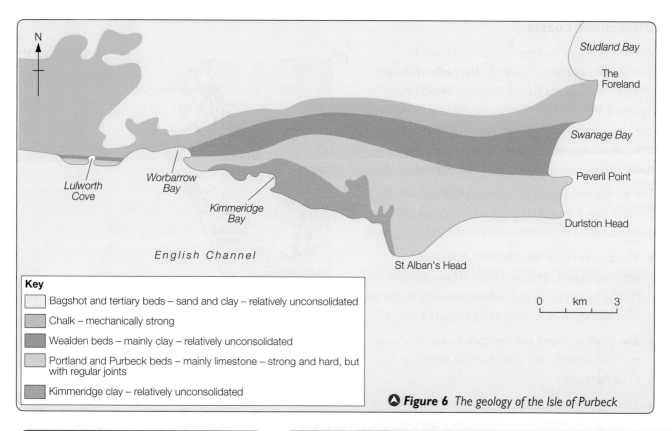

Key

◻ Bagshot and tertiary beds – sand and clay – relatively unconsolidated

◻ Chalk – mechanically strong

◻ Wealden beds – mainly clay – relatively unconsolidated

◻ Portland and Purbeck beds – mainly limestone – strong and hard, but with regular joints

◻ Kimmeridge clay – relatively unconsolidated

0   km   3

🔺 **Figure 6** *The geology of the Isle of Purbeck*

Dalmatian and Haff coasts

Dalmatian coasts are another type of concordant coastline. They have formed as a result of a rise in sea level. Valleys and ridges run parallel to each other. When the valleys flooded because of a rise in sea level, the tops of the ridges remained above the surface of the sea – as a series of offshore islands that run parallel to the coast.

The best example of a Dalmatian coastline is the one that gives this feature its name – the Dalmatian coast in Croatia (see Figure 7). Dalmatian coasts are also known as Pacific coasts, e.g. in southern Chile.

Haff coasts also consist of concordant features – long spits of sand and lagoons – aligned parallel to the coast. These are named after the Haffs, or lagoons, of the southern shore of the Baltic Sea, which are enclosed by sand spits or dunes.

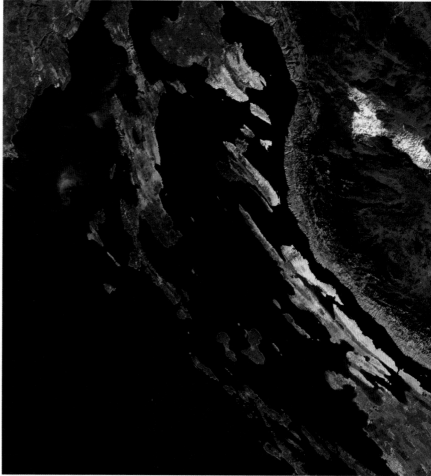

🔺 **Figure 7** *A satellite image of the Dalmatian coast in Croatia*

Discordant coasts

Whilst the Isle of Purbeck's southern coast is
concordant, its eastern coast is **discordant**. It runs
south from Studland Bay to Durlston Head (see
Figures 8 and 9). Here, more-resistant rocks (folded
into ridges) emerge at the coast as headlands and
cliffs, whilst less-resistant rocks form bays. Both
landforms can be seen clearly in both figures.

The geology and geological structure of the Isle of
Purbeck (see Figure 6) has influenced the coastal
morphology of its eastern coast in the following ways:

◆ The Bagshot and Tertiary beds consist of
unconsolidated sands and clays. These are less
resistant to erosion and, where exposed to the sea
at Studland, have formed a large bay as a result.

◆ The chalk is strong and resistant to erosion, so it
has formed cliffs and a headland at the coast
(The Foreland).

◆ The Wealden beds consist of unconsolidated clay,
which (like the Bagshot and Tertiary beds) is less
resistant to erosion and has led to the formation of
Swanage Bay, where it's exposed to the sea.

◆ The Purbeck and Portland beds consist mainly
of limestone. This is resistant and has led to the
creation of the headlands at Peveril Point and
Durlston Head. However, the limestone is also
jointed, which has created lines of weakness that
can be more easily eroded in places.

Key word

Discordant coast – The geology alternates between
bands of more-resistant and less-resistant rock, which
run at right angles to the coast.

◗ **Figure 8** *An aerial photo of the Isle of Purbeck's
discordant eastern coast*

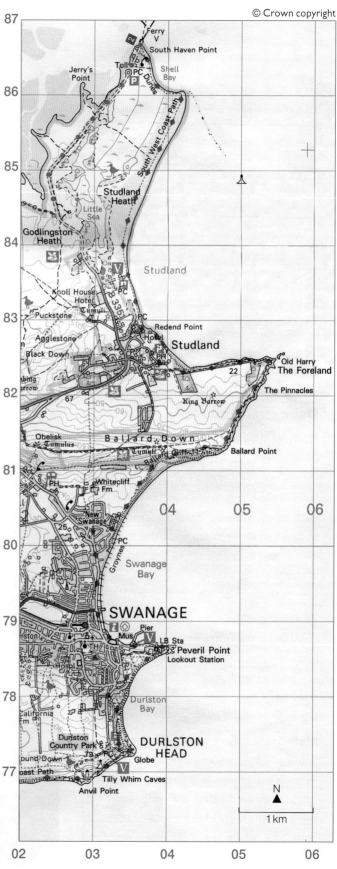

◔ **Figure 9** *An extract from an OS 1:50 000 map
showing the eastern coast of the Isle of Purbeck*

Headlands and bays

Headlands, such as The Foreland and Peveril Point (see Figures 8 and 9), jut out into the sea, with bays (such as Swanage Bay) lying between them. Headlands and bays commonly form when rocks of different strengths are exposed at the coast (see Figure 10). More-resistant rocks, such as chalk and limestone (or igneous and metamorphic rocks) tend to form headlands, whilst weaker rocks (such as shale and clays) are eroded to form bays.

Headlands and bays both affect incoming waves in different ways:

◆ Headlands force the incoming waves to refract or bend – concentrating their energy at the headlands. This increases the waves' erosive power, which leads to a steepening of the cliffs and their eventual erosion into arches and stacks (see page 111).

◆ By contrast, when waves enter a bay, their energy is dissipated (spread out) and reduced. This leads to the deposition of sediment (sand or shingle) – forming a beach.

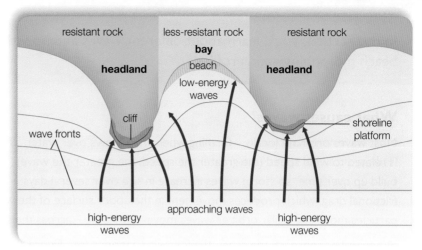

- When waves approach a headland, the depth of the water decreases. This causes the waves to get higher and steeper. Their velocity also reduces, and they become closer together. All of these factors increase their erosive power.

- However when waves enter a bay, the water is deeper. Therefore, they do not lose their velocity as rapidly – and they are also lower and less steep than those off the headland. This allows deposition rather than erosion to take place.

△ *Figure 10* *Headlands and bays*

 1 In pairs, produce a six-slide PowerPoint presentation of annotated photos of different UK coastlines to show the influence of geological structure on cliff profiles.

2 Using your evidence from question 1, and this section, create a spider diagram to show the factors which influence rates of coastal recession.

 3 Draw a field sketch of the coastal landscape shown in Figure 4. Annotate your sketch to show how the geology has influenced the development of this feature.

 4 In small groups, use Google Earth / Maps to identify one example each of concordant and discordant coastlines in the British Isles, and one example of each from overseas. Annotate their features.

On your own

5 Distinguish between the following pairs of terms: **(a)** geology and lithology, **(b)** relief and coastal morphology, **(c)** joints and bedding planes, **(d)** folds and faults, **(e)** strata and dip, **(f)** concordant and discordant coasts.

6 Study Figure 9, the OS map of the Isle of Purbeck. Identify features of its coastline that are related to its geology.

7 Use all of the material provided in this section about the Isle of Purbeck to write a 500-word report on the influence of geology and lithology on its coastal morphology.

Exam-style questions

AS 1 Explain the relationship between geology and coastal form along one named stretch of coast.
(6 marks)

A 2 Assess the extent to which rates of coastal recession and stability depend on lithology.
(12 marks)

In this section, you'll learn about different waves and how they influence beach morphology and profiles.

What causes waves?

Most waves originate locally – forming when wind blows over water. Their size is related to wind speed (the greater the speed, the greater the wave). They also build up over time, so storm waves increase in size over several days. Wind creates frictional drag, which produces movement in the upper surface of the water. Water particles move in circular orbit as waves move – or ripple – across the surface.

When waves approach the coast, the following happens (see Figure 1):

◆ The water becomes shallower and the circular orbit of the water particles changes to an elliptical shape.

◆ The **wavelength** (the distance between the **crests** of two waves) and the velocity both decrease, and the wave height increases – causing water to back up from behind.

◆ Force pushes the wave higher, so that it becomes steeper before **spilling** (or **plunging**) and breaking onshore.

◆ The water rushes up the beach as **swash**, and flows back as **backwash**.

⊙ *Figure 1 Waves approaching and breaking onshore*

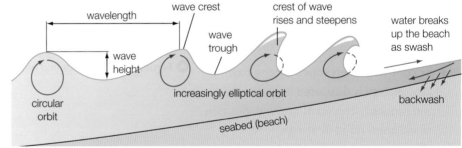

Wave period – the time interval from crest to crest in seconds. Sea waves have wave intervals between 1 and 20 seconds.

Swell waves

Not all waves originate locally. Some originate in mid-ocean, and maintain their energy for thousands of miles. The distance of open water over which they move is called the **fetch** – the greater the fetch, the larger the wave. On the UK coast, these mid-ocean waves appear as larger waves amongst smaller locally generated waves, and are called **swell waves**.

Different types of wave

Although waves vary, there are two main types: **constructive** and **destructive** (see Figure 2). These two types have significant impacts on processes and landforms at the coast.

⊙ *Figure 2 Constructive and destructive waves*

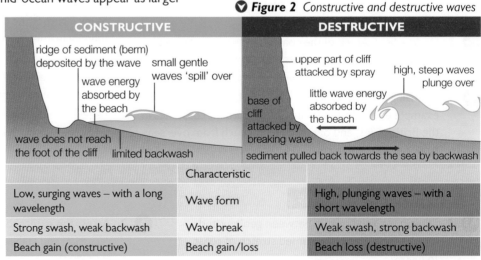

CONSTRUCTIVE	Characteristic	DESTRUCTIVE
Low, surging waves – with a long wavelength	Wave form	High, plunging waves – with a short wavelength
Strong swash, weak backwash	Wave break	Weak swash, strong backwash
Beach gain (constructive)	Beach gain/loss	Beach loss (destructive)

Beach morphology and profiles

Beaches consist of loose material, so their **morphology** (form or shape) alters as waves change. Seasonal changes in wave type create summer and winter **profiles** or gradients – sediment is dragged offshore by destructive waves during winter, and returned by constructive waves in summer. The material along a beach profile also varies in **size** and **type**, depending on distance from the shoreline (as Figure 3 shows).

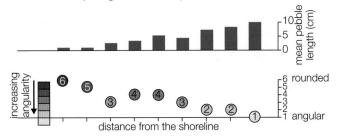

🔺 **Figure 3** *Pebble size and shape along a beach transect*

Beach profiles are steeper in summer, when constructive waves are more common than destructive. Constructive waves are less frequent (6-9 per minute), so wave energy dissipates and deposits over a wide area (weakening the backwash).

◆ The swash of a constructive wave deposits larger material at the top of the beach – creating a **berm** (usually of **shingle**).

◆ As the berm builds up, the backwash becomes weaker (draining by percolation, rather than down the beach). It only has enough energy to move smaller material, so the beach material becomes smaller towards the shoreline.

In winter, destructive waves occur at a higher frequency (11-16 per minute).

◆ Berms are eroded by plunging waves and high-energy swash.

◆ Strong backwash transports sediment offshore (depositing it as **offshore bars**).

◆ Sometimes, the backwash exerts a current known as a **rip**, or undertow – dragging sediment back as the next wave arrives over the top.

A 17 January midnight – 9.30am	Waves per minute	B 17 January 13.30 pm – 23.00pm	Waves per minute
00:00	9.1	13:30	14.0
00:30	9.0	14:00	14.3
01:00	8.7	14:30	14.3
01:30	8.6	15:00	15.4
02:00	8.8	15:30	16.2
02:30	8.2	16:00	16.2
03:00	8.2	16:30	15.8
03:30	8.6	17:00	15.8
04:00	8.5	17:30	16.2
04:30	8.8	18:00	15.8
05:00	8.8	18:30	16.2
05:30	9.2	19:00	15.8
06:00	9.7	19:30	15.0
06:30	9.5	20:00	15.0
07:00	9.5	20:30	15.0
07:30	9.8	21:00	14.6
08:00	10.0	21:30	14.0
08:30	10.2	22:00	14.6
09:00	10.2	22:30	14.0
09:30	10.5	23:00	15.0

🔺 **Figure 4** *Wave frequency data at Hornsea*

Exam-style questions

(AS) 1 Compare constructive and destructive waves. *(3 marks)*

(A) 2 Explain how different wave types result in different beach profiles. *(6 marks)*

Over to you

1 In pairs, use an atlas map of the UK (and a blank outline) to plot and name the UK coasts where high-energy waves are most likely.

2 Use Section 7.5 on 'Central tendency' to help you.
  a Using Figure 4, create a dispersion diagram for wave frequency for each of Columns A and B, and then calculate the range for each set of data.
  b Calculate the mean, median and mode for wave frequency in each of Columns A and B.
  c Compare the wave data for each period.
  d Explain the likely impact of each set of waves.

On your own

3 Distinguish between: **(a)** swash and backwash, **(b)** fetch and swell, **(c)** constructive and destructive waves, **(d)** beach morphology and beach profile, **(e)** berm and offshore bar.

4 a Draw annotated diagrams of summer and winter beaches.
  b Identify on each where largest sediment will be found, and why sediment size varies along the beach profile.

In this section, you'll learn about the processes of coastal erosion and the different coastal landforms created as a result.

An unbelievable storm

The residents of Riviera Terrace in Dawlish are used to their homes shaking when a storm hits the South Devon coast. They tend not to worry when high waves throw spray against their front windows. *'But this was different'* said one resident. *'It was like being in a car wash. The storm was unbelievable and waves were pounding against the terrace.'*

The winter storm that hit Dawlish in February 2014 (see Figure 1) was so powerful that the waves destroyed part of the sea wall – leaving a section of rail track dangling in mid air and cutting the rail connection between Devon, Cornwall and the rest of the UK for two months.

Cliff-foot erosion

The waves which destroyed the sea wall in Dawlish were spectacular. However, while not all waves are as powerful as those, they still attack the base of cliffs and erode the cliff foot using the following processes:

▲ **Figure 1** *The powerful storm of February 2014 – waves crash against the sea front and railway line in Dawlish*

Abrasion (also known as **corrasion**) – see Figure 2. When waves advance, they pick up sand and pebbles from the seabed. Then, when they break at the base of the cliff, the transported material is hurled at the cliff foot – chipping away at the rock. The size and amount of sediment picked up by the waves, along with the types of wave experienced, determines the relative importance of this erosive process.

cliff

waves advance

wave breaks at base of cliff, throwing sediment at it - known as **abrasion** or **corrasion**

▲ **Figure 2** *Cliff-foot erosion: abrasion*

Hydraulic action – see Figure 3. When a wave advances, air can be trapped and compressed (either in joints in the rock forming the cliff, or between the breaking wave and the cliff). Then, when the wave retreats, the compressed air expands again. This continuous process can weaken joints and cracks in the cliff – causing pieces of rock to break off. The force of the breaking wave can also hammer a rock surface. At high velocities, where bubbles form in the water and then collapse, they erode by hammer-like pressure effects.

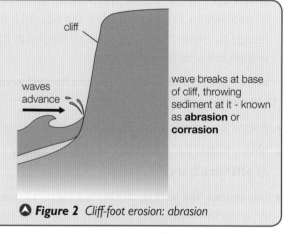

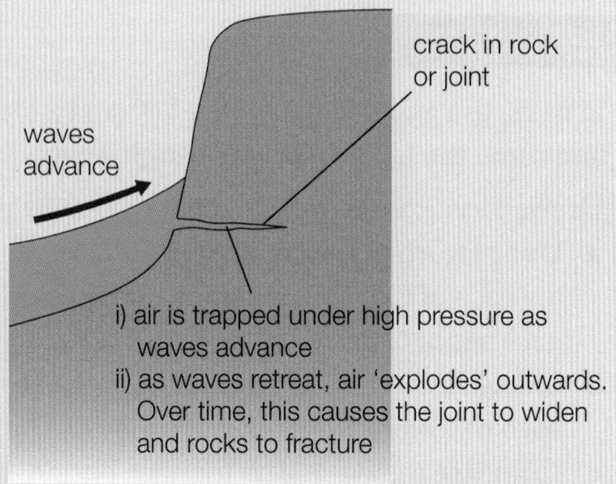

crack in rock or joint

waves advance

i) air is trapped under high pressure as waves advance
ii) as waves retreat, air 'explodes' outwards. Over time, this causes the joint to widen and rocks to fracture

▲ **Figure 3** *Cliff-foot erosion: hydraulic action*

Corrosion. When cliffs are formed from alkaline rock (such as chalk or limestone), or an alkaline cement bonds the rock particles together, solution by weak acids in seawater can dissolve them.

In addition, the process of **attrition** also takes place. This term refers to the gradual wearing down of rock particles by impact and abrasion, as the pieces of rock are moved by waves, tides and currents. This process gradually reduces the particle size and makes stones rounder and smoother.

Each of the above erosive processes can cause cliffs to become undercut and unstable (see Figure 4), which leads to their eventual collapse and retreat.

▲ **Figure 4** *Cliff undercutting and weakening*

How waves and lithology influence erosion

The rate and type of erosion experienced on a particular stretch of coast is influenced by the size and type of waves which reach that coast. Most erosion happens during winter storms (such as the one that hit Dawlish in February 2014) – when destructive waves are at their largest and most powerful. Hydraulic action and abrasion are powerful erosive forces, which attack differences in rock resistance (or weaknesses such as cracks, joints, bedding planes and faults). These geological factors are important in the development of coastal erosional landforms at a variety of scales.

Lithology (see Section 3.2) influences erosion. At a small scale, any geological weaknesses (such as joints, bedding planes or faults) are eroded more quickly, which can result in the formation of a range of different landforms (see pages 110–11). Bands of more-resistant rock between weaker joints and cracks erode more slowly. The selective erosion of areas of weakness – as opposed to more-resistant areas and types of rock – is called **differential erosion**.

At a medium and larger scale, areas of resistant rock generally form cliffs and headlands (see page 105), and areas of weaker rock form lowland areas with bays and inlets. The relief and structure of the coast – plus variations in wave type, exposure and fetch – work together to influence the coast. So, for example, headlands result in wave refraction (see page 105) – with wave energy and erosion concentrated on the headland.

Generally, erosion is faster where the rocks forming the coastline are weaker. For example, at Holderness (see pages 126–27) the weak boulder clays have eroded inland by 120 metres in just a century. By comparison, the resistant granites at Land's End have only eroded by 10 centimetres in the same period.

▼ **Figure 5** *Slumping on cliffs is South Cornwall, winter 2014*

109

Landforms created by coastal erosion

Figure 6 shows one of the most famous stretches of coast in the world. Many people travel along the Great Ocean Road (near Melbourne, Australia) just to see spectacular examples of the following landforms:

- **Headlands and bays**. Several headlands protrude out into the sea in the background of Figure 6, with sandy bays between them.

- **Cliffs**. The cliffs along this stretch of coast rise almost vertically from the shore to heights of up to 45 metres. The horizontal strata of the sedimentary sandstone which forms these cliffs are clearly visible.

- **Stacks and stumps**. These are isolated pillars of rock that lie just off the coast and are surrounded by water. The stumps are completely covered at high tide.

- **Shoreline platforms**. These are flat rocky platforms that extend out from the coast and surround the isolated stacks and stumps.

Wave-cut notches and shoreline platforms

When waves break against the foot of a cliff, erosion tends to be concentrated close to the high-tide line. This creates a **wave-cut notch**, like that shown in Figure 7, which begins to undercut the cliff. As the wave-cut notch gets bigger, the rock above it becomes unstable and eventually the upper part of the cliff collapses.

As these erosional processes are repeated, the notch migrates inland and the cliff retreats – leaving a remnant behind as a shoreline platform (sometimes known as a wave-cut platform); see Figure 8. The shoreline platform has a gentle slope of less than 4°, which is normally only completely exposed at low tide. Shoreline platforms rarely extend for more than a few hundred metres, because their width means that a wave will break earlier and its energy will be dissipated before it reaches the cliff – thus reducing the rate of erosion and limiting the further growth of the platform.

Cliffs

Figure 8 shows part of the chalk cliffs near Eastbourne. Constant wave action and erosion against the base of the cliff ensures that it maintains its steep profile as it retreats inland (see Figure 9).

Figure 6 *Coastal landforms on the Great Ocean Road in southern Australia. The stacks in the foreground are called the Twelve Apostles and are a big tourist attraction.*

Figure 7 *A wave-cut notch at Flamborough Head in Yorkshire*

Figure 8 *Cliffs and a shoreline platform near Eastbourne, West Sussex*

The steepest cliffs are found where rock strata are vertical or horizontal, or have almost vertical joints. The gentlest are found where rock dips towards or away from the sea (see Figure 2 on page 101).

Caves, arches, stacks and stumps

Caves, arches, stacks and stumps are all connected as part of a sequence of coastal landform development (see Figure 10):

◆ The erosion of rocks like limestone and chalk tends to exploit any lines of weakness – joints, faults and cracks.

◆ When joints and faults are eroded by hydraulic action and abrasion, this can then create **caves**. If the overlying rock then collapses, a **blowhole** will develop as the cave opens up at ground level. During storm high tides, seawater can be blown out of these blowholes with considerable and spectacular force.

◆ If two caves on either side of a headland join up, or a single cave is eroded through a headland, an **arch** is formed. The gap is then further enlarged by erosion and weathering – becoming wider at the base.

◆ Eventually, the top of the arch will become unstable and collapse – leaving an isolated pillar of rock, called a **stack**. The stack itself will continue to be eroded by the sea. As it collapses and is eroded further, it may only appear above the surface at low tide, and is now known as a **stump**.

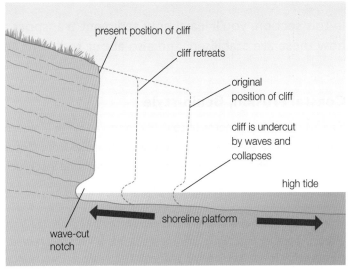

Figure 9 Cliff retreat

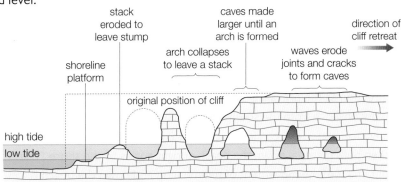

Figure 10 The formation of caves, arches, stacks and stumps at a headland

Exam-style questions

AS 1 Explain two processes of erosion that increase in importance during storms. *(6 marks)*

A 2 Assess the importance of different erosion processes in the development of cliff features. *(12 marks)*

Over to you

1 In pairs, discuss and suggest a possible rank order of the significance of abrasion, corrosion, hydraulic action or attrition **(a)** during a storm such as that at Dawlish (Figure 1), **(b)** on a quieter summer's day, **(c)** on chalk cliffs at Flamborough Head (Figure 7) in winter?

2 **a** Draw an annotated field sketch of the coastal landforms shown in Figure 6.
  b How active is the erosion on this stretch of coastline? Give evidence from the photo and annotate your sketch to justify your opinion.

3 Suggest how this landscape might change over the next few hundred years.

On your own

4 Distinguish between the following terms: **(a)** corrosion and corrasion, **(b)** abrasion and hydraulic action, **(c)** attrition and differential erosion.

5 Produce a flow chart to show the development of wave-cut notches, shoreline platforms and cliffs.

6 Using Google Maps / Earth, identify two stretches of coast in the UK – and two stretches in the rest of the world – where features in this section can be seen.

In this section, you'll learn how sediment transport and deposition create distinctive landforms, how these are stabilised, and also about sediment cells.

Coastal erosion, Benin-style

If an average of one metre of coastal erosion a year in Holderness sounds like a lot (see Section 3.4), Gilbert Adipeko from Cotonou, Benin, thinks differently. He remembers the night he lost much of his shorefront home – washed away by the sea.

> 'We were asleep when there was a deafening noise from the living room. I got up in a panic, and the whole room had disappeared under the waves,' said the 62 year old, still in shock, two months later. 'When I built my house in 1987, the sea was 250 metres away and there were four rows of houses in front of mine. Now they're all under the sea.'

Longshore drift

Most waves approach a beach at an angle – generally from the same direction as the prevailing wind. Along the coast of West Africa, winds blow onshore from the south-west throughout the year. As the waves advance, material is carried up the beach at an angle. The backwash then pulls material down the beach at right angles to the shore. The net effect of the movement of sediment up and down the beach is a **lateral shift** – and the process is known as **longshore drift** (see Figure 2). Where the removal of sediment is greater than the supply of new sediment, the beach is eroded.

Strong prevailing south-westerly winds and large waves create significant and sustained longshore drift in West Africa. As a result, a huge amount of sand is transported along the coast – from Ghana eastwards to Togo and Benin. However, despite that, not enough sand is deposited to replace that transported by longshore drift, so Benin's coastline is still being eroded at a rapid rate.

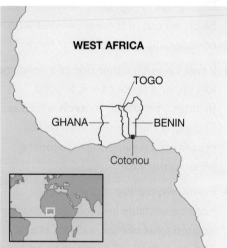

⌃ Figure 1 The coastline east of Cotonou in Benin has retreated 400 metres in just 40 years – an average of 10 metres a year

⌄ Figure 2 The process of longshore drift

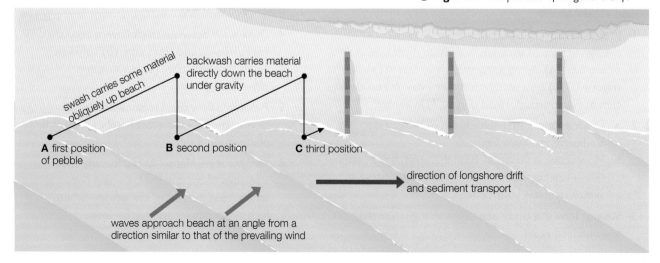

swash carries some material obliquely up beach

backwash carries material directly down the beach under gravity

A first position of pebble

B second position

C third position

direction of longshore drift and sediment transport

waves approach beach at an angle from a direction similar to that of the prevailing wind

Tides and currents

The angle at which waves approach a beach is a major factor in coastal sediment transport. However, tides and currents also affect the process of longshore drift.

◆ **Tides** are changes in the water level of seas and oceans – caused by the gravitational pull of the moon and, to a lesser extent, the sun. The UK coastline experiences two high and two low tides a day.

◆ The relative difference in height between high and low tides is called the **tidal range**. A high tidal range creates relatively powerful tidal currents, as tides rise and fall. Tidal currents can become particularly strong and fast in estuaries and narrow channels, and are important in transporting sediment.

Coastal depositional landforms

Deposition occurs along the coast when waves no longer have enough energy to transport sediment. Depending on how and where the sediment is deposited, a variety of landforms can be produced.

1 Spits

A spit is a long narrow feature, made of sand or shingle, which extends from the land into the sea (or part of the way across an estuary). Spits form on drift-aligned beaches (see Figure 5 on page 114). Sand or shingle is moved along the coast by longshore drift, but if the coastline suddenly changes direction (e.g. because of a river estuary), sediment will begin to build up across the estuary mouth and a spit will start to form (see Figure 4). The outward flow of the river associated with the estuary will prevent the spit from extending right across the estuary mouth. The end of the spit will also begin to curve round, as wave refraction carries material round into the more sheltered water behind the spit. This is known as a **recurved spit**.

The entrance to Poole Harbour (see Figure 3) is unusual in that it has two spits extending from both the northern and southern ends of the bay – forming a **double spit**. A salt marsh may develop behind a spit, where finer sediment settles and begins to be colonised by salt-tolerant plants (such as the one behind the left-hand spit in Figure 3).

BACKGROUND

Marine transport

There are four main ways in which coastal sediment is transported:

◆ **Traction** – relatively large and heavy rocks are rolled along the seabed.

◆ **Saltation** – smaller and lighter rocks 'bounce' along the seabed.

◆ Lighter sediment is carried in **suspension**.

◆ Dissolved sediment is carried in **solution**

▲ **Figure 3** *The double spit at the entrance to Poole Harbour*

▼ **Figure 4** *The formation of a spit*

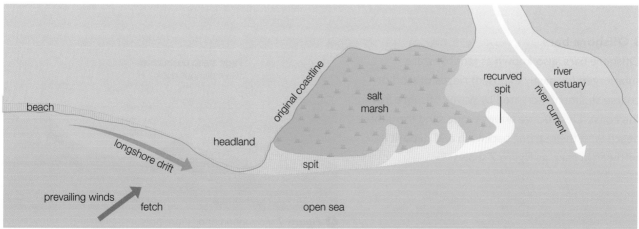

2 Different types of beaches

Beaches are commonly found in bays. Wave refraction creates a low-energy environment, which then leads to deposition (see page 105 on headlands and bays). A beach could consist of either sand or shingle – depending on factors like the nature of the sediment and the power of the waves. Beaches can also be **swash-aligned**, or **drift-aligned** (see Figure 5).

Sediment may be graded along a drift-aligned beach. Finer shingle particles are likely to be carried further by longshore drift – and also to become increasingly rounded as they move. The table in Figure 6 shows how shingle changes in size and shape (roundness) as it moves along the beach.

Sample site (see map)	Height of beach (metres above low-water mark)	Shingle diameter (cm)	Roundness index (%) *
1	5.5	8.4	18
2	7.5	8.5	20
3	8.0	7.7	26
4	11.5	8.1	38
5	11.0	6.1	23
6	7.5	5.8	15
7	10.0	6.2	21
8	10.5	7.2	30
9	10.0	7.5	27
10	11.0	6.2	35
11	14.0	6.5	44
12	12.5	5.8	41
13	18.0	4.8	62
14	13.5	5.0	71
15	15.0	5.8	65

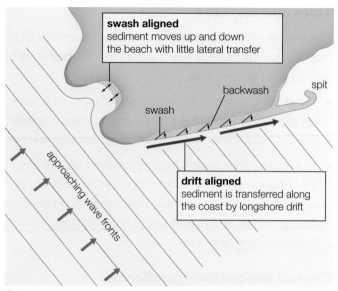

Figure 5 How swash-aligned and drift-aligned beaches form

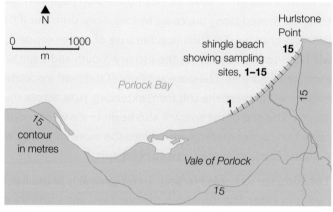

Figure 6 Sediment characteristics along a drift-aligned beach in Somerset
*Roundness index – This is the percentage of particles classified as rounded.

3 Offshore bars

Offshore bars, also known as sandbars, are submerged (or partly exposed) ridges of sand or coarse sediment – created by waves offshore from the coast. Destructive waves erode sand from the beach with their strong backwash and deposit it offshore in bars, as Figure 7 shows.

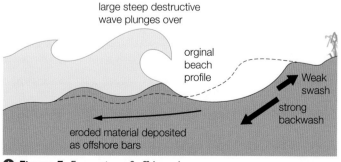

Figure 7 Formation of offshore bars

4 Barrier beaches (bars)

Where a beach or spit extends across a bay to join two headlands, it forms a **barrier beach** or **bar**, e.g. Loe Bar in Cornwall and Start Bay in Devon. The barrier beach at Start Bay (see Figure 8) is 9 km long and is formed from rounded shingle deposits (consisting mostly of flint and quartz gravel). Barrier beaches and bars can also trap water behind them to form lagoons, such as Slapton Ley in Figure 8.

Where a beach becomes separated from the mainland, it is referred to as a **barrier island**. Barrier islands vary in scale and form – are usually sand or shingle features – and are common in areas with low tidal ranges, where the offshore coastline is gently sloping. Large-scale barrier islands can be found along the Dutch coast, and in North America along the South Texas coast.

5 Tombolos

A tombolo is a beach (or ridge of sand and shingle) that has formed between a small island and the mainland. Deposition occurs where waves lose their energy and the tombolo begins to build up. Tombolos may be covered at high tide, e.g. at St Ninian's in the Shetland Islands (see Figure 9) and at Lindisfarne in Northumberland.

6 Cuspate forelands

A **cuspate foreland** is a triangular-shaped headland that extends out from the main coastline. It occurs where a coast is exposed to longshore drift from opposite directions. Sediment is deposited at the point where the two meet, which forms a natural triangular shape as it builds up. As vegetation begins to grow on the deposited sediment, it helps to stabilise the landform and protect it from storms that could erode it. Cuspate forelands can be small – extending out from the coast for just a few metres – or they can be larger features that extend up to 3 miles out from the coastline.

Dungeness in Kent (see Figure 10) is an example of a cuspate foreland. It has a pebble beach and an area of marshland. Dungeness has built up slowly since the first lighthouse was built there in the seventeenth century. Since then, several replacement lighthouses have had to be built, as the point of land has grown further out into the sea.

Stabilising depositional landforms

Many depositional landforms consist of sand and shingle – loose, unstable sediment that can be easily eroded and transported. Sandy beaches may be backed by **sand dunes**, like those at Studland in Dorset (see Figure 11), which consist of sand that has been blown off the beach by the onshore winds. Dunes can develop where sand is initially trapped by debris towards the back of the beach. Vegetation helps to stabilise the dunes as a result of **plant succession** (see page 116).

🔺 **Figure 8** *Start Bay barrier beach and Slapton Ley lagoon*

🔺 **Figure 9** *The tombolo linking St Ninian's Isle to the south-west Shetland mainland is the largest active sand tombolo in the UK*

🔺 **Figure 10** *Dungeness – a cuspate foreland*

▶ **Figure 11** *Sand dunes at Studland in Dorset*

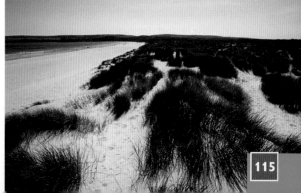

Plant succession

Bare ground (such as rock, sand dunes, salt marshes and mud flats) is gradually colonised by plants. The first colonising plants are called **pioneer species**. These species begin the process of plant **succession**, during which other species invade and take over until a balance is reached.

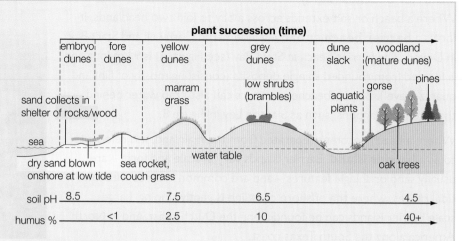

Pioneer species modify the environment by binding sand or soil with their roots and adding nutrients when they die and decay. Creeping plants, or those with leaf cover, help the sand or mud in dunes and salt marshes to retain moisture. These changes allow other species to colonise. The invaders also modify the environment by providing shade as well as improving the soil.

As the environment changes over time, different species colonise it until it becomes stable. The final community will be adjusted to the climatic conditions of the area, and is known as the **climatic climax community**.

- Embryo dunes are the first dunes to develop.
- As embryo dunes develop, they grow into bigger fore dunes – which are initially yellow in colour, but darken to grey as decaying plants add humus.
- Depressions between dunes can develop into dune slacks – damper areas where the water table is closer to, or at, the surface.

🔺 **Figure 12** Sand dunes and plant succession

Salt marshes

Salt marshes are areas of flat, silty sediments that accumulate around estuaries or lagoons (see Figure 13). They develop:

- in sheltered areas where deposition occurs
- where salt and fresh water meet
- where there are no strong tides or currents to prevent sediment deposition and accumulation.

Salt marshes are covered at high tide and exposed at low tide. They are common around the coast of Britain.

Sediment cells

Sediment moves along the coast in **sediment cells**. Within each cell, the sediment moves between the beach, cliffs and sea through the processes of erosion, transport and deposition. Any action taken in one place (e.g. erecting groynes) has an impact elsewhere in the cell. Each cell operates between physical barriers that prevent the sediment from moving any further along the coast (e.g. major headlands or river estuaries). The coastline of England and Wales is divided up into 11 major sediment cells (see Figure 14).

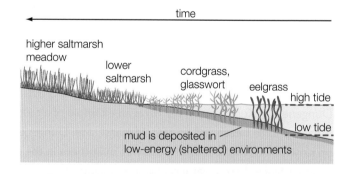

- As mud flats develop, salt-tolerant plants (such as eelgrass) begin to colonise and stabilise them.
- **Halophytes** (salt-tolerant species), such as glasswort and cordgrass, help to slow down tidal flow and trap more mud and silt.
- As sediment accumulates, the surface becomes drier and different plants begin to colonise, e.g. sea asters and meadow grass.
- Creeks (created by water flowing across the estuary at low tide) divide up the salt marsh.

🔺 **Figure 13** Salt marsh formation and plant succession

The sediment cell as a system

Sediment cells act as **systems** (see Section 3.1) – with **sources, transfers** (or flows) and **sinks**. Larger sediment is not transferred between cells, but finer sediment in suspension out at sea (e.g. some of the fine boulder clay eroded from Holderness) can be transferred.

The amount of sediment available within a sediment cell is called the **sediment budget**. Within each cell, depositional features build up which are in line – or **equilibrium** – with the amount of sediment available.

- If the sediment budget falls, waves continue to transport sediment (and erosion may therefore increase in some areas, because the sea has surplus energy). One change has led to another change – known as **positive feedback**.

- However if the sediment budget increases, more deposition is likely. The sea corrects itself, because it can only carry so much – and any surplus is deposited. This is known as **negative feedback** (the sea returns to a situation where it can handle the sediment supply).

The way in which systems attempt to balance in this way is known as **dynamic equilibrium**.

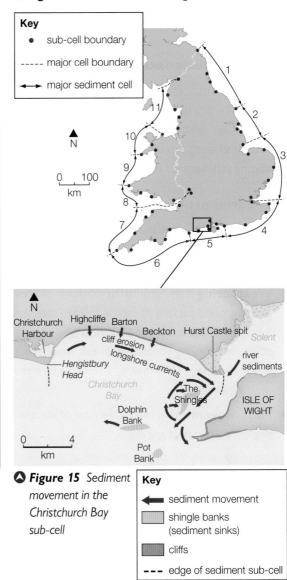

Figure 14 Sediment cells in England and Wales

Figure 15 Sediment movement in the Christchurch Bay sub-cell

Each major sediment cell can be divided into a number of sub-cells. One of these lies in Christchurch Bay (see Figure 15). Within this sub-cell, the processes of erosion, transport and deposition are all interlinked. Areas of deposition, like Dolphin Bank, form potential sources of sediment, as well as sinks.

Over to you

1 Create a systems diagram for a sediment cell, with inputs (sources of sediment), transfers (transportation) and sinks (stores of sediment).

2 a Explain the methods that were used to obtain the data in Figure 6.
   b Using Figure 6, analyse (i) the changes, and (ii) the statistical relationship between shingle diameter and roundness.
   c Use the student t-test (see Section 7.10) to analyse the significance of the statistical relationship between shingle diameter and roundness.

On your own

3 Explain why plant succession is important to the development of some coastal landscapes.

4 Using an annotated flow diagram, identify and explain the sequence of stages that follow plant succession to its climatic climax.

5 Explain how sediment cells aid our understanding of the development of some coastal landscapes.

Exam-style questions

1 Explain the characteristics of a 'drift-aligned' stretch of beach. (6 marks)

2 Assess the relative importance of depositional processes along a named stretch of coast. (12 marks)

In this section, you'll find out how weathering and mass movement influence coastal landscapes.

Reshaping the coast

A series of landslips (see Figure 1) affected the coast at Lyme Regis in Dorset during 2014 – in a sequence of movements that significantly reshaped the coastline. Less-dramatic mass movements, such as rock falls, also happen frequently all along this coast. These are important mechanisms for producing sediment for the sea to transport and redistribute.

Sub-aerial processes

Parts of the North Yorkshire coast suffer from rapid erosion (which you'll examine in more detail on pages 126–131). However, **sub-aerial processes** (weathering and mass movement) help to make things worse:

◆ **Weathering** is the gradual breakdown of rock, *in situ,* at or close to the ground surface. It can be divided into three different types: mechanical, chemical and biological (see below). By breaking rock down, weathering creates sediment which the sea can then use to help erode the coast. Weathering also helps to increase the rate of erosion of some coasts.

◆ **Mass movement** is the movement of weathered material down slope, as a result of gravity.

⬆ **Figure 1** *One of seven holiday chalets that had to be demolished and removed from this location at Lyme Regis in 2014, following a series of landslips*

Mechanical weathering

◆ **Freeze-thaw weathering** (also known as frost-shattering) occurs when water enters a crack or joint in the rock when it rains – and then freezes in cold weather. When water freezes, it expands in volume by about 10%. This expansion exerts pressure on the rock, which forces the crack to widen. With repeated freezing and thawing, fragments of rock break away and collect at the base of the cliff as **scree**. These angular rock fragments are then picked up by the sea, and used as tools in marine erosion. Despite the fact that the coast tends to be milder than places inland, freeze-thaw weathering is still important in coastal locations. For example, in 2001 (following a very wet autumn and a cold February) freeze-thaw weathering triggered several major rock falls along the south coast of England (see Figure 2). Chalk (a permeable and porous rock) was the main rock affected.

◆ **Salt-weathering**. When salt water evaporates, it leaves salt crystals behind. These can grow over time and exert stresses in the rock, just as ice does – causing it to break up. Salt can also corrode rock, particularly if it contains traces of iron.

◆ **Wetting and drying**. Frequent cycles of wetting and drying are common on the coast. Rocks rich in clay (such as shale) expand when they get wet and contract as they dry. This can cause them to crack and break up.

⬇ **Figure 2** *A major rock fall at the White Cliffs of Dover in February 2001, caused by freeze-thaw weathering*

Biological weathering

This occurs in several ways:

◆ Thin plant roots start to grow into small cracks in a cliff face. These cracks then widen as the roots grow thicker, which breaks up the rock (as shown in Figure 3).

◆ Water running through decaying vegetation becomes acidic, which leads to increased chemical weathering (see below).

◆ Birds (such as puffins) and animals (such as rabbits) dig burrows into cliffs.

◆ Marine organisms are also capable of burrowing into rocks (e.g. piddocks, which are similar to clams), or of secreting acids (e.g. limpets).

⊘ Figure 3 *Biological weathering*

Mass-movement processes

Mass movement can occur at a range of speeds from incredibly slow – less than 1 cm a year (e.g. soil creep) – to horrifyingly fast rock falls and slope failures, that can cause the landscape to change in a matter of minutes.

Mass movement at the coast is quite common, partly because constant undercutting of the cliffs makes them unstable and prone to collapse. More often, the sheer weight of rainwater – combined with weak geology – is the major cause of cliff collapse.

Different types of mass movement

Mass movement can be classified in different ways (see Figure 4). The resultant movement depends on a range of factors, including:

◆ the angle of the slope or cliff

◆ the rock type and its structure

◆ the vegetation cover

◆ how wet the ground is.

Chemical weathering

Carbonation Rainwater absorbs carbon dioxide from the air to form a weak carbonic acid. This reacts with calcium carbonate in rocks such as limestone and chalk – to form calcium bicarbonate, which is easily dissolved. The cooler the temperature of the rainwater, the more carbon dioxide is absorbed – increasing the effectiveness of carbonation in winter.

Nature of movement	Rate of movement	Type of mass movement
Flow	Imperceptible	Soil creep Solifluction
Flow	Slow to rapid	Earth flow/mudflow
Slide	Slow to rapid	Rock/debris fall Rock/debris slide Slump

⊘ Figure 4 *Classifying different types of mass movement*

Mass movement – flows

Soil creep

◆ Soil creep is the slowest form of mass movement – and is an almost continuous process.

◆ It is a very slow downhill movement of individual soil particles (see Figure 5).

Solifluction

◆ This movement averages between 5 cm and 1 metre a year.

◆ It occurs mainly in tundra areas, where the ground is frozen. When the top layer of soil thaws in the summer – but the layer below remains frozen (as permafrost) – the surface layer becomes saturated and flows over the frozen subsoil and rock.

Earth flows and mudflows

◆ An increase in the amount of water (e.g. as a result of heavy rain) can reduce friction – causing earth and mud to flow over underlying bedrock (see Figure 6).

◆ The difference between a slide and a flow is that in a slide the material remains intact (it moves 'en masse'). In a flow, the material becomes jumbled up.

Mass movement – slides

Rock falls

◆ Rock falls are most likely to occur when strong, jointed and steep rock faces/cliffs are exposed to mechanical weathering (such as freeze-thaw weathering).

◆ Rock falls occur on slopes over 40°.

◆ The material, once broken away from the source, either bounces or falls vertically to form **scree** (also known as **talus**) at the foot of the slope/cliff, as in Figures 2 (on page 118) and 7.

◆ Block falls are similar to rock falls. A large block of rock falls away from the cliff as a single piece, due to the jointing of the rock.

Rock/debris slides

◆ Rocks that are jointed, or have bedding planes roughly parallel to the slope or cliff surface, are susceptible to landslides.

◆ An increase in the amount of water can reduce friction – causing sliding.

◆ In a rock or landslide, slabs of rock/blocks can slide over underlying rocks along a slide or slip plane (see Figure 8).

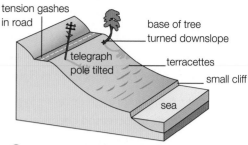

▲ *Figure 5* *Soil creep*

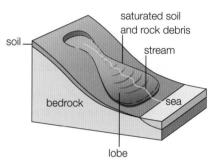

▲ *Figure 6* *A mudflow*

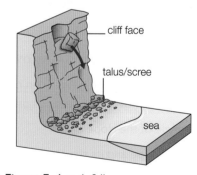

▲ *Figure 7* *A rock fall*

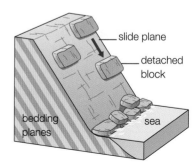

▲ *Figure 8* *A landslide*

Slumps

◆ Slumps often occur in saturated conditions.

◆ The difference between sliding and slumping is that there is a rotational movement in slumping (see Figure 9).

◆ Slumps occur on moderate to steep slopes.

◆ They are common where softer materials (clays or sands) overlie more-resistant or impermeable rock, such as limestone or granite.

◆ Slumping causes **rotational scars**.

◆ Repeated slumping creates a **terraced cliff profile** (see Figure 10).

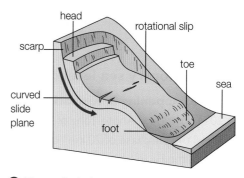

▲ *Figure 9 A slump*

◀ *Figure 10 A terraced cliff profile below Naish Farm Holiday Village in Christchurch Bay*

Over to you

1 In pairs, discuss and devise a flow diagram to outline the sequence of processes that probably led to the landslip at Lyme Regis.

2 In pairs, discuss and decide (**a**) the part played by weathering and its influence on the rate of coastal retreat, (**b**) which forms of weathering are likely to have the greatest impact in different parts of the UK?

3 Explain why evidence of past solifluction can be seen in some parts of the UK.

4 **a** Explain the significance of weathering and mass movement in the ways in which coastal landscape systems operate.

 b Explain how the methods by which sediment is supplied to the coastal system are likely to vary between areas of resistant and weak geology.

On your own

5 Distinguish between the following pairs of terms: (**a**) weathering and mass movement, (**b**) flows and slides, (**c**) slides and slumps, (**d**) soil creep and solifluction, (**e**) rockfalls and slumping.

6 Using Figure 4, identify the type of movement that produced the events in Lyme Regis in Figure 1.

7 Use the British Geological Survey website to investigate the geology of Christchurch Bay. Why do you think cliffs in this area are so susceptible to slumping?

Exam-style questions

AS 1 Assess the relative importance of different methods of mass movement along one stretch of coast. *(12 marks)*

A 2 Assess the significance of different weathering and mass movement processes in coastal landscape systems. *(12 marks)*

In this section, you'll find out how changes in sea level affect the coast.

Contemporary sea level change

What's happening in Kiribati?

In 2014, President Anote Tong of Kiribati finalised the purchase of 20 km² of land in one of the Fijian Islands – 2000 km from Kiribati. The inhabitants of Kiribati (a group of islands in the central Pacific Ocean) now own a refuge somewhere else.

The nation of Kiribati consists of 33 widely spaced islands, which stretch across an area of the Pacific Ocean nearly as wide as the USA. Kiribati's islands are very low-lying sand and mangrove atolls – only one metre or less above sea level in most places (see Figure 1). To visitors, Kiribati can seem like paradise, but it's been predicted that many of its islands could disappear under the sea in the next 50 years. In places, the sea level is rising by 1.2 cm a year (four times faster than the global average).

⬆ *Figure 1* *Tarawa Atoll, Kiribati – a vulnerable speck in the ocean*

Why are sea levels rising?

Global warming. Average global temperatures rose by 0.85°C from 1880 to 2012. During a similar period, 1870 to 2010, average sea levels rose by 21 cm. Sea levels are rising because the polar ice sheets (as well as glaciers worldwide) are melting, and because of **thermal expansion** (sea water expands as it warms).

No one knows exactly how far sea levels will rise. Climate scientists estimate that, by 2100, average sea levels will have risen by somewhere between 30 cm and 1 metre – with perhaps a 'best guess' of 40 cm. And that means that low-lying nations, like Kiribati, are at risk of disappearing under the waves.

What next for Kiribati?

The land purchased in Fiji by Kiribati will be used in the immediate future for agriculture and fish-farming projects – to guarantee the nation's food security. Rising sea levels in Kiribati are contaminating its ground water sources and affecting its ability to grow crops. In the future, if necessary, people could move from Kiribati to Fiji. The government has launched a 'migration with dignity' policy to allow people to apply for jobs in neighbouring countries, such as New Zealand. If the islands are submerged, Kiribati's population will become **environmental refugees** – people forced to migrate as a result of changes to the environment.

Longer-term sea level change

Sea level varies over time. It is measured relative to land, so the relative sea level can change if either the land or the sea rises or falls. The two types of sea level change are called:

◆ **eustatic change** – when the sea level itself rises or falls

◆ **isostatic change** – when the land rises or falls, relative to the sea

Eustatic change is global. In cold, glacial periods, precipitation falls as snow (rather than rain) and forms huge ice sheets that store water normally held in the oceans. As a result, sea levels fall. At the end of glacial periods – as temperatures rise – the ice sheets begin to melt and retreat. Their stored water then flows into the rivers and the sea again – and sea levels rise.

Isostatic change occurs locally. During glacial periods, the enormous weight of the ice sheets (which can be several kilometres thick) makes the land sink. This is called **isostatic subsidence**. As the ice begins to melt at the end of a glacial period, the reduced weight of the ice causes the land to readjust and rise. This is called **isostatic recovery**.

Eustatic changes occur relatively quickly, but isostatic changes take much longer. At the end of the last glacial period in Europe (about 8000 years ago), glacial meltwater caused a relatively rapid rise in sea level, which led to the formation of the English Channel and the North Sea (turning Britain into an island). Despite the melting of a huge amount of ice, the land only started to rise very slowly – and is still rising now.

In the UK, two different types of isostatic change have occurred since the last Ice Age (as Figure 2 shows):

◆ Land in the north and west – which was covered by ice sheets during the last Ice Age – is still rising as a result of isostatic recovery.

◆ However, land in the south and east (which the ice sheets never covered) is sinking. Rivers pour water and sediment into the Thames estuary. The weight of this sediment causes the crust to sink there and relative sea levels to rise. Therefore, south-east England faces increased flood risks as a result of the land sinking due to isostatic change, as well as a rising sea level caused by global warming.

◆ **Figure 2** *Isostatic change in the UK. The lines show how much parts of the UK are either rising or falling (the minus figures) in millimetres per year.*

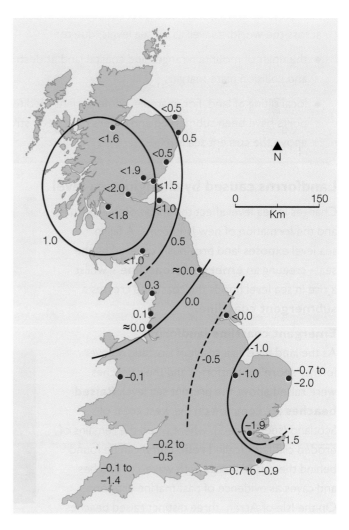

Sea level change due to tectonic activity

On Boxing Day 2004, an earthquake (measuring between 9.0 and 9.3 on the Richter Scale) caused a tsunami in the Indian Ocean that killed approximately 300 000 people. The Indonesian island of Sumatra was the worst hit, because it was the closest land to the earthquake's epicentre. The city of Banda Aceh was hit by 15-metre-high waves and flooded – just 15 minutes after the initial earthquake (see Figure 3). But this devastation was made even worse, because the earthquake caused the Earth's crust at Banda Aceh to sink – permanently flooding some parts of the city.

The 2004 earthquake was caused by an estimated 1600 km of fault line slipping about 15 metres along the subduction zone where the Indian Plate slides under the Burma Plate. The seabed rose several metres – displacing an estimated 30 km³ of water and triggering the tsunami. Not only that, but the raising of the seabed reduced the capacity of the entire Indian Ocean – producing a permanent rise in sea level of an estimated 0.1 mm.

Before

After

▲ *Figure 3* *The impact of the tsunami on Banda Aceh – before and after*

Past tectonic activity

Past tectonic activity has had a direct impact on some coasts across the world, as well as on sea levels, due to:

♦ the uplift of mountain ranges and coastal land at destructive and collision plate margins

♦ local tilting of land. For example, some ancient Mediterranean ports have been submerged and others have been stranded above the current sea level.

Landforms caused by changing sea level

Changes in sea level affect the shape of the coastline and the formation of new landforms. A fall in sea level exposes land previously covered by the sea – creating an **emergent coastline** – whilst a rise in sea level floods the coast and creates a **submergent coastline**.

Emergent coastline landforms

As the land rose as a result of isostatic recovery, former shoreline platforms and their beaches were raised above the present sea level. **Raised beaches** are common on the west coast of Scotland (see Figure 5), where often the remains of eroded cliff lines (called **relic cliffs**) can be found behind the raised beach, with wave-cut notches and caves as evidence of past marine erosion. On the Isle of Arran, three distinct raised beaches represent separate changes in sea level.

▼ *Figure 5* *A raised beach in Western Scotland*

Submergent coastline landforms

Rias (sheltered winding inlets with irregular shorelines) are one of the most distinctive features associated with a rise in sea level. They form when valleys in a dissected upland area are flooded. Rias are common in south-west England, where sea levels rose after the last Ice Age – drowning the lower parts of many rivers and their tributaries to form rias. The Kingsbridge estuary in Devon (see Figure 6) is one of these. It provides a natural harbour with the deepest water at its mouth.

Dalmatian coasts (also see Section 3.2) are similar to rias. In this case, the rivers flow almost parallel to the coast – rather than at right angles to it. The Dalmatian coast in Croatia gives this feature its name.

Fjords are formed when deep glacial troughs (see Section 2.8) are flooded by a rise in sea level. They are long and steep-sided, with a U-shaped cross-section and hanging valleys. Unlike rias, fjords are much deeper inland than they are at the coast. The shallower entrance marks where the glacier left the valley. Fjords can be found in Norway, New Zealand (see Figure 7) and Chile.

▲ **Figure 6** The Kingsbridge estuary

▶ **Figure 7** Milford Sound (a fjord) in New Zealand's South Island

1 In pairs, discuss and identify, using an atlas, which European coastlines are at risk from contemporary sea-level rises. Give examples.

2 To what extent are the risks from sea-level rise in south-east England (**a**) similar to, (**b**) different from the risks to Kiribati?

✦ 3 In pairs, devise a virtual online field trip that focuses on 'Isostatic change in the UK landscape'. Use Google Maps and an Internet search to identify the best UK locations to see and photograph (**a**) raised beaches, (**b**) relic cliffs, (**c**) any other evidence of eustatic change.

✦ 4 Use the phrase 'OS map Kingsbridge' to locate an OS map extract of Kingsbridge in south Devon.

   a Identify the OS map features of a ria coastline.

   b Research one other ria and its features in south-west England.

On your own

5 Distinguish between the following pairs of terms: (**a**) eustatic and isostatic change, (**b**) emergent and submergent coastlines, (**c**) rias and fjords, (**d**) relic cliffs and raised beaches.

6 **a** Summarise the problems facing Kiribati due to changes in sea level.

   b Use Internet research to update how well Kiribati is coping with climate change.

✦ 7 Search Google Maps for satellite images of Croatia's Dalmatian coast, and then annotate them to show evidence of past changes in sea level.

Exam-style questions

1 Explain the difference between eustatic and isostatic change. *(4 marks)*

Ⓐ 2 Evaluate the contribution of geologically recent eustatic changes to the UK's coastal landscapes. *(20 marks)*

In this section, you'll learn why and how rates of coastal erosion vary, plus the problems that this can cause.

Yorkshire is disappearing up to three times as fast

A dry spring and a soggy summer have been blamed for the sharply increased coastal erosion along Yorkshire's eastern flank, bordering the North Sea.

Engineers from the East Riding of Yorkshire District Council have been out surveying their stretch of collapsing cliffs, south of Bridlington. The results vary, but in places the county's coastline has retreated – that is, has eroded – by a startling 7 metres in just a year.

As a result, more properties teetering above the beach have been added to the list of homes no longer considered safe. Retired couples in Aldborough, 10 km south of Hornsea, face almost certain evacuation.

Adapted from *The Guardian*, September 2012

The Holderness coast ... going ... going...

For centuries, erosion has been a problem along the coast of the East Riding of Yorkshire, known as Holderness, which stretches between Bridlington and Flamborough Head in the north, and Spurn Head in the south. It has the fastest-eroding coastline in Europe. On average, it loses nearly 2 metres of coastline every year. Since Roman times, the Holderness coast has retreated by 4 km – and at least 29 villages have been lost to the sea (see Figure 1).

Why is erosion such a problem at Holderness?

Rates of coastal erosion vary, due to a range of physical and human factors – all of which can change in both the short- and the long-term.

However, there are three main reasons why the coastline at Holderness is retreating so rapidly:

◆ geology

◆ fetch

◆ longshore drift and beach material.

> *Figure 2* The geology of the Holderness coast

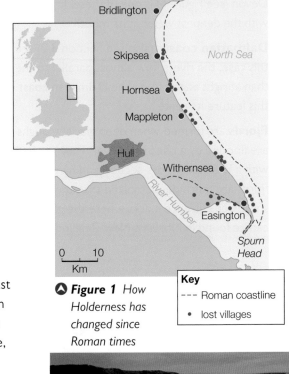

⬣ *Figure 1* How Holderness has changed since Roman times

Key
--- Roman coastline
• lost villages

⬣ *Figure 3* The chalk cliffs at Flamborough Head

⬣ *Figure 4* Holderness – low, unstable cliffs composed of boulder clay

Geology

◆ Most of the Holderness coast consists of **boulder clay** (see Figures 2 and 4). Boulder clay is also known as glacial till, or drift, and is a mixture of fine clays, sands and boulders deposited by glaciers after the last Ice Age. Boulder clay is structurally weak, and has little resistance to erosion. It produces shallow, sloping cliffs between 5 and 20 metres high.

◆ The chalk band that surrounds the boulder clay (see Figure 2) has created a headland at Flamborough Head (see Figure 3). Erosion along fault lines and bedding planes has created features such as cliffs, arches and stacks.

Fetch

One of the main factors affecting the rate of erosion is wave energy. This, in turn, depends on the fetch (how far the waves have travelled). Holderness is exposed to winds and waves from the north-east, with a small fetch of about 500–800 km across the North Sea. That isn't far – compared to the fetch of waves crossing the world's oceans – but the waves attacking the Holderness coast are also influenced by other factors, which help to increase their size and power:

◆ Currents (or **swell**) circulate around the UK from the Atlantic Ocean into the North Sea (see Figure 5). The Atlantic's fetch is 5000 km or more, so its currents add energy to the waves in the North Sea. Therefore, there are often powerful **destructive waves** at work along this coastline.

◆ Low-pressure **weather systems** and winter storms passing over the North Sea are often intense – producing locally strong winds and waves (see Figure 6). Low-pressure air weighs less, raising sea levels, which in turn produce much higher tides than normal (that reach the cliff base).

◆ Small, almost-enclosed seas (such as the Mediterranean and North Sea) often generate huge waves during storms. Waves move within the sea, but cannot disperse their energy – rather like water slopping up against the side of a washbasin.

◆ The **sea floor** is relatively deep along the Holderness coast, so waves reach the cliffs without first being weakened by friction with shallow beaches.

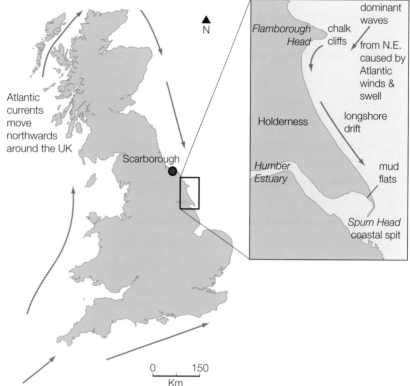

◢ **Figure 5** The fetch of the Holderness coast, and the additional swell circulating from the Atlantic

◣ **Figure 6** North Sea storm waves at Scarborough, just north of Holderness

127

Longshore drift and beach material

◆ The beaches at Holderness are its main problem. The boulder clay erodes to produce mainly clay particles, which are fine and easily transported out to sea in suspension, rather than accumulating on-shore as beach sand. Although there are beaches, they are narrow and they offer little friction to absorb wave energy. Plus, there is never enough sand to stop the waves from reaching the base of the cliffs at high tide.

◆ The tides flow southwards (see Figure 5 inset) – transporting sand south by **longshore drift** (see page 112), and leaving the cliffs at Holderness poorly protected against wave attack. Beaches south of Hornsea have reduced in width, because an imbalance exists between the input of sand (deposited by swash) and the removal of sand (by backwash).

Sub-aerial processes and coastal erosion

The cliffs at Holderness are affected by weathering and mass movement (collectively known as sub-aerial processes; see page 118).

◆ Chemical weathering is relatively ineffective at Holderness, except on the chalk cliffs at Flamborough Head. Mechanical and biological weathering are far more significant.
The main types of weathering experienced at Holderness are freeze-thaw and the alternate wetting and drying of the boulder clay, which makes it crumbly in dry periods.

◆ **Slumping** is the main form of mass movement

affecting the boulder clay cliffs at Holderness. The alternate wetting and drying of the clay causes expansion and shrinkage – producing cracks during long, dry periods. Subsequent rains then enter the cracked clay and percolate into the cliff, which becomes lubricated and much heavier. The weakened cliff cannot support the extra weight, and the clay slides downslope under gravity. The slumped material collects at the cliff base (see Figure 7) and is then removed by the sea – causing the cliff line to retreat.

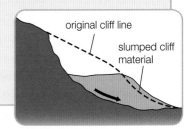

△ **Figure 7** *Slumping in action*

Human actions and coastal retreat

It's not just physical factors that affect coastal retreat. The **actions** that people and organisations (collectively known as **players**) take can impact on coastal retreat – and the outcome isn't always positive.

Key players on the Holderness coast

1 Central government agencies

The Environment Agency is responsible for coastal management (along with the local authorities). Its budget from central government has been cut since 2010.

2 Local government

The local authorities are jointly responsible for coastal management with the Environment Agency. However, in 2010, local-government funding was cut by central government, which restricted local councils to minimal increases in Council Tax.

3 Stakeholders in the local economy

◆ The tourist industry (including campsites) wants greater spending on coastal protection.

◆ Farmers want money spent to protect their farmland, which is of lower value than urban spaces.

◆ Residents at, for example, Hornsea and Mappleton want guaranteed coastal protection for their homes and businesses.

◆ Insurance companies are increasingly refusing to insure vulnerable properties.

4 Environmental stakeholders

English Nature and the RSPB want to protect Spurn Head (one of the UK's largest coastal spits), so a continuing flow of sand southwards by longshore drift is essential. One of the most-important assets of this spit is the protection that it gives to the mudflats of the Humber Estuary (one of the UK's most important birdlife reserves).

The impact of coastal management

The graph in Figure 8 shows erosion rates opposite the locations to which they relate. The gaps on the graph show where coastal defences are preventing erosion. However, the graph also shows that higher rates of erosion occur immediately to the south of those same coastal defences. For example, the sea wall, groynes and rock armour at Hornsea (see below) protect part of the coast, but they also interrupt the flow of beach material by longshore drift. The beach downdrift of Hornsea, at Mappleton, is then starved of material and its cliffs are exposed to wave attack. This is known as **terminal groyne syndrome**, and it affects many UK coasts.

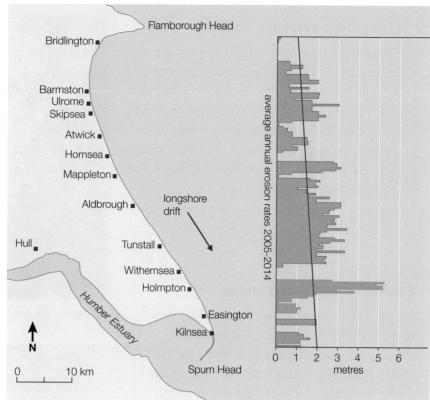

⬆ **Figure 8** *Erosion rates along the Holderness coast, 2005-2014*

- The first sea wall was built at Hornsea in 1870. It lasted for 6 years.

- In 1906, a stronger sea wall was built. It has now been extended five times.

- At the southern end, the defences were reconfigured in 1977. The T-shaped rock armour structure is designed to allow beach sediment to accumulate and pass behind it.

- The groynes built at Hornsea starved Mappleton (further south) of sediment. By the 1990s, nearly 4 metres of cliff were being eroded at Mappleton each year.

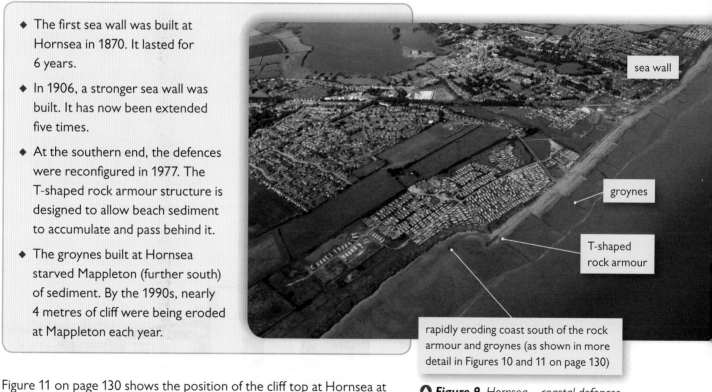

⬆ **Figure 9** *Hornsea – coastal defences*

Figure 11 on page 130 shows the position of the cliff top at Hornsea at different dates between 1854 and 2005. It illustrates that coastal retreat remains a continuing problem along this stretch of coast – in particular, immediately south of Hornsea's coastal defences, where the cliff top has retreated significantly since 1977-8 (when the rock armour was built).

Using map evidence and GIS

The photo in Figure 10 has been overlain with an OS map from 1852. It shows the section of coast in Figure 9 just south of the rock armour. You can clearly see how the coast has retreated since 1852. On its own, it doesn't prove that building coastal defences has increased the rate of coastal retreat – but if you look at it alongside Figures 8 and 11, the evidence becomes overwhelming.

▲ **Figure 10** *Coastal retreat south of Hornsea*

Economic and social losses

Margaret Fincham of Golden Sands Holiday Park, south of Withernsea, has lost 100 chalets to the sea in 15 years. She said *'If we hadn't lost the chalets, there could have been an extra 400 people visiting Withernsea and helping the local economy.'*

Withernsea isn't alone in suffering **economic** losses due to coastal erosion. It is predicted that 200 homes and several roads will fall into the sea between Flamborough Head and Spurn Point by 2100 (see Figure 12). Nationally, The Environment Agency suggests that 7000 homes will disappear due to coastal erosion by the same date.

Individuals lose out both **socially** and **financially** as a result of coastal erosion, and very little financial help is available for them. No compensation is paid out for the loss of private property or land caused by coastal erosion in England.

Between 2010 and 2012, the East Riding of Yorkshire Council used £1.2 million of direct money from Defra to trial different ways of helping people adapt to living on an eroding coastline. They gave some financial assistance to 36 households along the coast, supported 16 relocations and 43 property demolitions. The remainder of the money was used to help residents through the relocation and adaptation packages described on page 131.

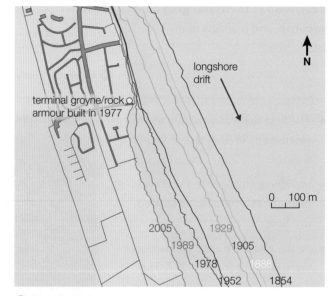

▲ **Figure 11** *Cliff top positions at Hornsea, 1854-2005*

▶ **Figure 12** *Homes and caravan parks are under threat on the Holderness coast*

East Riding Coastal Change Fund

This fund offers some limited help in terms of financial assistance, as well as free advice for those residents affected by coastal erosion. Its **relocation package** can fund:

◆ the demolition costs for a property

◆ some relocation costs (e.g. hiring a removal van) – up to a maximum of £1000.

◆ the expenses caused by relocating to a new home – up to a maximum of £200.

The fund also offers an **adaptation package**, which can help to pay for:

◆ rollback – the expenses incurred as a result of an individual's decision to replace a threatened coastal property with a new home inland. But it only covers things like planning application fees. Rollback was introduced particularly to address the risk to caravan parks, farms and homes in areas where coastal defences are not viable.

◆ assistance grants – to adapt properties which may be at risk from coastal erosion in the future (e.g. relocating septic tanks/waste pipes, and changing the access routes to some properties).

Over to you

1 In pairs, draw a large map of Holderness to show its main physical and human features. Add details about:

  • its geology and where most erosion is taking place

  • the influence of fetch, wave size, and tides

  • the direction of longshore drift and the features produced.

2 Still working in pairs, copy and complete the following table to classify human and physical factors that influence the rate of coastal erosion.

	Human	**Physical**
Short-term (i.e. a season, or a one-off event)		
Medium-term (i.e. from months to a few years)		
Long-term (i.e. many years or decades)		

3 Design a spider diagram to show (**a**) the economic and (**b**) the social impacts of erosion along the Holderness coast.

On your own

4 **a** Copy Figure 7 in Section 5.12 (the conflict matrix). Complete the players for the examples in this section about coastal erosion at Holderness.

  b Identify which groups at Holderness (**a**) agree and (**b**) disagree, and explain why this might be.

5 Google the phrase 'East Yorkshire coastal erosion Hull University' for its display of photo strips of the Holderness coast. Using strips TA47 and TA48, assess erosion rates at Hornsea.

 6 **a** Using Figure 11, take two transects from the cliff top to the shoreline. Then copy and complete the table below for each transect, in order to calculate the average rates of erosion per year for each of these seven time periods.

Time periods	Number of years	Erosion in metres	Erosion per year
1854-1888			
1888-1905			
1905-1929			
1929-1952			
1952-1978			
1978-1989			
1989-2005			

  b Plot the rate of erosion per year on a graph. How have rates of erosion changed?

  c Based on the current rate of erosion, estimate how much additional land is likely to be lost to the sea by 2025.

Exam-style questions

 1 Assess the impacts of coastal management along one stretch of coastline. *(12 marks)*

 2 Assess the relative importance of factors which have led to rapid coastal erosion along a stretch of coastline. *(12 marks)*

3.9 Coastal flooding

In this section, you'll learn about the significant and increasing risk of coastal flooding for some coastlines, as well as about some of its consequences.

Paying the price of increasing floods

As the waves pounded the road, seven-year-old Rima cried to her father 'Please don't let go of me. If you do, I'll drown.' Moments later, she slipped from his grasp and was swept away – one of 3363 victims of Cyclone Sidr.

This powerful storm struck Bangladesh in November 2007. Rima's father, Zafar, was carrying her as the family raced to a cyclone shelter. They were being pummelled by gushing water when he tripped – and Rima was flung from his arms.

Figure 1 *Some of the devastation caused by Cyclone Sidr in Bangladesh*

BACKGROUND

Bangladesh – a country at risk

- Bangladesh is the world's most densely populated country – with an estimated population of about 169 million in 2015.
- 46% of the country's population lives less than 10 metres above sea level.
- Bangladesh also lies on the floodplains of three major rivers – the Brahmaputra, Meghna and Ganges – which converge in Bangladesh and, together with 54 smaller rivers, empty into the Bay of Bengal through a series of estuaries.

The information in the above bullets helps to explain why Bangladesh's coastal regions are so densely populated, and also why so much of the country is threatened so frequently by both river and coastal flooding. Almost every year, huge areas of the country flood as Himalayan snowmelt adds to monsoon rains and high tides in the Bay of Bengal. Coastal flooding is a major problem – between March and May, violent thunderstorms produce strong southerly winds of 160 km per hour, which bring in six-metre-high waves from the Bay of Bengal to swamp coastal areas.

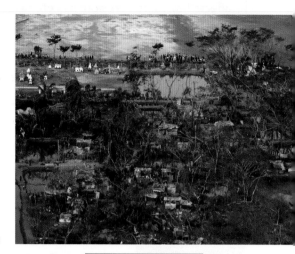

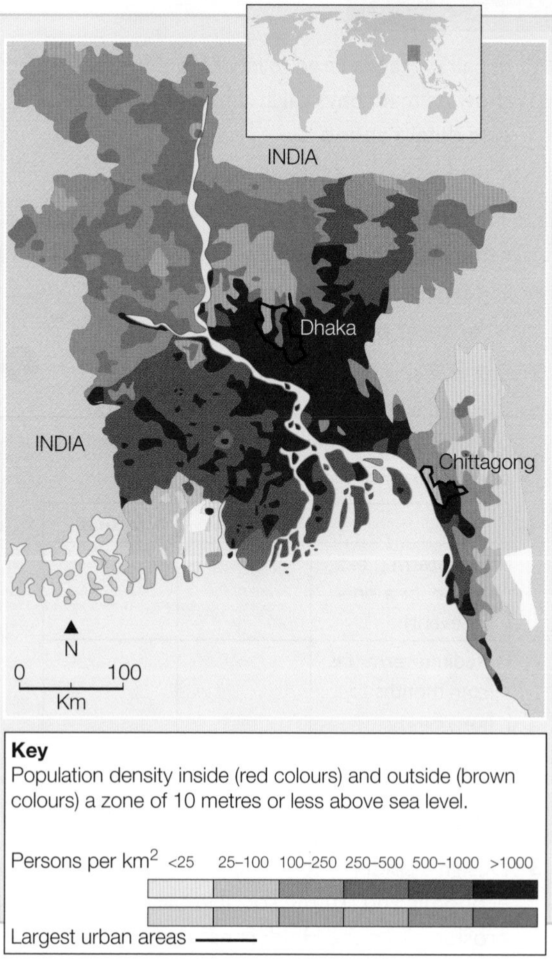

INDIA
Dhaka
INDIA
Chittagong
N
0
100
Km
Key
Population density inside (red colours) and outside (brown colours) a zone of 10 metres or less above sea level.
Persons per km² <25 25–100 100–250 250–500 500–1000 >1000
Largest urban areas

Figure 2 *Population density at risk of flooding in Bangladesh*

Increasing the flood risk

There is not a great deal that Bangladesh can do about those physical factors that make it prone to flooding, such as its height above sea level or the number of major rivers passing through the country on their way to the sea. However, human actions (such as those explained opposite) are increasing the risk of coastal flooding.

1 Subsidence

According to the journal *Nature Climate Change*, some of Bangladesh's estuarine islands have sunk by as much as 1.5 metres in the last 50 years or so. Isostatic readjustment is partly responsible, but the main reason is clearance and drainage of more than 50 large islands in the Ganges-Brahmaputra river delta. These islands used to be forested, but have now been cleared and are being used to grow rice to feed the country's large population. In the 1960s and 1970s, large earth embankments were built around these islands to protect them against tidal and storm-surge inundations (see Figure 3). However, this human action has also prevented the natural deposition of sediment that used to maintain the islands' height. Now these islands are fast submerging and millions of people living on them are at increased risk of flooding if the embankments give way.

2 Removing vegetation

Mangrove forests are found along the tropical and sub-tropical coasts of Africa, Australia, Asia and the Americas – but the largest remaining tract of mangrove forest in the world is found in the Sundarbans region of Bangladesh, on the edge of the Bay of Bengal (see Figure 4).

Mangroves are one of the most productive and complex ecosystems on the planet. They are essential to marine, freshwater and terrestrial biodiversity, because they stabilise coastlines against erosion, collect nutrient-rich sediments, and provide a nursery for coastal fish. Not only that, but they also provide protection and shelter against extreme weather events (such as storm winds and floods, as well as tsunamis). They absorb and disperse tidal surges associated with these events.

The Sundarbans forest helped to take the sting out of Cyclone Sidr in November 2007, but recent satellite studies show that 71% of Bangladesh's mangrove-forested coastline is now retreating by as much as 200 metres a year. The causes of this are erosion, rising sea levels – and human actions that deliberately remove the vegetation. Globally, half of all mangrove forests have been lost since the mid-twentieth century. Converting the forest into lucrative shrimp farms accounts for 25% of this loss. This is happening in Asia (including Bangladesh) and also Latin America.

▲ **Figure 3** *The clearance and drainage of land for cultivation, together with the building of large earth embankments, has caused some estuarine islands to shrink and subside*

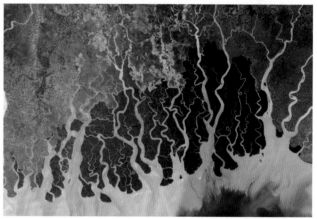

▲ **Figure 4** *In this satellite image, the Sundarbans mangrove forests show up as dark green, surrounded to the north by lighter-green farmland*

Sri Lanka – protecting mangrove forests

A report published twelve months after the devastating 2004 Indian Ocean tsunami, compared two coastal villages hit by the wall of water. It showed that two people died in the settlement protected by dense mangrove and scrub forest, while up to 6000 people died in the village that had removed the protective vegetation.

Partly as a result of this, Sri Lanka has become the first nation in the world to protect all of its mangrove forests. Its scheme will protect all 8800 hectares of remaining forest, as well as replanting mangroves that have already been felled (and funding micro-loans for villagers in exchange for them protecting local mangrove forests).

Storm surges

Cyclone Sidr swept in from the Bay of Bengal on 15 November 2007 – bringing with it a **storm surge** that reached up to 6 metres high in places. Storm surges causing devastating results are common in the Bay of Bengal (see Figure 5). But what are they and how do they form?

Storm surges are changes in sea level caused by intense low-pressure systems – **depressions** and **tropical cyclones** – and high wind speeds. For every drop in air pressure of 10 mb, the sea level rises by 10 cm. During tropical cyclones, the air pressure may be 100 mb lower than normal, which will raise the sea level by 1 metre. This rise in sea level is intensified in areas where the coastline is funnel-shaped. During high tides and in low-lying areas – such as much of Bangladesh – the results can be deadly (see Figure 5).

Date of cyclone	Max wind speed (km/hr)	Storm surge height (metres)	Death toll
11 May 1965	161	3.7 – 7.6	19 279
15 December 1965	217	2.4 – 3.6	873
1 October 1966	139	6.0 – 6.7	850
12 November 1970	224	6.0 – 10.0	300 000
25 May 1985	154	3.0 – 4.6	11 069
29 April 1991	225	6.0 – 7.6	138 882
19 May 1997	232	3.1 – 4.6	155
15 November 2007 (Sidr)	223	4.5 – 6.0	3363
25 May 2009 (Aila)	92	-	190

Figure 5 *Bangladesh – cyclones, storm surges and their effects*

Bangladesh – the impacts of Cyclone Sidr

Cyclone Sidr brought heavy rain with it, as well as strong winds of up to 223 km/hr and a huge storm surge reaching up to 6 metres in height. It was a category 4 storm. According to Bangladesh's Meteorological Department, Sidr's eye crossed the coast near the Sundarbans mangrove forests around 9.30 pm on 15 November. As Figure 6 shows, the coastal districts and offshore islands suffered the highest number of deaths and the worst effects.

Rima's mother (see page 132) says that the river embankment on one side of her village held firm against the powerful waves of the storm surge. However, an embankment on the other side of the village had been lowered by the local authorities to make it into a paved road – allowing the storm surge into the village.

▶ **Figure 6** *Deaths caused by Cyclone Sidr in Bangladesh*

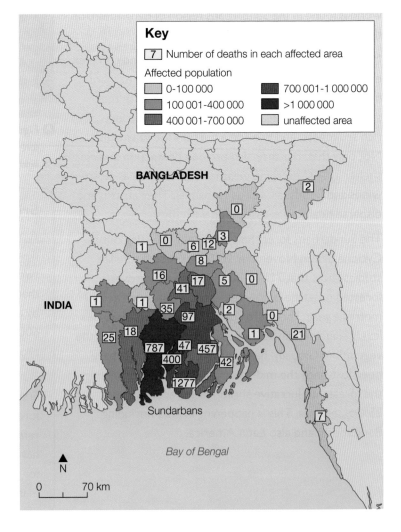

Key

7 Number of deaths in each affected area

Affected population
- 0-100 000
- 100 001-400 000
- 400 001-700 000
- 700 001-1 000 000
- >1 000 000
- unaffected area

BANGLADESH

INDIA

Sundarbans

Bay of Bengal

N

0   70 km

Rima's family weren't the only ones to suffer. 'I have never seen such a catastrophe in my 20 years as a government administrator' said Harisparasad Pal, an official from Barguna District. 'Village after village has been shattered. Millions of people are living out in the open.' The situation in coastal districts like Barguna was described as the worst in decades – and that's saying something, given that Bangladesh is used to dealing with major floods and storms every year.

◆ The storm surge breached many coastal and river embankments – causing severe flooding in low-lying areas.

◆ The high winds and floods damaged housing, roads, bridges and other infrastructure.

◆ Electricity supplies and communications were knocked out, and roads and waterways became impassable.

◆ Drinking water was contaminated by debris, and many freshwater sources (such as wells) were inundated with salt water.

◆ The sanitation infrastructure was destroyed, raising the risk of disease.

Figure 7 gives you an idea of the scale of the immediate **short-term** impacts.

The total cost of Cyclone Sidr to Bangladesh was estimated to be US$1.7 billion. Most of the destruction and social and environmental losses were caused by the severity of the cyclone; the failure of the extensive embankment system; and the consequent flooding of many villages. But the casualties were lower than expected, because of improved disaster-prevention measures (including an improved forecasting and warning system and the use of cyclone shelters).

Damaged crops	685 528 hectares
Damaged houses	1 518 942
Dead and missing	4234
People injured	55 282
Cattle and poultry killed	1 778 507
Damaged educational institutions	16 954
Damaged roads	8075 km
Damaged bridges/culverts	1687
Damaged electricity lines	703 km
Tube wells affected	901
Cost of damage to roads, embankments, sluice gates and riverbank protection	US$ 29.6 million

Figure 7 *Some of the impacts of Cyclone Sidr (from Bangladesh's Department of Disaster Management)*

Coastal flooding and storm surges in developed countries

From mid-December 2013 to early January 2014, the UK experienced a spell of extreme weather – as it was hit by a succession of major storms. The storms were driven by a powerful jet stream bringing low-pressure weather systems across the Atlantic. One of the most significant storms occurred during 5-6 December 2013 – bringing with it a storm surge that affected the coast of the UK (as well as other countries in Northern Europe).

What caused the storm surge?

◆ Intense low pressure (976 mb deepening to 968 mb). This was similar to the storm which caused the 1953 storm surge that killed 307 people in the UK and 1800 people in the Netherlands.

◆ Sea shape and coastline. The North Sea is open to the Atlantic Ocean (see Figure 5 on page 127) and tapers towards the south – in a dangerous funnel shape. This allows strong northerly winds to push storm surges towards cities like Amsterdam and London, as well as low-lying coastal areas including Norfolk and much of the Netherlands. Some of these areas are at, or below, sea level.

◆ Sea depth. The North Sea gets shallower as well as narrower towards the south, which has the effect of increasing the height of tides and storm surges.

◆ High seasonal tides.

◆ Strong northerly winds pushed the storm and the surge further south – increasing the height of the surge and tides.

The impacts of the 2013 storm surge

The storm of 5-6 December 2013 had impacts across the whole of Northern Europe, as Figure 8 shows.

Key

Hazards
- ● storm surge
- ★ strong wind
- ■ floods

Damage
- 🏚 houses destroyed
- 🏠 houses affected
- 🛫 travel disrupted
- 🏃 evacuations

Alerts
- 〰️ wind
- 〰️ coastal event

Impacts in the Netherlands
- Strong winds, together with a storm surge that was predicted to be up to 2 metres high in the Netherlands. However, on the East Frisian coast (in Germany, but right on the Dutch border), it reached 3.74 metres above mean sea level.
- Unlike in the 1953 storm surge, no-one died.
- The Netherlands constructed a series of *Delta Works* in response to the 1953 storm surge. *Delta Works* consist of a series of dams and storm-surge barriers, designed to protect the country from flooding by the North Sea. The Eastern Scheldt storm-surge barrier is the largest of the Delta Works and one of the biggest construction projects in the world. It cost 2.5 billion Euros to build, and costs a further 17 million Euros a year to operate. It was opened in 1986, and was closed against the storm surge on 5-6 December 2013.

Impacts in the UK
- Strong winds (gusts of over 200 km/hr in Scotland).
- Coastal flooding (1400 homes flooded) and forced evacuation – mostly along the coast of Eastern England, but also in North Wales.
- At Hemsby (Norfolk), cliff erosion resulted in several properties collapsing into the sea.
- Bridges were shut, and rail services in eastern counties were disrupted.
- Two people died.
- Hundreds of thousands of properties were protected by flood defences, and the Thames Barrier was closed to protect London.
- Insurers calculated the cost of the damage at £100 million.

Figure 8 *Storm and storm surge across Northern Europe 5–6 December 2013*

Increasing flood risk

More storms

It's not really clear whether climate change will mean more hurricanes (cyclones), but warmer ocean-surface temperatures and higher sea levels are expected to make them more intense – with stronger winds (2-11% stronger) and more rain (projected to increase by about 20%).

The connection between climate change and hurricane frequency isn't straightforward. Globally, the number of tropical storms that form each year ranges from 70-110. About 40-60 of these reach hurricane force. But records show large changes year-on-year in the number and intensity of these storms (see Figure 9). Note that this graph is just for Atlantic storms.

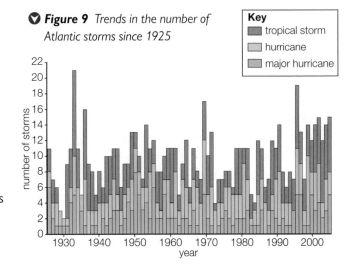

Figure 9 *Trends in the number of Atlantic storms since 1925*

Key
- ■ tropical storm
- ■ hurricane
- ■ major hurricane

Changes in frequency and intensity vary around the world. In the North Atlantic, the long-term average number of storms per year is 11, with about six becoming hurricanes. More recently though (from 2000-2013), the average is 16 tropical storms per year – including eight hurricanes. This increase correlates with the rise in North Atlantic ocean-surface temperatures.

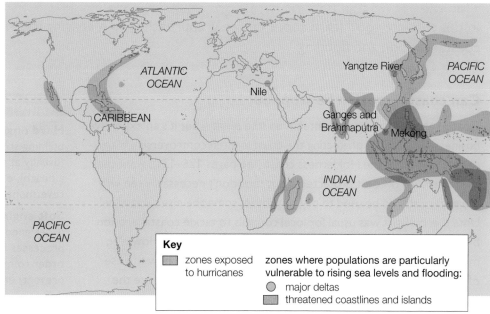

Key

	zones exposed to hurricanes	zones where populations are particularly vulnerable to rising sea levels and flooding:
○		major deltas
		threatened coastlines and islands

The strongest hurricanes have increased in intensity over the past two or three decades in both the North Atlantic and Indian Oceans, but in the Pacific and Indian Oceans there are no significant trends.

Recent predictions indicate that in some areas – as warmer oceans lead to more powerful, intense, hurricanes – fewer storms will actually develop.

▲ Figure 10 *Rising sea levels and coastal flooding*

More flooding

Countries with low-lying coasts already suffer from coastal flooding, but climate change is likely to increase the risk. The IPCC has predicted that, by 2100, hundreds of millions of people will have been forced to abandon many coastal zones worldwide, because of rising sea levels. Figure 10 shows those areas of the world where populations are vulnerable to rising sea levels and coastal flooding.

As sea levels continue to rise, storm surges will also become higher. This effect is already being seen – such as in 2012, when rising sea levels intensified the impact of Hurricane Sandy in the eastern USA (causing $65 billion worth of damage; much of it related to coastal storm-surge flooding).

Preparing for the future

Rising sea levels, together with more-intense and possibly more-frequent storms, can all lead to increasing flood risks. There are two possible approaches to dealing with the risk:

◆ **Adaptation** – making changes to lessen the impact of flooding. This can include things like building sea walls (see page 138); building storm-surge barriers (such as the Eastern Scheldt storm surge barrier; see Figure 8); reinstating mangrove forests (see the example of Sri Lanka on page 133).

◆ **Mitigation** – making efforts to reduce the emission of greenhouse gases and so reduce the impacts of climate change.

Over to you

1 a In pairs, list the impacts of Cyclone Sidr and then classify them into economic, social and environmental.

 b Repeat activity (**a**) for the December 2013 storm surge in the North Sea.

 c Which impacts were the worst, and why?

2 Draw a flow diagram to show stages in the development of a storm surge.

On your own

3 Define the following terms: (**a**) flood and storm surge, (**b**) mitigation and adaptation, (**c**) mangrove and Sundarbans, (**d**) flooding and inundation.

4 Write a 750-word report: 'The flood threat is less about flooding, and more to do with a country's level of development'.

Exam-style questions

AS 1 Explain the physical and human causes of one flood in a developing country. *(8 marks)*

A 2 Evaluate the influence of a country's level of development in determining the impacts of coastal flooding. *(16 marks)*

In this section, you'll learn about different approaches to managing the risks associated with coastal erosion and flooding.

Can coastal erosion be stopped?

Coastal erosion can be prevented – up to a point – but it's an expensive business. The cost of protecting the coast is often controversial. Many people, e.g. along the Holderness coast (see pages 126–131), want their own stretches of coast protected – but they don't necessarily see why their taxes should be used to pay to protect someone else's coastline! Until the 1990s, it was usual for local councils to tackle coastal erosion by designing **hard-engineering** structures (see Figure 1). However, most of those structures are very expensive to build, so now the use of **soft-engineering** techniques is more popular. But what are the advantages and disadvantages of the different methods?

Key words
Hard engineering – This involves building structures along the coast (usually at the base of a cliff or on a beach), e.g. sea walls, groynes and revetments.
Soft engineering – This approach is designed to work with natural processes in the coastal system, in order to manage (but not necessarily prevent) erosion.

Type of structure	Advantages	Disadvantages	Cost
Groynes Timber or rock structures built at right angles to the coast. They trap sediment being moved along the coast by longshore drift – building up the beach.	The built-up beach increases tourist potential and protects the land behind it. Groynes work with natural processes to build up the beach. Not too expensive.	Groynes starve beaches further along the coast of fresh sediment, because they interrupt longshore drift. This often leads to increased erosion elsewhere. Groynes are unnatural and rock groynes can be very unattractive.	£5000 to £10 000 each (at 200-metre intervals).
Sea walls Made of stone or concrete at the foot of a cliff, or at the top of a beach. They usually have a curved face to reflect waves back into the sea.	Effective prevention of erosion. They often have a promenade for people to walk along.	They reflect wave energy, rather than absorbing it. They can be intrusive and unnatural looking. They are very expensive to build and maintain.	£6000 a metre.
Rip rap (rock armour) Large rocks placed at the foot of a cliff, or at the top of a beach. It forms a permeable barrier to the sea – breaking up the waves, but allowing some water to pass through.	It is relatively cheap and easy to construct and maintain. It's often used for fishing from, or for sunbathing by tourists.	The rocks used are usually from somewhere else (e.g. granite), so they don't fit in with the local geology and can look out of place. It can be very intrusive. The rocks can be dangerous for people clambering over them.	£100 000 to £300 000 for 100 metres.
Revetments Sloping wooden, concrete or rock structures – placed at the foot of a cliff or the top of a beach. They break up the waves' energy.	They are relatively inexpensive to build.	They are intrusive and very unnatural looking. They can need high levels of maintenance.	Up to £4500 a metre
Offshore breakwater A partly submerged rock barrier, designed to break up the waves before they reach the coast.	An effective permeable barrier	It is visually unappealing. It's a potential navigation hazard.	Similar to rock armour – depending on the materials used.

⬥ **Figure 1** *Hard engineering – advantages and disadvantages*

Hard engineering and Holderness

The Holderness coast is 85 km long. Only 9.2 km are protected by hard-engineering structures (maintained by the East Riding of Yorkshire Council). An additional 2.15 km are protected by other bodies. The rest of the coastline is unprotected.

Most of the defences on the Holderness coast consist of a mixture of nineteenth-century structures, together with more-recent upgrades, extensions and alterations.

Hard engineering – the impacts

Decisions taken by councils and coastal authorities to use hard-engineering methods to protect particular places on the coast, can then lead to problems elsewhere (see Figure 2).

Cost-benefit analysis (CBA)

A cost-benefit analysis is carried out before a coastal-management project is given the go-ahead. Costs are forecast (e.g. a sea wall – its design, building costs, maintenance, etc.) and then compared with the expected benefits (e.g. value of land saved, housing protected, savings in relocating people, etc.). Costs and benefits are of two types:

♦ **Tangible** – where costs and benefits are known and can be given a monetary value (e.g. building costs).

♦ **Intangible** – where costs may be difficult to assess but are important (e.g. the visual impact of a revetment).

A project where costs exceed benefits is unlikely to be given permission to go ahead.

Hornsea

Defences: Concrete sea walls, groynes, rock armour (see above).

Impact: The groynes trap sediment and maintain the beach at Hornsea, but Mappleton downdrift has been starved of sediment as a result. There, rapid wave attack has eroded the cliffs, so that by the 1990s, nearly 4 metres of cliff were being eroded each year.

Mappleton

Defences: Two rock groynes (costing £2 million) were built in 1991 (see right), with the aim of preventing the removal of the beach by longshore drift. Rock armour was also used.

Impact: At Cowden, 3 km south of Mappleton, the resultant sediment starvation caused increased erosion of the cliffs (from 2.5 to 3.8 metres a year between 1991 and 2007).

Withernsea

Defences: A straight sea wall was built in 1875. However, over time, wave energy eroded (**scoured**) the base of the wall – causing it to collapse. So, in the 1990s (following a **cost-benefit analysis**) the straight wall was replaced by a curved wall – at a cost of £6.3 million (£5000 per metre).

Impact: The waves are now noisier when they break against the wall, and the promenade is smaller. The views from sea-front hotels have also been restricted. Some tourists find the rip-rap at the base of the sea wall unattractive.

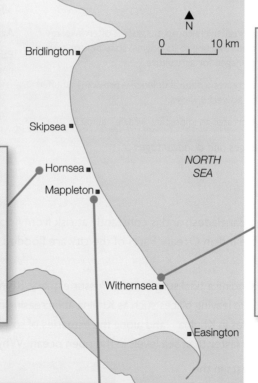

Bridlington

Skipsea

Hornsea

Mappleton

NORTH SEA

Withernsea

Easington

N

0   10 km

Figure 2 *Some of the impacts of hard engineering*

Soft engineering

A range of soft-engineering techniques attempt to work with natural processes to protect coasts (see Figure 3). These can also be used to manage changes in sea level, e.g. by allowing low-lying coastal areas to flood (creating marshes).

Method	Advantages	Disadvantages	Cost
Beach nourishment The addition of sand or pebbles to an existing beach to make it higher or wider. The sediment is usually dredged from the nearby seabed.	Relatively cheap and easy to maintain. It looks natural and blends in with the existing beach. It increases tourist potential by creating a bigger beach.	It needs constant maintenance, because of the natural processes of erosion and longshore drift.	£300 000 for 100 metres.
Cliff regrading and drainage Cliff regrading reduces the angle of the cliff, to help stabilise it. Drainage removes water to prevent landslides and slumping.	Regrading can work on clay or loose rock, where other methods won't work. Drainage is cost-effective.	Regrading effectively causes the cliff to retreat. Drained cliffs can dry out and lead to collapse (rock falls).	Details unavailable
Dune stabilisation Marram grass can be planted to stabilise dunes. Areas can be fenced in to keep people off newly planted dunes.	It maintains a natural coastal environment. It provides important wildlife habitats. It is relatively cheap and sustainable.	It is time consuming to plant marram grass. People may respond negatively to being kept off certain areas.	£200 to £2000 for 100 metres.
Marsh creation A form of managed retreat, by allowing low-lying coastal areas to be flooded by the sea. The land then becomes a salt marsh.	It is relatively cheap, because it often involves land reverting to its original state before it was managed for agriculture. It creates a natural defence – providing a buffer to powerful waves. It creates an important wildlife habitat.	Agricultural land is lost. Farmers or landowners need to be compensated.	The cost is variable – depending on the size of the area left to the sea.

⊙ Figure 3 *Soft engineering – advantages and disadvantages*

Managing future threats

Adaptation 1

Khulna is a river port in south-west Bangladesh and is constantly at risk from flooding, despite being 125 km inland from the Indian Ocean. Parts of the city are flooded on average at least 10 times a year.

Cyclone Aila swept inland in 2009, sending a tidal surge up the Pussur estuary. Rising sea levels, as a result of climate change, are leaving places such as Khulna at increasing risk from storm surges. However, high tides in Khulna – and along the estuaries of south-west Bangladesh – are rising six times faster than sea levels in the open ocean. Why?

⊙ Figure 4 *The tidal surge created by Cyclone Aila damaged nearly 2000 km of embankments in Khulna and surrounding districts*

◆ The destruction of mangrove forests in the Sundarbans is partly to blame (see page 133).

◆ But it's also thought that the embankments built to protect people from the rising tides are making the problem worse – by constricting and funnelling tidal flows, pushing water further inland, and increasing the tidal range.

In the past 50 years, Bangladesh has built 4000 km of coastal embankments. About 30 million people live on

polders (land enclosed by embankments) or planned polders.
In 2014, Bangladesh was planning to upgrade 600 km of embankments in the Sundarbans delta region (using $400 million from the World Bank).

In 2014, the *New Scientist* reported that John Pethick, a British geomorphologist, said that Bangladesh's upgraded embankments will put millions of lives at risk. He argued that the flood defences in the Sundarbans delta are doing more harm than good. He suggested that some polders should be abandoned, and that embankments should be positioned further back from estuaries (to reduce the funnelling effect). But this argument is controversial. Local people don't want to give up land to rising tides, and – despite warnings that a global rise in sea level (in conjunction with the funnelling effect) could result in a 2.5-metre rise in high tides by the end of this century – the Bangladeshi government and the World Bank seem intent on going ahead and upgrading the embankments.

Adaptation 2

Embankments may not provide the most **sustainable** solution to managing future threats – but there are other ways.

Down the coast from Khulna, in India's Mahanadi Delta in the state of Odisha, people are trying a different approach. The Mahanadi Delta is also prone to cyclone disasters and, again, the loss of mangroves is part of the story. 50 years ago, coastal villages in Odisha had an average width of 5.1 km of mangroves protecting them. Today, that figure is an average of 1.2 km. in 1999, during 'super cyclone' Kalina, villages that still had four or more kilometres of mangroves, recorded no deaths. However, in areas where the protective belt was less than 3 km wide, death rates rose sharply.

The NGO Wetlands International, the Indian Government and Odisha's Integrated Coastal Zone Management Project (see Section 3.11), are trying to reverse decades of mangrove destruction. They are helping villagers to plant mangroves along the coast, and also on the banks of all tidal rivers along Odisha's coast.

According to Daniel Alongi of the Australian Institute of Marine Science, the faster sea levels rise, the faster mangroves accumulate sediment in their roots. They can keep up with a rise of 25 mm a year – eight times the current global rate. No sea wall can do that.

⬆ *Figure 5 Mangroves in India at low tide with their roots on display*

Exam-style questions

(AS) 1 Assess the effectiveness of hard-engineering approaches designed to protect the coast from erosion. *(12 marks)*

(A) 2 Evaluate the effectiveness of coastal-management strategies along a stretch of coast. *(20 marks)*

Over to you

 1 a Using Google Maps, find Hornsea and Mappleton (plus another stretch of the Holderness coast of your choice). Select a 100 m × 100 m (i.e. one hectare) stretch of **each** coast.

b Count the number of properties within each hectare. Assume an average value of £250 000 per house, plus £25 000 per hectare for farmland. Then calculate the land value of each hectare.

c In pairs, decide which of the coastal-management strategies discussed in Figures 1, 2 and 3 would be best for each hectare, and why.

 2 Consider each of the methods used to manage Hornsea, Mappleton and Withernsea (see Figure 2). Then use the 'Environmental Impact Assessment' technique from Section 7.2 to assess the impact of coastal management at each place. Which place has (a) the most and (b) the least favourable score, and why?

On your own

3 Distinguish between: (a) hard and soft engineering, (b) tangible and intangible costs, (c) environmental impact and cost-benefit analysis.

4 Using examples from this section, explain how the **actions** of different players can have unforeseen consequences on the coast.

In this section, you'll find out how coasts in different parts of the world are being managed in a more holistic and sustainable way.

Odisha's coastal zone

Odisha's coastal zone, on India's north-east coast, has a wide range of coastal and marine flora and fauna (including 1435 km^2 of mangrove forest). It is rich in mineral deposits and has huge potential for offshore wind, tidal and wave power. Cultural and archaeological sites also dot the coast. Coastal fishing employs large numbers of people as fishermen, as well as those employed to process the fish caught.

However, Odisha's coastal zone is under stress from:

◆ rapid urban industrialisation

◆ marine transport, fishing and **aquaculture**

◆ tourism

◆ coastal and seabed mining

◆ coastal erosion

◆ offshore oil and natural gas production

◆ an increase in the frequency and intensity of severe weather events, such as cyclones

◆ and rising sea levels.

In an attempt to manage some of these problems, an Integrated Coastal Zone Management (ICZM) project has been implemented – with the aim of managing the coast and its resources in a **sustainable** way.

In Odisha, many different organisations have an interest in managing the coast, and these have consulted with others who have a stake in its future. Some of the main organisations involved in the consultations have been listed in the stakeholders box on page 143.

In addition to the inter-organisational consultations, a wide range of public consultations have also been held – including with individual villages about issues including:

◆ the assessment and control of coastal erosion

◆ the development of eco-tourism

◆ planting or replanting mangroves

◆ building cyclone shelters.

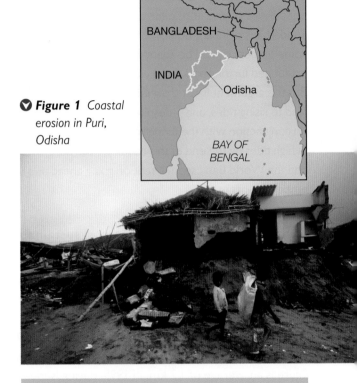

▼ *Figure 1 Coastal erosion in Puri, Odisha*

BANGLADESH

INDIA

Odisha

BAY OF BENGAL

Key word

Aquaculture – The farming of aquatic organisms, such as shellfish, for food.

BACKGROUND

Integrated Coastal Zone Management (ICZM)

The move to adopt an **Integrated Coastal Zone Management** strategy means that complete sections of coast are now being managed as a whole – rather than by individual towns or villages. This is because we now know that human actions in one place will affect other places further along the coast – because sediment moves along the coast in **sediment** (or **littoral**) **cells** (see page 117).

ICZM is a process that brings together all of those involved in the development, management and use of the coast. The aim is to establish sustainable levels of economic and social activity; resolve environmental, social and economic challenges and conflicts; and protect the coastal environment.

Greenpeace India (an environmental pressure group) has also been involved in meetings about income generation and the management of marine resources, acting with some of the villages included in the ICZM project.

Back to Holderness

The East Riding of Yorkshire Council developed an ICZM, which was launched in 2002. It involved over 80 organisations and was called '*Towards a Sustainable Coast*'. It was based on the UK government's principles for coastal management in England, which included:

* taking a holistic approach
* adopting a long-term perspective
* pursuing adaptive management
* seeking specific solutions and flexible measures
* working with natural processes
* providing participatory planning.

The ICZM was used to develop the Flamborough Head to Gibraltar Point **Shoreline Management Plan (SMP)** – published in 2011 – and to deliver the East Riding Coastal Change Pathfinder, 2010-12 (see page 131). Flamborough Head and Gibraltar Point are the northern and southern limits of a major sediment cell on England's east coast (see Figure 14 in Section 3.5).

What is the SMP?

The Flamborough Head to Gibraltar Point SMP sets out the policy for managing the coastline and responding to coastal erosion (and flood risks) over the next 100 years. It assesses potential erosion and flood risks, and then identifies sustainable coastal defence and management options, which take into account the influences and needs of the human, natural and historic environments (including existing defences and adjacent coastal areas).

SMPs are recommended for all sections of the coastline in England and Wales by Defra (the government Department for Environment, Food and Rural Affairs). Four options are considered for any stretch of coastline – as outlined on the right.

The East Riding of Yorkshire Council worked with a number of players and stakeholders in developing the SMP, including:

* **National government agencies** – Environment Agency, Natural England
* **Local government** – Lincolnshire County Council, North East Lincolnshire Council, East Lindsey District Council
* **Stakeholders in the economy** – The National Farmers Union
* **Environmental stakeholders** – English Heritage

ICZM Project Odisha, players and stakeholders

1 **Central (Federal) government**
   * Archaeology Department of Culture
   * Water Resource Department
   * Fisheries Department

2 **State and local government**
   * Odisha State Disaster Management Authority
   * Odisha State Pollution Control Board
   * Wildlife Wing of Forest and Environment Department (State)
   * Paradeep Municipality (local)

3 **Stakeholders in the local economy**
   * Odisha Tourism Development Corporation
   * Handicraft and cottage Industries

Options for coastal action

1 Hold the line. This involves maintaining the current position of the coastline (often using hard-engineering methods).

2 Advance the line. This involves extending the coastline out to sea (e.g. by encouraging the build-up of a wider beach, using beach-nourishment methods and groyne construction).

3 Managed retreat/strategic realignment. This involves allowing the coastline to retreat, but in a managed way. It can involve the deliberate breaching of flood banks built to protect low-quality farmland from flooding (creating salt-marsh environments).

4 Do nothing/no active intervention. This involves letting nature take its course and allowing the sea to erode cliffs and flood low-lying land (whilst letting the existing defences collapse).

What's the plan for Holderness?

Figure 2 shows how the Holderness coast will be managed up to 2025. Beyond that, plans are in place but they may change in a few areas (e.g. the coastline adjacent to the gas terminals at Dimlington and Easington – depending on whether these sites are still in use by then).

CBA and EIA

In order to make the decision about what and where to protect, a **CBA** (see page 139) and an **Environmental Impact assessment** or EIA (see Section 7.2) are carried out. For each different area shown on Figure 2, the economic assessment (CBA) identified whether:

◆ the benefits clearly outweighed the costs

◆ the benefits marginally outweighed the costs

◆ the costs clearly outweighed the benefits.

The CBA concluded the following:

◆ Along the undefended parts of the coast, the 'Do nothing' policy has no costs. However, there will be some economic losses (land, buildings, etc.).

◆ The benefits outweigh the costs of continuing to protect Bridlington, Hornsea and Withernsea

◆ The economic benefit of holding the line at Mappleton is similar to the cost.

◆ Because of the current importance of the gas terminals at Dimlington and Easington, the benefits clearly outweigh the costs.

◆ Spurn Point will be allowed to evolve – requiring minimal costs.

An EIA decides whether environmental quality will improve, or worsen, as a result of the different options for managing the coast. The decision under the SMP is to 'hold the line' for current defences at Dimlington and Easington gas terminals. An EIA for coastal protection works recommended the current protection scheme of a rock revetment made up of large granite boulders (which is approximately 1 km long).

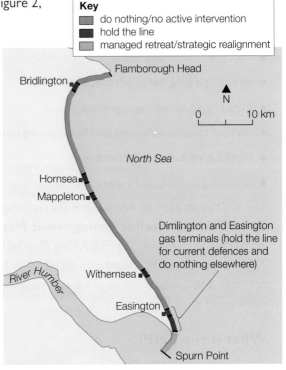

Key
- do nothing/no active intervention
- hold the line
- managed retreat/strategic realignment

Flamborough Head
Bridlington
North Sea
Hornsea
Mappleton
Dimlington and Easington gas terminals (hold the line for current defences and do nothing elsewhere)
River Humber
Withernsea
Easington
Spurn Point

N
0   10 km

▲ **Figure 2** *Coastal management options for Holderness up to 2025*

Wider issues

Decisions about whether to defend the coast or not are complex judgements, based on a range of factors – not just a CBA and EIA. Figure 4 (opposite) summarises some of the main factors considered.

Winners and losers

From Odisha to Holderness, many of the world's coastal zones are vulnerable to flooding or erosion (or face threats from a range of factors). Different players are involved – local authorities, homeowners, environmental pressure groups, to name but a few. As decisions are made about how to manage the issues they face, some people are bound to come out on top, whilst others lose out.

▼ **Figure 3** *Easington gas terminal is protected by a rock revetment*

Over the next few decades, countries face difficult decisions about the best way
to manage the coast. In the UK: farmland and isolated houses are likely to remain
unprotected; residents, councils and businesses often disagree about the best approach;
conflicts arise when coastal defences in one place have a negative impact elsewhere along
the coast – and where there are delays in the implementation of coastal protection.

▼ **Figure 4** *Managing the coast*

Engineering feasibility
This considers the following:
• Is it the right method (i.e. what would
  work best – sea walls, revetments, etc.)?
• Is it achievable?
• Is it within budget?
• What are the risks?

Environmental sensitivity
• Does the coastline include sites which are protected, e.g. National
  Nature Reserves, Sites of Special Scientific Interest (SSSI)?
• Flamborough Head and Spurn Head are both defined as Heritage
  Coasts (because of the value of their landscapes). Hornsea Mere
  is a SSSI, and a Special Protected Area. Flamborough Head is also
  a Special Area of Conservation.

Managing the coast

Land use and value
• What is the coast used for and how much is the land worth?
• Much of the Holderness coastline consists of agricultural land.
  Where the coastline is not protected it will continue to be lost
  to the sea.
• Agricultural land is classified from Grades 1-5 (1 being excellent,
  5 being very poor). It is estimated that by 2025 approximately
  160 hectares of grade 3 and 4 land will be lost to erosion.

Political, social and economic reasons
• Agriculture is a key employer in the area – many jobs
  depend on it.
• Tourism is another key industry along the coast, and is
  a major contributor to the local economy.
• Most coastal villages will not be at risk of erosion
  over the lifetime of the SMP – but some individual
  properties are. Approximately 37 homes are at risk of
  disappearing into the sea by 2025.
• Politically, costs have to be acceptable to the
  government of the day – and often something has to
  be seen to be 'done'. 'Do nothing' may be a reasonable
  option, but it's rarely acceptable to those affected.

Impacts on coastal processes
• Doing nothing means that coastal processes continue
  uninterrupted. By allowing Flamborough Head to continue to
  erode, sediment continues to be supplied to other parts of the
  coastline.
• Holding the line at Bridlington, Hornsea and Withernsea
  means that erosion is prevented there – thus interrupting the
  sediment supply further south. The same happens at Mappleton,
  Dimlington and Easington.
• Defended areas are likely to become promontories and beaches
  may become narrower.

Over to you

1 As a class, discuss whether the policies of 'Hold the line'
  and 'Do nothing', when combined, are suitable options
  for Holderness.

2 a In pairs, brainstorm and list a set of up to ten criteria
    to decide whether different coastal-management
    options are sustainable or not. Include economic,
    social, environmental, cultural and political factors.
  b Apply your criteria from (a) to (i) the ICZM in Odisha,
    (ii) the SMP between Flamborough Head and Gibraltar
    Point.
  c Write a 750-word explanatory report on your findings.

3 To what extent are the aims of the ICZM in Odisha
  and the SMP for Holderness (a) similar, (b) different?

On your own

4 Distinguish between the terms: 'Integrated Coastal
  Zone Management' and 'Shoreline Management Plan'.

5 Explain the differences between the players and
  stakeholders in Odisha and those in Holderness and
  East Riding.

6 Using all of the criteria/factors in the spider diagram
  (Figure 4), assess the success of coastal management
  along the Holderness coast.

Exam-style questions

1 Assess the effectiveness of holistic management
  strategies used to protect a stretch of coast from
  erosion. *(12 marks)*

2 Evaluate the success of policies which are designed
  to manage coasts holistically. *(20 marks)*

Having studied Coastal Landscapes and Change, you can now consider the three synoptic themes embedded in this chapter. 'Players' and 'Attitudes and Actions' were introduced on page 95. This page focuses on 'Futures and Uncertainties', as well as revisiting the four Enquiry Questions around which this topic has been framed (see page 95).

3 Futures and Uncertainties

People approach questions about the future in different ways. They include those who favour:

- **business as usual**, i.e. letting things stand. This might involve doing nothing, or only doing what's absolutely necessary when it's unavoidable.

- **more sustainable strategies** towards coastlines, particularly managing landscapes faced with different threats.

- **radical action** when faced with climate change, e.g. adapting to or mitigating threats from climate change (see Sections 3.9 and 3.11).

Working in groups, select two coastal landscapes from this chapter and then discuss the following questions in relation to each one:

1 How far has management of coastal landscapes created unintended consequences?

2 Is there evidence to suggest that radical actions are needed to manage the threats to coastal landscapes from climate change?

Revisiting the Enquiry Questions

These are the key questions that drive the whole topic:

1 Why are coastal landscapes different and what processes cause these differences?

2 How do characteristic coastal landforms contribute to coastal landscapes?

3 How do coastal erosion and sea level change alter the physical characteristics of coastlines and increase risks?

4 How can coastlines be managed to meet the needs of all players?

Having studied this topic, you can now consider answers to these questions.

Discuss the following questions in a group:

3 Consider Sections 3.1-3.2. How far does geology and lithology help to create unique coastal landscapes?

4 Consider Sections 3.3-3.6. What is unique about particular named coastal landscapes? Explain your judgments.

5 Consider Sections 3.7-3.9. What unique contributions does sea level change make to coastal landscapes?

6 Consider Sections 3.10-3.11. Which is best – hard engineering, soft engineering, or holistic approaches to coastal management?

Books, music, and films on this topic

Books to read

1. *Coast: The Journey Continues* by Christopher Somerville (2006)

This book from the BBC TV series looks at the different coastlines around the British Isles and how their different properties determine their appearance and character.

2. *The Beach Book: Science of the Shore* by Carl Hobbs (2012)

This book assesses some of the processes that take place around coastlines, and how such processes shape beaches. It also looks at how different players are involved in managing beaches.

3. *Coastal Flooding impacts and adaptation measures for Bangladesh* by Saquib Ahmad Khan and Ali Hossain (2012)

This is a factual book which assesses the impact of coastal flooding on large populations in Bangladesh, and how the country's approach to dealing with this needs to change to cope with the growing threats to the coast and its people.

Music to listen to

1. 'Ocean Rising' by Justin Sullivan (2003)

This song considers sea levels rising around the world.

2. 'After the Storm' by Mumford and Sons (2009)

The lyrics of this song describe how a flood can cause separation between people.

Films to see

1. *Global Flooding over the next 100 years – National Geographic* (2015)

A documentary film which looks at potential floods that could occur over the next 100 years, their possible causes, and where some of the worst hit areas could be.

2. *Flood* (2007)

A disaster film that shows the terrifying impact a flood can have on a coastal region, in this case London and the Thames Estuary, and how prepared flood defences can prove to be inadequate.

3. *Extreme Engineering: Venice Flood Gates* (2004)

Part of a TV series looking at Extreme Engineering, the coastal city of Venice was studied as the programme assesses the MOSE project that manages the threats the city faces from flooding.

Chapter overview – introducing the topic

This chapter studies globalisation, including the processes that have led to increased globalisation, plus its characteristics, impacts and consequences.

In the Specification, this topic has been framed around three Enquiry Questions:

> 1 What are the causes of globalisation and why has it accelerated in recent decades?
>
> 2 What are the impacts of globalisation for countries, different groups of people and cultures and the physical environment?
>
> 3 What are the consequences of globalisation for global development and the physical environment and how should different players respond to its challenges?

The sections in this chapter provide the content and concepts needed to help you answer these questions.

Synoptic themes

Underlying the content of every topic are three synoptic themes that 'bind' or glue the whole Specification together:

> 1 Players
>
> 2 Attitudes and Actions
>
> 3 Futures and Uncertainties

Both 'Players' and 'Attitudes and Actions' are discussed below. You can find further information about 'Futures and Uncertainties' on the chapter summary page (page 192).

1 Players

Players are individuals, groups and organisations involved in making decisions about globalisation processes. They can be individuals and national/international organisations (e.g. inter-governmental organisations), national and local governments, businesses (from small operations to the largest TNCs), pressure groups and non-governmental organisations.

Players that you'll study in this topic include:

- Section 4.3 – How **international political and economic organisations** (e.g. the **WTO**, **IMF** and **World Bank**) have contributed to globalisation through finance and trade. You'll also see how **national governments** join trading blocs such as the EU and ASEAN, and adopt liberalisation policies, to encourage globalisation.

- Section 4.4 and 4.8 – How **TNCs** are important in globalisation, particularly through cultural diffusion.

- Section 4.10 – How **pressure groups** oppose aspects of globalisation (e.g. immigration, free communication), and how censorship by national governments challenges freedoms associated with globalisation. Such groups might include those promoting, for example, the Paralympic Games (see Section 4.8).

2 Attitudes and Actions

Actions are the means by which players try to achieve what they want. For example, those promoting globalisation see it for its business opportunities. However, others see inequalities resulting from shifts in wealth ownership – part of a debate between winners and losers.

Actions largely concern the following issues:

1 **Economic versus environmental interests.** The differing views of pro- and anti-globalisation groups often lead to disputes (see Section 4.8). Environmental pressure groups seeking greater protection for threatened environments often play a significant role in these, as do those promoting the Transition movement (see Section 4.11).

2 **Economic versus social issues.** Globalisation depends upon workers being able to migrate internationally to where there is work. However, immigration does not always sit well with host populations (see Section 4.10).

> In groups, consider and discuss the following questions in relation to the UK:
>
> 1 How have UK governments adopted policies to encourage globalisation since 1980?
>
> 2 What have been some of the social impacts?
>
> 3 How and why has the UK's membership of the EU become increasingly contentious?

In this section, you'll learn about global economic shifts and some of their characteristics.

Globalisation contained

Container ships keep getting bigger! In July 2013, Maersk took delivery of the world's largest container ship. With 18 000 containers, Maersk claimed that freight could now be transported more cheaply – since the new ship used 20% less fuel per container than one carrying 10 000 containers. However, Maersk's number one position didn't last long!

◆ In January 2015, China Shipping Container Lines' huge new ship *Globe* arrived at Felixstowe Port in Suffolk, from China, on its maiden voyage – with 19 000 containers (see Figure 1).

◆ A month later, the Mediterranean Shipping Company's ship *Oscar* became the biggest container ship of all (for now) – with 19 224 containers!

These super-sized container ships were all built in South Korea, and they all sail between ports in Asia and Europe – the world's largest trade route (see Figure 2). The huge distances between these ports makes size important to keep costs down. In 1990, the average container ship held just 4000 containers, and there were many shipping companies. Now, fewer but much larger shipping companies dominate global trade. Containerised shipments have shifted the balance of economic power from Europe towards Asia.

The journey back to China

Almost everything on board the *Globe* was made in China (or elsewhere in Asia) by European- or US-owned companies. Relocating (or **out-sourcing**) production to Asia, exploits cheaper Asian labour costs. However, the return journey to China often only carries plastic waste from Europe for recycling or incinerating. So these huge ships bring high-value goods to Europe and take back low-value waste in return.

Amazon primed for take off

Many of the products in the *Globe*'s containers will probably be ordered by shoppers through Amazon. Amazon is a product of the digital age, and – as an **e-tailer** (an electronic online retailer) – it has re-shaped the retail industry. Many people now buy online instead of in a physical shop. Quick delivery times allow Amazon to take advantage of global connections to reduce the costs of storing large numbers of items in warehouses for long periods. In 2014, Amazon established a base in Shanghai, because its Chinese sales are now rising as China's wealth increases.

⯆ **Figure 1** *The mega-ship CSCL Globe arriving at Felixstowe in January 2015. Goods bought in Europe but made in Asia.*

The cargo on board the *Globe* included 4000 containers destined for Britain, including:

◆ Soft furnishings and sofas (109 containers)

◆ White goods, e.g. washing machines (62)

◆ Footwear (104)

◆ Bags (16)

◆ Children's toys (15)

When full, it could carry 156 million pairs of shoes, 300 million tablet computers or 900 million cans of baked beans!

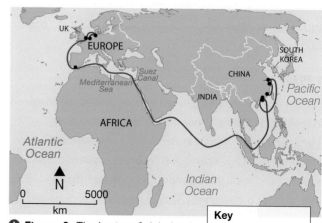

⯅ **Figure 2** *The basics of globalised transport – linking Asian producers and European consumers*

Key
╲ main container ship route
• shipping port

148

Amazon uses Internet technology to open up and exploit new markets:

◆ It has a media machine selling music, movies and books through its 11 websites across the world (the green 'Global marketplaces' in Figure 3).

◆ It offers manufacturing companies, sellers, writers and musicians access to a global market for their sales through its warehouses

◆ It has grown from being an online bookseller to become the world's leading e-tailer – with customers in 180 countries.

◆ Its 'Prime' product offers online TV and films.

◆ Its tablets (Kindles) give its customers access 24 hours a day, anywhere in the world.

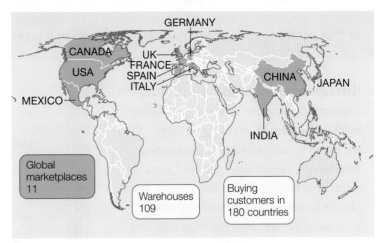

▲ **Figure 3** *Amazon's global business in 2015*

Operating in most countries, Amazon now works in a world without borders. As a result, national governments find it hard to keep track of its sales in each country, in order to calculate the level of business tax that it owes in each location (and it's relatively easy for global online companies like Amazon to register their sales in those countries with lower tax rates).

The throw away society

Items purchased through Amazon are cheaper than in shops, because of lower operating costs and bulk buying (known as **economies of scale**). Economies of scale and a race to deliver ever-cheaper goods – called '**the race to the bottom**' – mean that people in the twenty-first century buy and throw away more. Just like the cargo of return waste on the *Globe*, over 30% of what is purchased will be thrown away within a year – and that doesn't include the packaging. People know the price of each item, but its *value* and what it costs in terms of human rights (in sweatshops) and environmental impacts (using unrecyclable packaging, or rare metals) are often ignored. The costs are huge, as Figure 4 shows.

> These ships are bringing in the goods that Europe used to make. Whole sectors of global trade are now being dominated by companies operating out of China. The real cost of the goods that Maersk is bringing in should include the environment, the markets destroyed in developing countries, and the millions of jobs lost.

▲ **Figure 4** *A comment by Caroline Lucas (Green Party MP)*

Over to you

1 In pairs, list the benefits and problems of **(a)** the increasing size of container ships, and **(b)** online shopping.

2 **a** Research five common consumer items in your class (e.g. mobiles) and list: when they were bought; their average lifespan; the number of repairs they've undergone; when their owners expect to replace them.

   b Write a short report on the 'throw away society' among your friends.

3 In pairs, prepare a table to summarise the economic, social and environmental impacts of e-tailing.

On your own

4 Explain what is meant by the terms 'container revolution', 'economies of scale', 'race to the bottom' and 'throw away society'.

5 Research where the headquarters of Maersk, Mediterranean Shipping Company, and China Shipping Container Lines are located, and where their ships are built and registered.

6 Using Figure 3 and the information obtained from question 5, explain how far these companies are 'global'.

'Imagine there's no country …'

Imagine a British tourist, wearing a T-shirt made in Guatemala, scrolling through the menu list on her American-designed smartphone (which was made in China). She selects the music of a Canadian singer, sends photos home to the UK, while relaxing in a hardwood chair imported from Bali. Her meal arrives – it's an Indian dish, served on the veranda of a Cuban hotel, managed by a Spanish leisure chain. Globalisation indeed!

What is globalisation?

Globalisation is the process by which people, culture, finance, goods and information transfer between countries with few barriers. In some ways, it's like the global trade that occurred for centuries – between wealthy countries (who invested and manufactured) and poorer countries (who supplied raw materials and the basic labour to produce them). But modern globalisation is actually very different from that original model – the manufacturing process itself has now shifted to poorer countries, and modern globalisation is much wider in scope. It's no longer just about raw materials – known as **commodities** – or goods; it also concerns people, capital, culture and information technology.

Of course, the physical distances between places remain unchanged, but communication technologies have been revolutionised, which has massively reduced the time it takes to trade and communicate globally. In little over a hundred years, the world has moved from communication taking months (by ship or on horseback), to days (by rail or telegraph), to hours (by plane), to seconds (with a 'click') – as Figures 1 and 2 show. It's called **time-space compression**, and it has led people to refer to a '**shrinking world**'.

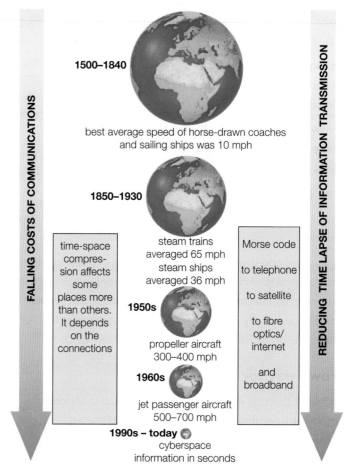

FALLING COSTS OF COMMUNICATIONS

REDUCING TIME LAPSE OF INFORMATION TRANSMISSION

1500–1840
best average speed of horse-drawn coaches and sailing ships was 10 mph

1850–1930
steam trains averaged 65 mph
steam ships averaged 36 mph

time-space compression affects some places more than others. It depends on the connections

Morse code
to telephone
to satellite
to fibre optics/ internet
and broadband

1950s
propeller aircraft 300–400 mph

1960s
jet passenger aircraft 500–700 mph

1990s – today
cyberspace information in seconds

Figure 1 *A shrinking world*

Figure 2 *The explosion in global digital communications by 2013 (a massive sevenfold increase since 2008)*

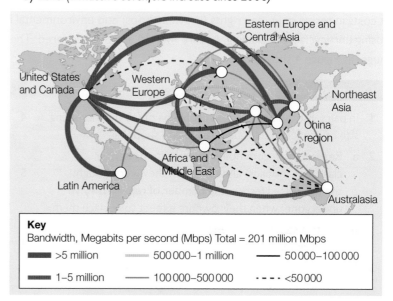

Eastern Europe and Central Asia
United States and Canada
Western Europe
Northeast Asia
China region
Africa and Middle East
Latin America
Australasia

Key
Bandwidth, Megabits per second (Mbps) Total = 201 million Mbps

— >5 million
···· 500 000–1 million
— 50 000–100 000
— 1–5 million
— 100 000–500 000
- - - - <50 000

How globalisation works

◆ Those who make decisions to invest and manufacture overseas – and also who help to determine consumer tastes and opinions – come mainly from North America, Japan and Europe, as well as oil-rich billionaire investors from Russia, the Middle East and Nigeria.

◆ China, India and South East Asia have become manufacturers for the world.

◆ India also provides financial and IT support services for Higher Income Countries.

◆ Global brands use much the same advertisements worldwide, and also dictate where products are made.

Outsourcing and relocation processes are fluid and often change. The fashion industry, for example, has moved much of its manufacturing from India and China to Vietnam, Bangladesh and Lesotho (in Southern Africa), because costs there are even lower. Meanwhile, much of sub-Saharan Africa remains detached and isolated, with little economic influence.

The processes and impacts of globalisation

Financial
◆ Global capitalism is spread by large TNCs – some with incomes even larger than the GDPs of many countries.

◆ The cheaper labour in developing economies helps supply consumers in wealthier nations with goods.

◆ Trillions of dollars are exchanged globally by electronic means every day (in payments, loans, share purchases and debt).

Political
◆ Some TNCs, like News Corp, seek to influence how people think. News Corp owns Sky TV and the *Sun* and *Times* newspapers; UK politicians often seek its support to influence voters' opinions.

◆ International political organisations have expanded to promote economic growth, e.g. the EU, the G8 / G20.

◆ TNCs and international political organisations can influence national governments.

◆ Many trade barriers (e.g. tariffs and quotas) have been reduced or removed to liberalise world trade.

Population
◆ Those with skills in management, finance and IT move around the world to where they are most in demand.

◆ Economic migrant labour flows to areas with higher incomes and higher rewards.

Communication and information
◆ Lower transport costs allow increasing long-distance tourism.

◆ Cheaper global phone networks, increasing mobile phone usage, and fast fibre-optic connections allow exchanges of information and ideas by email, social media and online news websites.

◆ A 'global village' is emerging, interested in universal sport, music and films, with no political or social boundaries.

Over to you

1 Identify globalisation processes in your class, by investigating:

  a where two student colleagues and their families were born

  b where **(i)** two items of their clothing were made, **(ii)** two current music tracks they like are from

  c where their parents are **(i)** working, and **(ii)** for whom, **(iii)** where those organisations are based

 2 a Using Figure 2, describe the flows of data and communication in 2013 **(i)** of over 5 million mbps, **(ii)** of 1-5 million mbps.

  b Explain the contrasts in flows between **(i)** and **(ii)**.

On your own

3 Write a 300-word view about how far it matters that TNCs like News Corp can attempt to exert political influence.

Exam-style questions

 1 Explain how changes in technology have speeded up the process of globalisation. (*6 marks*)

A 2 Explain how technology has contributed to the process of globalisation. (*6 marks*)

In this section, you'll learn that decision-making by political and economic players are important factors in globalisation.

Pakistan's fishermen

In 1995, Pakistan joined the World Trade Organisation (WTO). To comply with the WTO's trade rules, Pakistan opened up its fishing grounds to foreign competition. Until then, it had enforced a 200-mile exclusion zone around its coast – designated so that only Pakistani fishing boats could fish there. However, after joining the WTO, deep-sea trawlers owned by TNCs were allowed to fish in Pakistan's coastal waters. Huge trawlers from India and elsewhere now take most of the catch, while Pakistan's own fishing communities are left in poverty (see Figure 1).

In 2013, the charity ActionAid published a report on Pakistan's fishing industry, called 'Taking the fish'. ActionAid found that Pakistan's fish stocks were falling to dangerously low levels. Coastal villagers felt that their right to fish had been taken away, and thousands were forced to give up fishing altogether.

The global players

Pakistan's story is a common one. Events that affect Pakistan's local fishing communities are actually a consequence of global decision-making. Increasingly, decisions that affect people locally are being made by global organisations – which seem to have no accountability, and are invisible in most people's lives. These are the **players** in globalisation. Who are they, and how have they gained their power?

After the end of the Second World War, in 1945, the USA and many war-ravaged and economically weak Western European countries felt threatened by the territorial advances being made by communism. By 1949, the Soviet Union, many of the countries of Eastern and Central Europe, as well as mainland China, were all communist states (with a strong and growing communist influence in Korea and Vietnam as well). The accepted view in the USA and Western Europe was that the best way to combat the spread and influence of communism was economic development. The USA saw the job of rebuilding Europe as the priority (the reasons for which can be seen in Figure 2), with subsequent aid programmes designed to stimulate economic growth in South-East Asia. That view determined much of American foreign policy for the next half century.

Siddique Malah, local Pakistani fisherman

> Now all the fish are caught by trawlers and I'm in such poverty that I had to pull my 15-year-old son out of school.

> I can't give good food to my family, or the other basic needs of life.

Hassan Dablo, local Pakistani fisherman

⊘ **Figure 1** *The impact of global trade on local communities in Pakistan*

▶ **Figure 2** *This view of Nuremburg, Germany, in 1945 shows just how much economic investment was needed to rebuild post-war Europe. The USA and many Western European governments thought that promoters of communism might exploit this situation to expand its foothold westwards.*

To achieve the desired economic development, three global organisations were established to promote it, as well as to restore and maintain international financial stability after the end of the Second World War. Each organisation still remains fundamental to global decision-making today.

- The **World Bank** and the **International Monetary Fund** (IMF) were both established at the 1944 Bretton Woods Conference in the USA. Both organisations are based in Washington, D.C.

- The International Trade Organisation, which became the **WTO** in 1995, is based in Geneva, Switzerland.

▶ **Figure 3** *Greek police at an anti-austerity protest after 2008*

1 The IMF

Like the World Bank, the IMF lends money for development purposes. However, its primary role is to maintain international financial stability. In return for loans, it tries to force countries to privatise (or sell off) government assets in order to increase the size of the private sector and generate wealth. Many observers believe that this policy has forced poorer countries to sell off their assets to wealthy TNCs.

The IMF also exists to stabilise currencies – and therefore countries – in order to maintain economic growth, such as its involvement in the Greek debt crisis in the years after the 2008 financial crash. From 2008 onwards, Greece was forced to cut back on its government expenditure – known as austerity. Many protests occurred in Greece against these cutbacks forced on the country by other Eurozone countries and the IMF.

2 The World Bank

The World Bank was formed to finance economic development. It uses bank deposits placed by the world's wealthiest countries to provide loans for development in countries that agree to certain conditions concerning repayment and economic growth. Its first loan was to France for post-war reconstruction. It also focuses on natural disasters and humanitarian emergencies.

3 The WTO

Since 1945, governments have been keen to use trade as a way of generating economic growth in the world's poorer regions. The WTO (and its predecessor the ITO) believes in Free Trade without subsidies or tariffs (duties) – known as **barriers**. Removing barriers is known as '**trade liberalisation**'. The WTO advocates for trade liberalisation, and seeks to encourage all trade between countries free of **tariffs**, **quotas** (i.e. set amounts which could not be exceeded), or **restrictions** on trade (e.g. by preferring to trade with some countries over others). By 2016, it had 162 member states.

International trading blocs

Increasingly, countries are grouping together as members of **trading blocs** (see Figure 4 on page 154), to promote free trade between them. There are now a number of these blocs, e.g. the EU and NAFTA. Most are located in particular geographical regions, and support trade for their members by:

- removing tariffs between member states

- creating barriers for non-member states by placing tariffs on imports. This increases the price of imports and helps to protect their own industries.

This approach has been advantageous for many countries, and has resulted in rapid economic growth, e.g. Asia's Newly Industrialising Countries (NICs). However, non-members are excluded, which prevents their development. Some blocs even subsidise their producers, such as EU farmers, in order to protect them from the influence of the global market (although this is against WTO trade rules). However, the size of the US and EU economies gives them both significant clout in the WTO, whose policing of this issue has been toothless.

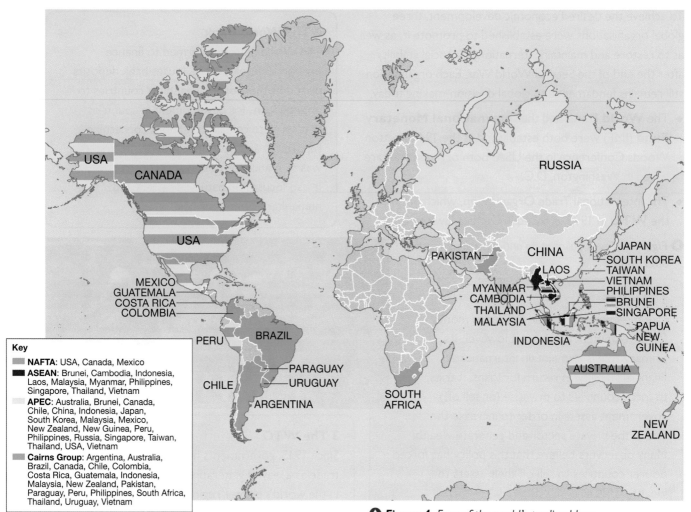

Key

▬ **NAFTA**: USA, Canada, Mexico
■ **ASEAN**: Brunei, Cambodia, Indonesia, Laos, Malaysia, Myanmar, Philippines, Singapore, Thailand, Vietnam
▢ **APEC**: Australia, Brunei, Canada, Chile, China, Indonesia, Japan, South Korea, Malaysia, Mexico, New Zealand, New Guinea, Peru, Philippines, Russia, Singapore, Taiwan, Thailand, USA, Vietnam
▬ **Cairns Group**: Argentina, Australia, Brazil, Canada, Chile, Colombia, Costa Rica, Guatemala, Indonesia, Malaysia, New Zealand, Pakistan, Paraguay, Peru, Philippines, South Africa, Thailand, Uruguay, Vietnam

△ **Figure 4** *Four of the world's trading blocs*

Individual national governments

1 The UK

Just as important to the globalisation process is the willingness of individual national governments to promote international strategies for growth. In the 1980s, the Conservative government, led by Margaret Thatcher, was the first to embrace globalisation strategies fully. Some industries were left to close if their profitability depended on government **subsidies**, and the government also refused to artificially support industries facing competition from cheaper overseas products (e.g. the coal-mining industry, which was decimated by cheap foreign coal imports during the 1980s and 1990s).

> **Key word**
>
> **Subsidies** – Grants given by governments to increase the profitability of key industries.
>
> **Foreign Direct Investment (FDI)** – Investment made by an overseas company or organisation into a company or organisation which is based in another country.

Instead, the Conservative government developed two strategies for growth:

◆ It gave tax breaks – i.e. subsidies – to companies investing in areas such as London Docklands. Almost all companies establishing themselves in London's Canary Wharf development since the late 1990s have been given life-long tax breaks. This is a highly attractive benefit, which has encouraged a number of large overseas financial institutions to relocate to London.

◆ It also gave grants and subsidies to encourage foreign companies to locate new manufacturing plants in the UK. Nissan's Washington, Tyne and Wear, factory and Toyota's factory in Burnaston, Derbyshire, were each subsidised in order to attract investment from their Japanese parent companies – known as **Foreign Direct Investment** (FDI). By 2015, the UK was the fourth largest recipient of FDI.

2 China

After decades of economic and political isolation, the Chinese government declared an '**open door**' policy to international business in 1978. It needed Western technology and investment to develop its economy – and from then on its government welcomed foreign businesses setting up in China.

Companies from Europe and the USA quickly saw the advantages of out-sourcing and relocating into one of southern China's four '**special economic zones**', later known as '**Export Processing Zones**'. These zones offered tax incentives and huge pools of cheap labour. Since then, China's economy has grown rapidly and (in 2001) it became a member of the World Trade Organisation. By 2005, around 50% of Chinese exports came from foreign companies with connections in these zones.

China's rapid economic growth has altered the flow of FDI. China is still the world's largest recipient of FDI, but as growth has shifted into Asia, traditional flows of FDI have changed. Now, countries such as China and India control flows of FDI, together with governments and companies in Brazil, Russia and South Africa (collectively known as the BRICS countries). These five countries now invest heavily in the USA, EU, sub-Saharan Africa (see Figure 5) and South America. Therefore, China is now a major player in both the inflow and outflow of FDI. Investment flows from the BRICS to other countries increased twenty-fold (to $145 billion) between 2000 and 2012 – and now account for over 10% of the global total.

Key word

Special Economic Zones (SEZ) – Set up by national governments to offer financial or tax incentives to attract FDI, which differ from those incentives normally offered by a country. China now uses the term '**Export Processing Zones**'

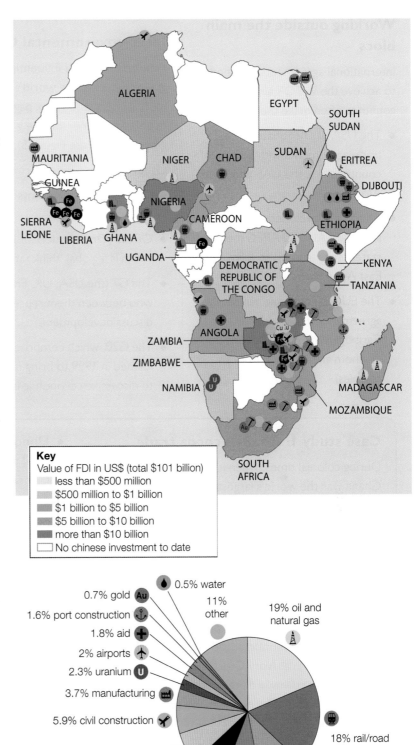

Key
Value of FDI in US$ (total $101 billion)
- less than $500 million
- $500 million to $1 billion
- $1 billion to $5 billion
- $5 billion to $10 billion
- more than $10 billion
- No chinese investment to date

0.5% water
0.7% gold Au
1.6% port construction
1.8% aid
2% airports
2.3% uranium U
3.7% manufacturing
5.9% civil construction
6.9% copper Cu
7.4% iron ore Fe
9.2% hydroelectric dams
11% other
19% oil and natural gas
18% rail/road
10% other mining

🔺 **Figure 5** FDI flows from China into Africa

155

Working outside the main blocs

International agreements are needed to achieve the WTO's aims, but countries often make their own deals:

◆ The USA has negotiated free-trade agreements with a range of countries, e.g. the UAE and South Africa.

◆ China has developed deals with some ASEAN and APEC member states (see Figure 4) to form a new East Asian bloc.

◆ The EU has economic partnership agreements with 69 African, Caribbean and Pacific nations (most of whom are former European colonies).

Inter-governmental Organisations (IGOs)

Amongst the most influential players are partnerships constructed by governments of the world's wealthiest countries to develop policies of mutual interest (e.g. tax payments by TNCs). Some develop informal economic partnerships, e.g. China's engagement with sub-Saharan African governments (see Figure 5).

Other IGOs are more formalised and have significant influence, including:

◆ OPEC, which represents 40% of global oil producers and therefore influences oil prices.

◆ OECD (Organisation for Economic Co-operation and Development), which is a global 'think tank' of the 34 wealthiest nations.

◆ The G7 (the USA, UK, France, Canada, Germany, Italy and Japan) – who between them represent 50% of global GDP. It meets annually to discuss development.

◆ The G20, which comprises 19 individual countries plus the EU. It was formed in 1999 to bring together developed and developing economies to discuss key economic issues (e.g. debt relief).

Case study 1: Ghana's cocoa trade

During colonial times, when it was ruled by Britain, Ghana was the world's largest producer of cocoa. The British government set the price that Ghanaian farmers would receive. Since independence in 1957, three factors now dictate global cocoa prices:

◆ **Commodity traders**. In financial centres, such as London, traders buy cocoa in advance for TNCs like Cadbury, on what is known as the **futures market**. This guarantees the supply, price and delivery date of the product months ahead. However, other cocoa growers, such as Ivory Coast, also supply TNCs – which puts downward pressure on prices by giving the commodity traders alternative sources to negotiate with.

◆ **Overseas tariffs**. Although WTO rules seek tariff-free international trade, the EU sets tariffs for processed cocoa – but none for raw cocoa beans. Ghana could gain extra income and employment by processing its cocoa beans into powder or chocolate before export. The value-added would be higher, but tariffs would then be applied on entry to the EU – driving up the price for anyone who buys the product. This forces Ghana to export raw cocoa beans instead.

◆ **Unequal power**. Ghana joined the WTO in 1995. Until then, its government had subsidised Ghanaian farmers to encourage food production for its growing urban population. But the WTO imposed a joining condition that Ghanaian farmers should not be subsidised (even though the USA and EU freely pay subsidies to their own farmers). As a result, Ghanaian farmers cannot compete with imported and subsidised American or European foods. Some Ghanaian rice producers and tomato growers (see Figure 6) have given up altogether.

⬆ **Figure 6** *Tomato farmers in Ghana now find themselves unable to sell their own crop, because cheap subsidised imports from the EU have flooded the market*

Case study 2: Vietnam calling

Agreeing to WTO rules can help countries to forge new trade links. In 2014-15, the EU and ASEAN negotiated new trade deals. One of the largest took place in August 2015 – with Vietnam. It removed all import duties and quotas on items traded between EU countries and Vietnam. Over 31 million jobs in Europe depend on exports, so having easier access to a growing and fast-developing market like Vietnam, with its 90 million consumers, is great news. Before the deal, the value of EU trade with Vietnam in 2014 was already US$30 billion.

◆ Vietnam exports the following items to the EU: telephones, electronic goods, footwear, clothing, coffee, rice, seafood and furniture.

◆ The EU exports the following items to Vietnam: electrical machinery, aircraft, vehicles and pharmaceuticals.

Vietnam's labour costs are lower than China's, so Western consumers can now benefit from lower prices – the race to the bottom continues!

⊘ Figure 7 *This Vietnamese woman is likely to be kept busy as a result of trade deals between the EU and Vietnam*

Case study 3: Cotton in Guatemala

In the 1980s, 75% of Guatemala's cotton crop was exported. The income generated was used to buy pesticides, machines and equipment for future crops. However, if Guatemala had processed its raw cotton into finished clothes, and then exported these instead, its export earnings would have been greater (a similar situation to that of Ghana's cocoa crop). More significantly, only 1% of the land devoted to cotton production would have been needed to generate the same income – leaving Guatemalans able to grow other crops and open new markets. Instead, WTO policies made this difficult. Guatemala was tied to exporting raw cotton, until competition from other countries ended production completely in 2005. TNCs now import raw cotton into Guatemala and use its workforce to produce cheap T-shirts for export.

Exam-style questions

AS 1 Explain the term 'liberalisation of trade'. *(4 marks)*

A 2 Assess the extent to which the globalisation of trade can bring problems as well as benefits. *(12 marks)*

Over to you

1 Make a larger copy of the diagram on the right. Read through this section and add **(a)** the names of all organisations or players who contribute to decision-making about globalisation at different scales, **(b)** an example of something that they do to bring this about.

[Diagram: concentric circles labelled from outer to inner: Global, International, National, Local]

2 Distinguish between the work of the IMF and that of the World Bank.

3 Draw three spider diagrams – one each for Pakistan, Ghana, and Guatemala. In pairs, discuss and label the gains and the losses for each country from their global connections (use different colours for each).

4 Compare the experiences of these three countries with those of China and Vietnam. What lessons can be learnt if globalisation is to prove more successful?

On your own

5 Distinguish between the following pairs of terms: **(a)** colonialism and globalisation; **(b)** trade liberalisation and subsidies; **(c)** tariffs and quotas; **(d)** Special Economic Zones and trading blocs; **(e)** FDI and investment.

6 Outline the similarities and differences between the adoption of globalisation by the UK and China.

7 Research where Chinese FDI is affecting one sub-Saharan African country (see Figure 5). Produce a six-slide PowerPoint presentation to show how this country wins and loses through its FDI.

Transnational corporations: corporate colonialism?

It's common to think of **transnational corporations (TNCs)** as a recent phenomenon. However – in the eighteenth and nineteenth centuries – a large part of India was controlled by The East India Company (from its headquarters in London). The Company controlled trade routes and ruled 20% of the world's population.

TNCs are vital players in globalisation. Each of the 2015 Top Ten TNCs (shown in Figure 1) earns more revenue in a year than the GDP of some countries! In 2016, there were over 60 000 TNCs. The top 200:

◆ employed just 1% of the global workforce, but

◆ accounted for 25% of the world's economic activity.

However, a new world order is emerging. In 2006, six of the Top Ten TNCs were based in the USA. By 2015, only two of the Top Ten TNCs were American (Walmart and Exxon Mobil) and three Chinese TNCs had overtaken Ford, General Motors and ConocoPhillips.

Rank	Company	Country of origin	Turnover 2015 (US$ billion)	Profit 2015 (US$ billion)	Revenue compared to the GDP of whole countries
1	Walmart Stores	USA	485.6	16.3	More than Sweden
2	Sinopec	China	446.8	5.1	More than Norway
3	Royal Dutch Shell	Netherlands	431.3	14.8	More than Norway
4	PetroChina (CNPC)	China	428.6	16.3	More than Norway
5	Exxon Mobil	USA	382.6	32.5	Equal to Austria
6	BP	UK	358.6	3.7	Almost equal to UAE
7	State Grid	China	339.4	9.7	More than Colombia
8	Volkswagen	Germany	268.5	14.5	More than Chile
9	Toyota Motors	Japan	247.7	19.7	More than Finland
10	Glencore (mining)	Switzerland	221.0	2.3	More than Ireland

▲ **Figure 1** *Fortune magazine's Top Ten TNCs in 2015*

Key word

Corporation – A business which exists separately from its owners. Its owners are shareholders, who mostly appoint directors to run the business.

TNCs and global production

TNCs are vital to the spread of globalisation, because their expansion involves the free flow of capital, labour, goods and services. Three factors have led to this: motive, means and mobility.

1 Motive

Under capitalism, there is one motive – profit. Those companies that become dominant do so by:

◆ controlling and minimising their costs (e.g. raw materials, labour)

◆ increasing their revenues by expanding their markets and merging with or taking over their competitors.

Corporations manage this through:

◆ **achieving economies of scale** by expanding their capacity, e.g. Amazon (warehouse size) and Maersk (ship size), in order to reduce their unit costs.

◆ **developing new markets**. New markets are essential, and they depend on creative product design and product desirability. This process can involve expanding the market to new customers, or creating frequently updated models that existing customers will want to buy (e.g. mobiles and computer software).

◆ **horizontal integration**, in which a company expands at one level in the production process. For example, Apple acquired Emagic and Logic Pro (for music downloading technologies) to help it expand its services.

- **vertical integration**, in which a company controls and owns every stage of production from exploration to sales. The US oil giant Exxon Mobil (see Figure 1) owns and controls oil wells, tankers, refineries and petrol stations!

- **diversifying their product range**. Many companies expand their product range in order to future-proof their sales – if one product fails, another will succeed.

2 Means

None of the motives could become reality without the means. Banking and the free flow of **capital** (money) around the world are the mechanisms for company growth. Global, relatively unrestricted flows of finance connect banks, businesses and countries in complex webs. These flows between countries are variable:

- from year to year (as Figure 2 shows)

- over the medium term (for example, they slowed during the 2008 financial crisis, when many businesses reduced their levels of investment, but had largely recovered by 2014)

- over the longer term (although the USA is a constant player, in recent years the Top Ten countries both investing and receiving FDI from overseas have changed. Some countries that received FDI in the 1980s and 1990s are now providers of FDI to others).

A kind of **reverse colonialism** has happened – Hong Kong, Singapore, China, South Korea, Malaysia, India and Brazil are now established High- or Middle-Income Countries, and are all net providers of overseas investment. For example:

- in 2015, there were over 800 Indian-owned businesses in the UK – employing more than 110 000 people

- Russia invested US$87 billion overseas in 2013

- China has bought Japanese household brands, and is investing heavily in infrastructure in Africa (see Figure 5 in Section 4.3), as well as in the UK (where it already owns several well-known brands, such as Pizza Express) and the USA.

3 Mobility

Mobility has been fundamental to the spread of globalisation. This includes:

- faster and cheaper transport, e.g. the use of ever-larger container ships (see Section 4.1)

- rapid communication systems, using fibre optics, satellite and digital technology

- new production technology, such as '**just in time**' and flexible production systems, which provide cheaper products and fast turn-around from orders to delivery

- global production networks, which link sources of raw materials, finance, manufacturing, markets and sales. These are run from company headquarters, almost always in High Income Countries.

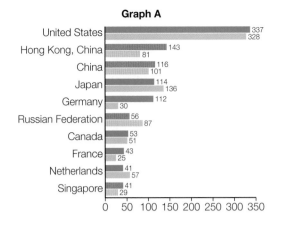

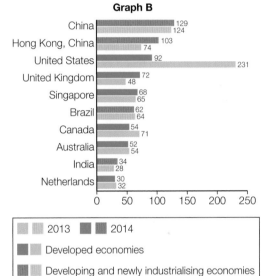

 Figure 2 *The Top Ten countries for FDI in 2013 and 2014: Graph A shows the source countries and Graph B shows the recipients. The figures at the end of each bar are in US$ billion.*

Key word

Just in time – The means by which the time gap between production and delivery to the customer is sharply reduced – cutting warehousing and storage costs.

However, mobility flows mean that production and sources of materials can be flexible and can lead to controversial decision-making, as the news headlines in Figure 3 show:

Thousands of jobs at risk as India's Tata Steel seeks British exit

Reuters, March 2016

Kraft takes over Cadbury in multi-billion dollar deal to create global confectionery leader – no guarantees for 4500 workers at Bourneville factory

Telegraph, January 2010

The world's largest sports shoe manufacturer, which supplies Nike and Adidas, has been forced to halt production at a huge Chinese factory, after thousands of workers continued to protest yesterday over pay and benefits

FT.com, April 2014

🔺 **Figure 3** *TNCs: some of the impacts of TNCs and the decisions that they make*

Key word

New economy – Where GDP is earned more through expertise and creativity in services such as finance, media, law, technology and management, than from the manufacture of goods. Also known as the '**knowledge economy**' (see Section 5.2).

Off-shoring – When a company does work overseas, either itself or using another company.

Outsourcing – When work is contracted out to another company (it's known as off-shoring when that company is overseas).

The big brands

Some TNC brands really influence our lives. These days, Disney is not just about theme parks and films; it is a globally recognised brand (see Figure 4). In the Forbes list of the top global brands in 2016, Disney ranks eleventh and is easily the world's biggest leisure brand. In 2014, its turnover (or revenue) was US$48.8 billion. It employs 180 000 workers of its own, as well as having 40 000 suppliers in 70 countries. From a small animation studio in California in the 1950s, Disney has expanded rapidly to take advantage of the media and technology revolution to become a truly global company.

New economy – new heroes?

Disney is typical of the new economy, where ideas are as important as goods. Disney's ideas originate within the USA, but its merchandise (e.g. toys and T-shirts) is produced overseas. Merchandise alone earns Disney US$37 billion each year. Between 2013 and 2015, one film alone – *Frozen*, of course! – earned the company almost US$2 billion!

Disney operates a just-in-time system for merchandise production, which has both benefits and problems:

- For Disney, the company can wait to judge the success of a venture before investing in merchandise. It uses overseas manufacturers – called **off-shoring** or **outsourcing** – and demands quick delivery times. This avoids having to operate its own expensive production lines.

- However, outsourcing is not problem-free. Overseas workers often earn low wages (e.g. in Vietnam). Overseas factories have also used toxic substances banned in the USA (e.g. between 2007 and 2013, some toys manufactured in China had to be recalled from shops because of unsafe levels of toxic lead used in their paint).

TNCs like Disney are aware of the poor publicity that can result from such production methods, so they increasingly monitor their suppliers.

The Walt Disney Company is involved in many global activities, such as:

- over 250 linked satellite and cable TV companies
- 6 film/TV production and distribution companies, including Pixar and Lucasfilm (which makes the *Star Wars* films)
- 12 publishing companies, 15 magazines and newspapers
- 728 shops worldwide, plus galleries and toy companies
- 5 record labels and music publishing companies
- 2 theatrical production companies
- 14 theme parks and resorts, and a cruise line
- sports franchises and teams, plus the sports broadcaster ESPN
- multimedia (producing CD-ROMs, e-games, infotainment)
- property development and human resources agencies

🔺 **Figure 4** *Disney – the world's second largest media company and one of the world's best-known brands*

Cultural globalisation and glocalisation

Is the world becoming 'Disneyfied'? Disney owns 40 Spanish-speaking radio stations, together with foreign language television channels and a Chinese radio station in Hong Kong. It also supplies reading materials to help teach English in Chinese and African schools. It was one of the first global TNCs to **glocalise**, by tailoring its products to specific market areas. For example:

◆ *Mulan* marked a decision to enter the Chinese market

◆ *The Hunchback of Notre Dame* was launched to rebrand Disneyland Paris

◆ *The Lion King* was aimed at African markets, *Aladdin* at the Middle East, and *Rescuers Down Under* and *Finding Nemo* at Australia.

Disney's influence now spreads widely, as Figure 5 shows. Retail/leisure locations the world over are becoming clones of Disneyland – merging consumerism, popular culture and entertainment.

> **Key word**
>
> **Glocalise / glocalisation** – When a company re-styles its products to suit local tastes

△ Figure 6 *Disneyland Hong Kong – the latest to join the Disney stable of global entertainment in an expanding market*

Influences on urban planning: 'cloned towns'	Media influences	Influencing governments
• Shopping malls like Disneyland on suburban out-of-town developments • Disney-themed fast-food outlets • Crowd monitoring with CCTV cameras • Resort tourism with everything on site	• 24-hour TV channels in North Africa, the Middle East, Europe, Australia, Malaysia and Cuba • Disney owns shares in European and Brazilian commercial TV channels • Chinese state TV uses Disney's ESPN for sports coverage • Internet/iPod services for TV and film downloads	• The US government enforces copyright for US companies like Disney • The French government paid US$2 billion towards EuroDisney's construction (providing 30 000 jobs) • The Hong Kong and Tokyo governments paid US$1 billion towards their respective Disneylands (see Figure 6)

△ Figure 5 *The 'Disneyfication process' – Disney's approach is seeping into people's everyday lives*

Over to you

1 In pairs, complete a table showing the benefits and problems of the ways in which goods are produced for:
   a Disney **b** its workers.
2 On a world map, locate and show (using different symbols) where Disney's:
   a main decisions are made
   b goods are produced
   c products are consumed
   d influence can affect governments.
   You may find it useful to use its website.
3 How far is Disney a truly global corporation?
4 In pairs, discuss and feed back the extent to which you feel Disney simply offers the world what it wants.

On your own

5 Distinguish between the following pairs of terms: **(a)** globalisation and glocalisation, **(b)** outsourcing and off-shoring, **(c)** FDI and reverse colonialism, **(d)** horizontal and vertical integration of products.
6 Draw a spider diagram with three spokes – one for each of motive, means and mobility. Show how each of these three terms **(a)** helps a TNC to expand, **(b)** leads to increased globalisation.
7 Using Figure 3, list the advantages and disadvantages of attracting FDI from TNCs.

Exam-style questions

AS 1 Explain two ways in which TNCs promote globalisation. *(4 marks)*

A 2 Assess the role played by TNCs in the globalisation process. *(12 marks)*

In this section, you'll learn how globalisation affects some places and organisations more than others.

A two-speed world

Globalisation has opponents as well as supporters:

◆ Those in favour of globalisation claim that increased connectedness – i.e. becoming **'switched on'** – improves many countries' economic development, which in turn leads to a higher standard of living for their citizens.

◆ Those against globalisation argue that it leads to corrupt practices, and also that some countries and people are left behind – i.e. are left **'switched off'**. Globalisation thus creates a two-speed world.

Degrees of globalisation

As countries become more globalised, certain characteristics – or **indicators** – change. These include:

◆ **flows** (e.g. a higher volume of international trade, more migration of people, more FDI)

◆ **technologies** (e.g. increased Internet usage, flows of information, telecommunications)

◆ **movements** (e.g. increased international air traffic)

◆ **media** (e.g. the spread of global advertising, publishing, music, TV and film)

These indicators improve connections, which are the basis of globalisation. The more connected that places become, the smaller the differences between them will be. Distances between places will 'shrink' (see Section 4.2), and the best-connected places will become 'switched on'.

How degrees of globalisation vary

Countries and cities can be ranked according to their degree of globalisation. Data from a broad range of indicators are used to compile an index which measures how globalised a country has become. The KOF and Kearney Indexes are two such measurements:

Measuring globalisation 1 – the KOF Index

Each year, the Swiss Economic Institute produces an annual Index of Globalisation, known as the **KOF Index**. KOF is an acronym for the German word *Konjunkturforschungsstelle*, which means 'business cycle research institute'. The results for 2015 are shown in Figure 1.

Rank	The 15 most globalised countries and their scores in 2015	Rank	The 15 least globalised countries and their scores in 2015
1	Ireland 91.3	177	Myanmar 33.01
2	Netherlands 91.24	178	Sao Tome and Principe 32.86
3	Belgium 91	179	French Polynesia 32.82
4	Austria 90.24	180	Burundi 32.26
5	Singapore 87.49	181	Tonga 31.89
6	Sweden 86.59	182	Sudan 31.54
7	Denmark 86.3	183	Comoros 31.15
8	Portugal 86.29	184	Afghanistan 30.62
9	Switzerland 86.04	185	Bhutan 29.3
10	Finland 85.64	186	Equatorial Guinea 27.49
11	Hungary 85.49	187	Eritrea 27.13
12	Canada 85.03	188	Laos 26.91
13	Czech Republic 84.1	189	Kiribati 26.00
14	Spain 83.71	190	Somalia 25.39
15	Luxembourg 83.56	191	Solomon Islands 25.26

⬥ **Figure 1** *The world's most and least globalised countries in 2015 (based on the KOF Index)*

The KOF Index score for each country is calculated using specific interactions, grouped as follows:

◆ **Economic globalisation** – e.g. cross-border transactions and the volume of FDI.

◆ **Social globalisation** – e.g. cross-border contacts (telephone calls, letters and tourists, foreign residents) and information flows (Internet/TV/Press). It also measures the presence of McDonald's and Ikea as indicators of 'global affinity'.

◆ **Political globalisation** – e.g. the number of foreign embassies in the country, the country's membership of different international organisations (e.g. the WTO), and its participation in UN peacekeeping activities.

Each set of different indicators in each group is scaled (or weighted), because some indicators are more significant than others (e.g. the volume of FDI). The three sets of indicators are then each aggregated into one value for each of economic, social and political interactions, which are then ranked. The Index rankings and scores shown in Figure 1 represent the final average rank for all three interactions.

The Index shows that 13 of the Top 15 countries are European, which contrasts with tables of GDP (where the USA leads) or manufacturing output (where China leads). The KOF Index measures *international* interactions. European countries are small compared to China and the USA – so every country has embassies, its own Internet systems, investments – all recorded as transactions. By contrast, China and the USA have large domestic economic markets, so *internal* connections are important – but these do not count in the KOF values.

Measuring globalisation 2 – the A T Kearney Index

By contrast, the 'Carnegie Endowment for International Peace' think-tank also publishes a globalisation index, called the **A T Kearney Index**. It identifies four main indicators from which to calculate its Index:

◆ **Political engagement** – e.g. a country's participation in international treaties and organisations, as well as peacekeeping operations.

◆ **Technological connectivity** – e.g. the number of Internet users, hosts and servers.

◆ **Personal contact** – e.g. through telephone calls, travel, remittance payments (payments sent home from overseas by migrants).

◆ **Economic integration** – e.g. the volumes of international trade and FDI.

The A T Kearney Index 'top performers' for 2015, across the four indicators, are shown in Figure 2. This Index is slightly different to the KOF Index, because it uses more holistic indicators (e.g. the number of web servers, rather than internet communications) and also volumes of trade as well as FDI. It also includes countries that are players on the political stage, so the participation of the UK and Ireland in international treaties (such as the HIPC initiative; see page 165) is recognised. The overall ranking is worked out using a complex points and weighting system for the four individual indicator rankings.

🔻 **Figure 2** *The 'top performers' in 2015, according to the A T Kearney Index*

Overall ranking	Country	Economic ranking	Personal ranking	Technological ranking	Political ranking	Best scores for
1	Singapore	1	3	12	29	Trade, FDI and personal
2	Switzerland	9	1	7	23	Phone contacts
3	USA	58	40	1	41	Internet hosts and servers
4	Ireland	4	2	14	7	Economic and personal
5	Denmark	8	8	5	6	Ranked highly across the board
6	Canada	23	7	2	10	Technological then personal
7	Netherlands	21	11	6	5	Political and technological
8	Australia	18	36	3	27	Internet users
9	Austria	15	4	13	2	Political and personal
10	Sweden	19	12	9	9	Internet users and political
11	New Zealand	35	15	4	24	Secure Internet servers
12	UK	25	14	8	4	International treaties and technological

Barriers facing sub-Saharan Africa

Many African countries have received FDI for years (see Figure 5 in Section 4.3). The two case studies investigate progress in Zambia and Tanzania. Both are former British colonies, and demonstrate the physical, political, economic and environmental reasons why some countries are struggling to 'switch on'. For example, mobile connections in sub-Saharan Africa are growing rapidly (they are much cheaper than landlines), but the costs of 3G and 4G connections make Internet usage more expensive than just sending a mobile text.

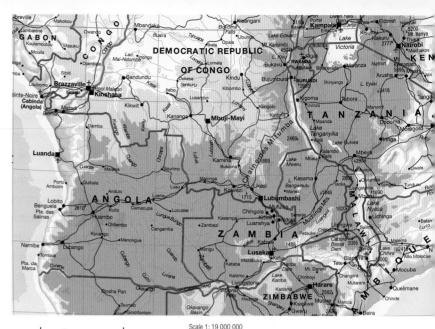

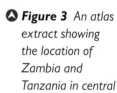

1 Zambia

Zambia is the world's eighth largest producer of raw and part-processed copper. However, it's a **landlocked** country (it has no coast), see Figure 3, so it relies on good political relations with its neighbours to access ports on the Tanzanian or Angolan coasts by rail:

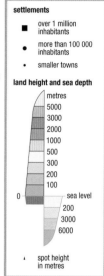

Figure 3 *An atlas extract showing the location of Zambia and Tanzania in central southern Africa*

◆ In the 1970s, a 1860km rail link (the TanZam railway) – developed using Chinese investment – took Zambia's copper to the Tanzanian coast. Although by 2000 this rail link was in poor repair, further Chinese investment has now upgraded it – and, as a result, copper exports have risen since 2008.

◆ The recently developed Benguela rail link, also paid for by the Chinese, now carries copper for export to the Angolan coast (see Figure 4). Copper remains Zambia's biggest export, although its value has fallen in recent years as fibre-optic cables increasingly replace it in telecommunications.

Since 2000, privatisation and debt cancellation have reduced Zambia's debt, and US$20 billion of FDI has been invested in the copper industry, so it can part-process the ore and add value to it – increasing the country's income. Expanding copper towns, like Kitwe, have started a process of helping localities in Africa to 'switch on' – and Zambia's development indicators have improved as a result (see Figure 5).

Figure 4 *The new Benguela rail link between Zambia and the Angolan coast*

Indicator	2004	2014
External debt (% of GNI)	159	31
GDP (US$ billions)	7.3	27.07
GNI per capita (US$)	500	1080
Life expectancy at birth (years)	38	58
Primary school enrolment (%)	86	91
HDI	0.47	0.56
KOF Index	55.71	52.16
Mobile phone users as a % of the adult population	n/a	75
Internet users as a % of the adult population	n/a	23

Figure 5 *Some development indicators for Zambia*

Key word

Human Development Index (HDI) – A single index figure, published by the UN each year, which expresses the level of education, health, and GDP indicators for every country.

2 Tanzania

Tanzania has very fertile volcanic soils, and 80% of its working population is employed in agriculture. However, during the 10-year period from 2006 to 2016, the global market price of one of its main crops – raw cotton – fluctuated from just US$0.40 per pound weight to US$2.00. When prices are high, Tanzanian farmers feel encouraged to grow cotton (see Figure 6). However, due to global overproduction, cotton prices frequently fall. When this happens, the country is less able to pay for imported manufactured goods. There is no guarantee of income and GDP fluctuates, so Tanzania is struggling to 'switch on'.

Key word

Heavily Indebted Poor Countries (HIPC) initiative – The HIPC consist of 38 of the least developed countries with the greatest debts – which, since 1996, have been eligible to have their debts with the IMF and the World Bank either cancelled or rescheduled.

◆ Until 2001, Tanzania had serious debt problems. Then the global **Heavily Indebted Poor Countries (HIPC)** initiative led to the cancellation of many of its debts. Now the income gained from growing cotton and other export crops is helping the country to invest in its schools and health care – and improve its development indicators (see Figure 7).

◆ Tanzania also has growing investment links with India, China (via the TanZam railway; see opposite), Japan and the United Arab Emirates (UAE), in order to export its farm produce and mineral resources.

▼ **Figure 6** Cotton growing in Tanzania

Indicator	2004	2014
External debt (% of GNI)	67.9	39.9
GDP (US$ billions)	12.1	49.2
GNI per capita (US$)	340	930
Life expectancy at birth (years)	46	61
Primary school enrolment (%)	86	83
HDI	0.42	0.49
KOF Index	36.12	38.39
Mobile phone users as a % of the adult population	n/a	75
Internet users as a % of the adult population	n/a	26

⚫ **Figure 7** Some development indicators for Tanzania

Over to you

1 Study Figures 1 and 2. Explain **(a)** how the rank order has been reached, **(b)** reasons for differences between the two indices, **(c)** the advantages and disadvantages of each as ways of ranking globalisation.

2 **a** Study the human and physical features on the map of Zambia and Tanzania in Figure 3. Identify and explain features of their physical and human geography which may have influenced their lack of connectedness with the rest of the global economy until recently.

 b In pairs, draw and complete two spider diagrams, each with four legs – physical, political, economic and environmental – and decide in what ways these factors might lead in future to Zambia and Tanzania becoming more 'switched on'.

On your own

3 Distinguish between: **(a)** switched on and switched off; **(b)** an indicator of development and an index; **(c)** HIPC and HDI.

4 Suggest possible reasons why the four separate indicators in Figure 2 show such wildly different rank orders for the same countries (except Denmark!).

5 Using Zambia and Tanzania as examples, explain how far globalisation seems to help or hinder sub-Saharan African countries.

Exam-style questions

AS 1 Explain why Internet usage in sub-Saharan Africa is low compared to the rest of the world. *(4 marks)*

A 2 Using examples, explain why some countries are more globalised than others *(8 marks)*

Silk, cotton and spices

Long before Europe's industrial age, ancient trade routes (known as 'The Silk Road') brought valuable silk overland to Europe from China, and sea routes brought expensive spices, tea and textiles (cotton in particular) from India and South East Asia. However, in the nineteenth century, Europe's industrialisation and the development of mechanised production techniques, shifted a lot of textile production to Europe and North America, and Asia's significance as a textile producer (rather than just a provider of raw materials) declined. Then, in the late twentieth century, production began to shift again – when cheaper labour in Asia, and faster shipping times, provided an incentive to relocate a lot of textile production back to Asia, as well as the production of many other goods (such as electronic items).

The shift begins

In the 1970s and 1980s, the **global shift** began – that is, the movement of manufacturing from Europe and the USA to many Asian countries. This shift led to the economic re-emergence of the Asian region:

◆ Pacific Rim countries, such as Japan, Hong Kong and South Korea – closely followed by China and India – became major players in the globalised economy.

◆ By 2013, the value of two-way trade between the Americas and Asia was nearly double that between the Americas and Europe.

Three factors helped to accelerate this global shift:

◆ Individual Asian countries, such as India, began to allow overseas companies access to their markets, with a new open-door policy.

◆ TNCs began to seek new areas for manufacturing (e.g. China) and for outsourcing services (e.g. call centres and software development in India).

◆ FDI began to flow into the emerging or re-emerging Asian countries.

Early 1990s After the start of the Chinese open-door policy in 1978, FDI in China begins to rise

Mid-late 1990s The collapse of communism in the Soviet Union and Eastern Europe leads to stagnated investment, which only picks up again after 2004, when many former communist countries join the EU

Mid-2000s After the start of the Indian open-door policy in 1991, India begins to attract significant FDI

2008 When the global financial crisis begins, there is a major collapse in FDI across all regions

Key
— Latin America & Caribbean
— Europe & Central Asia
— East Asia & Pacific
— Middle East & North Africa
— Sub-Saharan Africa
— South Asia

▲ **Figure 1** *Foreign Direct Investment (FDI) into different global regions, and its links to international events*

Figure 1 shows the flow of FDI into different regions. Notice that the flow of FDI into China and India did not begin immediately. Although restrictions on receiving FDI were relaxed in 1978 and 1991 respectively, the flow did not increase significantly for another 10-15 years. Similarly, it was over a decade after the collapse of communism before FDI in Eastern Europe took off, after many former communist Eastern European countries joined the EU in 2004.

FDI in China and South East Asian countries has focused mainly on manufacturing. However, India's close political links with the UK – together with its adoption of English as its business and second language, and its good technical universities – has given it an edge in software development, as well as in providing call-centre and other support services. At the same time, faster electronic communication and shipping, trade liberalisation (see Section 4.3), and the HIPC initiative (see Section 4.5) have also helped Low Income Countries to develop.

The global shift – China

The global shift into China has mostly focused on manufacturing. China has been the world's largest recipient of FDI since 2000, and its share of global trade by value rose from 3% in 2001 to 10% by 2013. Rapid industrialisation in China has also been accompanied by rapid urbanisation – particularly in the large cities near the coast, shown in Figure 3. By 2015, China had 150 cities with populations of over 1 million – up from 30 cities in 2000! However, such rapid growth has brought both costs and benefits.

The benefits of growth

1 Investment in infrastructure

There has been an impressive expansion of Chinese infrastructure. By 2016:

◆ China had developed the world's longest highway network

◆ its rail system had reached 100 000 km in length (linking all cities and provinces)

◆ its high-speed rail (HSR) system was the world's longest – having doubled in length in ten years – with high-speed lines linking Beijing with Guangzhou and Shenzhen, as well as with Shanghai

◆ Shanghai's Maglev (a magnetic levitation train; see Figure 3) had become the world's fastest commercial train service – the 30 km journey between Shanghai's airport and the CBD takes just eight minutes and reaches 431 km/h (268 mph)!

◆ 82 airports had been built since 2000 – taking China's total to 250 – with eight of the world's top 12 airports by freight tonnage, e.g. Shanghai.

2 Reductions in poverty

Over 300 million Chinese people are now considered to be middle class – nearly as many as the entire population of the USA! By 2022, an estimated 45% of the Chinese population will be classed as urban middle class. Sales of consumer items have also rocketed, e.g. the Chinese bought more TVs and laptops than Americans in 2013.

Poverty in China has reduced significantly. Between 1981 and 2010, China reduced the number of people living in poverty by 680 million. It has also reduced its extreme-poverty rate (those earning US$1 a day or less) from 84% in 1980 to just 10% in 2016. Although 20% of the Chinese population still live on less than US$2 a day (particularly in rural areas), many cope with low incomes because of payments sent home by urban family members (known as **remittance payments**).

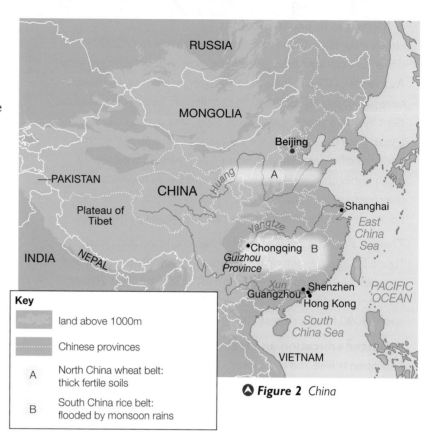

Key

	land above 1000m
	Chinese provinces
A	North China wheat belt: thick fertile soils
B	South China rice belt: flooded by monsoon rains

⬢ *Figure 2* China

⬢ *Figure 3* Shanghai's Maglev train, which links the airport and the CBD

3 Increases in urban incomes

Urban incomes in China have increased sharply since 2000 – driven by both economic growth and slower population growth. As a result of China's (now relaxed) one-child policy, employers have had to pay higher wages to recruit staff. Urban incomes have risen by 10% a year since 2005; by 2014, they averaged US$9000 a year. Even though there are still variations in average incomes between different urban industries (see Figure 4), they are still much higher than workers would receive if they had remained in the countryside – plus their terms and conditions include a 40-hour week, with higher overtime rates as well as paid holidays.

There is a big and growing rural-urban divide in China (see Figure 5). In 2013, per capita net annual income after taxes and rent (known as **disposable income**) for the poorest 20% of rural households was the equivalent of £412 – compared with over £9000 for the richest 20% in cities.

4 Better education and training

Education is free and compulsory in China between the ages of 6 and 15. 94% of Chinese over the age of 15 are now literate – compared to just 20% in 1950. In 2014, 7.2 million Chinese graduated from university – 15 times higher than in 2000. This growth in higher education has helped to create a skilled workforce for the Chinese economy's expanding knowledge and service sectors. However, again, there is a big rural-urban divide – with per capita spending on secondary education varying widely from £2200 in Beijing to just £300 in Guizhou.

The costs of growth

1 The loss of productive farmland

Despite increased food production, China's industrialisation has led to an increasing loss of farmland since 2000. Over 3 million hectares of arable farmland (the size of Belgium) has been polluted with heavy metals; 12 million tonnes of grain were polluted in 2014. The increased use of fertilisers and pesticides has also led to farmland near rivers (used for drinking water) being taken out of production.

2 An increase in unplanned settlements

China's rapid industrialisation has created an urgent need for more urban housing, which has resulted in a big increase in **informal homes**. Land prices have rocketed and made decent housing unaffordable, particularly near city centres (see Figure 6). Two types of informal housing have emerged, both illegal under Chinese law:

◆ Expanded housing in villages located on the edges of cities. Villagers add extra storeys to their houses and then rent the extra space to migrant workers.

◆ Farmland (owned collectively under communism) is privately developed for housing without permission.

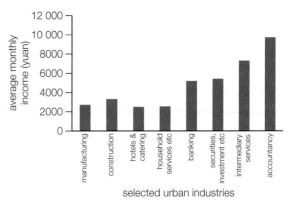

Figure 4 *Average monthly incomes (in yuan) for selected urban industries in China in 2014 (the exchange rate was about 6 yuan to £1)*

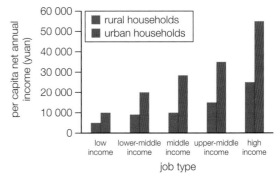

Figure 5 *Variations in annual disposable income (in yuan) between urban and rural areas in China by job type*

Figure 6 *Poor-quality housing in Shanghai. At the time this photo was taken, the land had already been earmarked for redevelopment for 7 million square metres of much needed low-cost housing. Another city, Shenzhen near Hong Kong, has the highest percentage of informal housing in China – half of its housing was classed as informal in 2010.*

3 Pollution and health problems

Chinese economic growth has caused environmental problems that affect human health:

◆ China's air pollution, caused mostly by coal-fired power stations, is so bad that the capital, Beijing, has frequent pollution alerts (see Figure 7).

◆ 70% of China's rivers and lakes are now polluted. The water in 207 of the Yangtze's tributaries is not even fit for irrigating farmland – let alone drinking.

◆ 100 cities suffer from extreme water shortages, and 360 million Chinese do not have access to safe drinking water. Tap water in Chongqing contains 80 out of 101 forbidden toxins under Chinese law.

▲ **Figure 7** *Air pollution in Beijing, caused by the burning of fossil fuels to power the country's economic growth*

In 2015, one US climate research organisation calculated that Chinese air pollution kills an average of 4400 people every day (or 1.6 million people each year). Their data showed that a third of China's population breathes in air that would be considered unhealthy by US or European standards. Air pollution causes asthma, lung cancer and heart problems.

4 Land degradation

Despite having 22% of the world's population, China only has 6.4% of its land and 7.2% of its farmland. Rapid urbanisation and industrialisation are reducing this further, and over 40% of China's farmland is now suffering degradation. The rich black soils in the north (the source of China's grain; see Figure 2) are eroding, while in the south the soils are suffering from acidification caused by industrial emissions. Land clearance has also led to deforestation and over-intensive grazing.

5 Over-exploitation of resources and resource pressure

China has abundant oil and coal, as well as key metals such as iron ore. But its resources cannot keep up with its demand, so the Chinese government has sought additional resources in Africa and Latin America (see Figure 8). Amazonian rainforest has been cleared in Ecuador, the Cerrado savannah has been converted to soy fields in Brazil, and oil fields are under development in Venezuela's Orinoco belt – all for China's consumption.

6 Loss of biodiversity

In 2015, the environmental charity WWF (in a research report about China's demands on nature) found that China's terrestrial vertebrates had declined by 50% since 1970. The WWF researchers tracked over 2400 populations of nearly 700 vertebrate species in China and discovered that almost half had vanished in the 45 years since 1970. The main cause was habitat loss and the degradation of natural environments by economic development.

▲ **Figure 8** *The Peruvian President visiting a copper mine in Peru owned by Chinese TNC Chinoco in 2013*

The global shift – Leicester

Some regions in High Income Countries also face social and environmental problems as a result of the global shift. Leicester (with a population of 330 000 in 2011) is an East Midlands city once dominated by the textile industry. Its story is typical of many long-established industrial areas:

◆ In the 1920s, over 30 000 people worked in Leicester's textile mills.

◆ By the 1960s, one factory supplying knitwear for Marks and Spencer (now M&S) employed 6500 workers on its own.

◆ The demand for extra factory workers brought Indian, Pakistani (and, later, Ugandan Asian) families to Leicester. These families set up home in the cheaper inner-city wards of Spinney Hills and Belgrave (see Figure 11).

◆ However, by the 1970s, overseas competition meant that cheaper clothes were available from Asia, and many manufacturing jobs were lost in Leicester. Industries closed, causing **deindustrialisation**.

Leicester still has a small textile industry using local designers (see Figure 9), but most items are now made in Asia. In 2016, Leicester's designers and specialist textile workers were helping to keep the textile industry alive there, although with far fewer local employees. Liam Green's HYPE company (see Figure 9) and the Internet businesses BooHoo and ASOS are examples. But British manufacturing continues to decline. By 2015, just 12% of M&S's clothing was made in the UK.

The impacts of the global shift on the UK's industrial cities

1 Dereliction and contamination

Many textile companies in Leicester were forced to close when business declined – and a lot of the previous industrial land was left abandoned or **derelict** (see Figure 10).

Industrial dereliction also scarred a number of other UK inner cities. For example, **Sheffield** suffered when its steelworks closed, while most of Glasgow's shipyards fell into disrepair as work shifted to the Far East. Much of the derelict industrial land was contaminated from the previous dumping of chemical waste (e.g. in dyes), or from manufacturing domestic gas from coal (e.g. the site of London's O2 Arena), as well as from other industrial waste disposal.

2 Unemployment, depopulation and deprivation

Although the populations of most UK inner cities are now increasing, the 1970s and 1980s experienced major declines. For example, as traditional industries closed, the population of Newcastle-upon-Tyne fell by 12% in the 1970s and another 6% in the 1990s. Many inner-city areas became run down and the housing was low cost. As a result, many people on low incomes or unemployment benefit moved to these cheap areas, which became pockets of deprivation.

In Leicester, areas of deprivation (see Figure 11) often coincide with the previous industrial areas (see Figure 12), as well as with wards containing large ethnic populations (particularly British Indians; see Figure 11) – i.e. the very people who originally came to the UK to answer the urgent need for extra factory workers in the boom years.

My company has grown from selling printed T-shirts from a bedroom to processing 350 website orders a day from a 4000 square metre warehouse in Leicester. We supply chain stores, but depend on our online presence. Blank T-shirts come from Dubai, Pakistan and China, but all cut-and-sew pieces and knitwear are made locally, and they use a local printer and dye house.

⬆ *Figure 9 Liam Green, the co-founder of HYPE (a fashion business based in Leicester) speaking in 2013*

⬆ *Figure 10 The impact of deindustrialisation – one of Leicester's former textile mills, now semi-derelict*

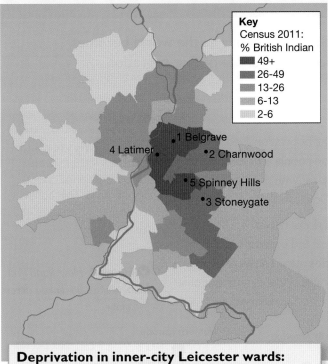

Deprivation in inner-city Leicester wards:

1 Belgrave – 6.7% unemployed (in the top 20% of England's most deprived wards)
2 Charnwood – 8.1% unemployed (top 20%)
3 Stoneygate – 7.0% unemployed (top 20%)
4 Latimer – 6.5% unemployed (top 5%)
5 Spinney Hills – 7.9% unemployed (top 5%)

 Figure 11 *Levels of deprivation and ethnic concentrations (primarily British Indian) in Leicester*

 Figure 12 *A land use GIS dataset of Leicester, compiled in 2012*

Turning around some inner-city areas has proved difficult. Many such areas have gained reputations for crime (see Section 6.5), mainly property-related (theft or burglary) or linked to anti-social behaviour. In fact, crime rates have fallen sharply since 2000 – particularly violence, burglary, robbery and car theft. Some reasons are technological, e.g. car immobilisers. But the decline in crime may be due to regeneration, employment opportunities, and gentrification (Sections 5.3 and 6.9).

Over to you

1 In pairs, draw a mind map to show factors which led to rapid economic growth in China and India in the 1990s.

2 **a** Use an enlarged photocopy of Figure 2 to annotate the economic and social benefits of the global shift for China.
 b In pairs, list benefits of growth for China. Score each one to show its importance to China between 1 (unimportant) to 5 (very important). Add up a total score.
 c What, in your view, are the three most important benefits? Compare your decision with others'.

3 **a** In pairs, research an image for each of the six 'Costs of growth' and produce a six-slide PowerPoint presentation to show the environmental costs as a result of China's economic growth'.
 b Write a 400-word report on 'The above costs outweigh the benefits of growth'.

4 **a** Using Figures 11 and 12, outline the patterns of land use, ethnicity and deprivation in Leicester.
 b Assess the impacts of deindustrialisation on Leicester.

On your own

5 Define the terms: global shift, deindustrialisation.

6 Study Figures 4 and 5. Explain to what extent there is a divide between rich and poor in China and between urban and rural.

7 'China's story is one of economic gain, but environmental disaster'. In 500 words, explain to what extent you agree with this statement.

Exam-style questions

AS 1 Explain the impacts of the global shift on one country that you have studied. *(6 marks)*

 2 Assess the impacts of the global shift on one named country. *(12 marks)*

It's an urban world

In 2008, the world reached a new milestone – over half of the population lived in urban settlements of over 10 000 people. By 2050, an extra 3 billion people are expected to be living in urban areas. As a result, the numbers of **million cities** (with populations of over 1 million) and **mega cities** (with populations of over 10 million) are also growing rapidly.

Figure 1 shows how the most important cities are linked in a global network. These **world cities** represent **hubs** in the global economy – where political and economic decisions are made, and where most investment takes place. **Hub cities** attract flows of economic migrants, as well as of capital. They are the best-connected cities, and their international airports provide gateways to the rest of the country as well as to secondary cities.

Key word

World cities (or 'hub cities') – Cities with a major influence, based on: finance, law, political strength, innovation and ICT.

🔽 *Figure 1 How world – or 'hub' – cities are linked across the globe*

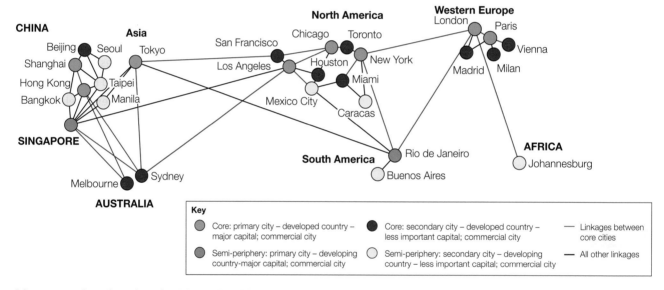

Hyper-urbanisation in New Delhi

New Delhi is experiencing **hyper-urbanisation** – its rapid population growth is outstripping the ability of the authorities to provide for basic needs, e.g. sanitation. As India's fastest-growing city, it's predicted to grow by 40% between 2010 and 2020. The main causes of this rapid population growth are:

◆ a high birth rate and low death rate, which together produce a high rate of natural increase

◆ one of the world's fastest rates of rural-to-urban migration.

Every year, hundreds of thousands (even millions) of people leave rural areas and move to urban areas to live (particularly large cities). These rural migrants consist of:

◆ the rural poor, who lack opportunities in their villages and who hope for a better future in the city

◆ the rural rich, who move to cities to invest in urban property and provide a better education and standard of living for their families.

Figure 2 explains some of the factors that **pull** (or attract) people into cities, as well as other factors that **push** people away from rural areas – plus some key background factors. In New Delhi's case, increased FDI has created many new jobs, with its financial district in particular becoming a global finance hub (persuading TNCs like Coca-Cola and Microsoft to base their Indian operations there). However, many of New Delhi's recently arrived rural migrants end up living in slums and struggling to find work. Many live on the city edge and travel for work. City girls are vulnerable to assault on public transport routes, and few can afford the modern metro.

Social challenges caused by rapid growth

Unfortunately, size is not always good news. Today, half of the world's urban dwellers live in poverty. Many countries (such as India) have persuaded TNCs to invest there by offering them low tax rates, so the reduced revenue received by the government is unable to pay for badly needed infrastructure (such as piped water supplies). Rapid economic growth may provide a range of new jobs, but it also creates a number of **social challenges**:

◆ It challenges governments to provide services and basic needs such as housing and education.

◆ In most countries, private companies are more likely than governments to provide housing, water, healthcare, energy and sanitation systems to meet demand from growing urban populations. More often than not, such companies will target high earners first – so the wealthier areas of cities will have piped safe water, while poorer areas won't.

◆ Sprawling shantytowns are the products of uncontrolled urban growth, and are common in most cities in Low and Middle Income Countries.

◆ Even in wealthy cities (such as London), the number of homeless people rises as accommodation becomes more unaffordable.

Environmental challenges caused by rapid growth

The World Health Organisation (WHO) surveyed 1600 cities worldwide for air quality. They measured 'particulates' (particles created by diesel emissions, domestic stoves, forest fires and coal-burning industries). New Delhi was judged the worst – with 153 micrograms of particulates per cubic metre. By contrast, London was 2516th, with 16 micrograms.

New Delhi's growing population and prosperity is predicted to increase the number of vehicles on its streets from 4.7 million in 2010 to 26 million by 2025. Air pollution is India's fifth largest killer – causing respiratory and cardiac problems. This is added to other environmental problems, such as sewage pollution, chemical dumping from factories, and fuel spillages from poorly maintained vehicles.

International migration

Globalisation has meant that many people now move around the world more freely. Such movements increase **interdependence** between regions. For example, in 2014, the flow of people around the world as migrants (rather than as holidaymakers or on business trips) resulted in 231.6 million migrations (see Figure 3). These migrations involved travel from the global 'South' (i.e. the developing world) to the global 'North' (i.e. the developed world), and vice versa, as well as travel within those regions.

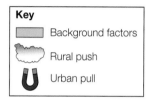

▲ **Figure 2** Urban growth factors

Key
	the number of migrants (in millions)
82.3	
36%	the percentage of total global migrations represented

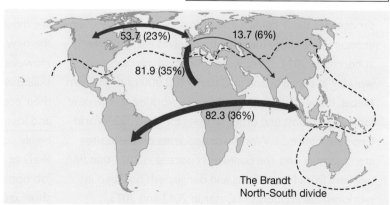

▲ **Figure 3** The global flow of people in 2014. The arrows indicate the flow, not the direction.

London – a destination for elite migrants

London's position as a major world city has arisen from its knowledge economy (see Section 4.4). Flows of skilled, wealthy migrants (experts in finance and investment) – known as **elite migrants** – have moved there. London is now a global property hotspot (see Section 6.9), and in 2013 the *Financial Times* reported that foreign buyers were involved in 82% of property deals in Central London.

Overseas buyers, particularly from Qatar and Russia, have invested in a number of landmark London properties:

◆ Qatari Investments have bought into The Shard (see Figure 4), Canary Wharf, Harrods, Stratford's East Village (formerly the 2012 Olympic Village), and the One Hyde Park development.

◆ A third of all foreign purchases of residential property in London between 2004 and 2014 went to Russians. Unfortunately, this high demand has led to property price inflation, and much London housing has now become too expensive for many native Londoners. In 2011, an apartment in the One Hyde Park development (see Section 6.9) became London's most expensive at £136 million, when it was sold to a Ukrainian oligarch.

Low-wage migrants

By contrast, some locations – such as the United Arab Emirates (UAE) and Qatar – have benefited from flows of cheap labour. The UAE (population 9 million) and Qatar (2 million) both have labour shortages, because of their small populations, so they recruit many manual workers from overseas. Migrant workers make up 90% of the UAE's workforce (including 1.75 million Indians and 1.25 million Pakistanis). Their entry visas are tied to particular jobs, so there is always the risk of being deported when their contract ends. Movements on this scale are referred to as **mass low-wage economic migrations**.

In both countries, low-wage migrants – often working for US$4 a day – have led to the rapid building of infrastructure for high-end international tourism and services (as these oil-rich emirates seek to diversify their economies, and the migrants themselves escape poverty at home). In 2014, there were 1.4 million migrants working in construction in Qatar – 400 000 of them from Nepal. Many of these migrants were working on the new football stadiums and other facilities for the 2022 World Cup (see Figure 5). Workplace accidents on these sites are common, and *The Guardian* reported in 2014 that 964 workers from Nepal, India and Bangladesh had died in work-related accidents in Qatar in 2012 and 2013.

⬤ *Figure 4* *The Shard in London – owned by Qatari investments*

⬤ *Figure 5* *Low-wage migrant workers in Qatar helping to build the new facilities for the 2022 football World Cup*

The costs and benefits of migration

Voluntary international migrations – as opposed to forced migrations (e.g. as a result of the Syrian Crisis) – are mostly economic. The **host locations** gain, because the new migrants fill gaps in their labour markets. However, the **source locations** lose, because young skilled workers are lost from their own workforces, so their economies suffer. Figure 6 summarises the gains and losses for each location. Despite public perceptions, highly educated and professional people are more likely to migrate, because they are most able to seek job opportunities overseas – and can fill urgent skills shortages, e.g. as nurses or doctors.

The host location...	The source location...
• receives skilled foreign workers • fills skills shortages in key areas • also gains lower-skilled workers for jobs which are difficult to fill (the dirty, difficult and dangerous jobs, known as the 3Ds) • can sustain a growing leisure sector, through gaining workers for restaurants, hotels and bars • can sustain the lifestyle of its middle classes, by gaining workers to provide childcare, cleaning and elderly care • can often balance an ageing population with young adults (the most common age group for migrants), particularly if they have children – increasing the birth rate • can also experience pressure on housing, healthcare and school places.	• experiences reduced unemployment, as people emigrate to find jobs elsewhere • loses its most skilled and dynamic workers (the brain-drain); the losses occur in key areas (e.g. healthcare), as well as those with entrepreneurial and business skills • earns remittance payments (see Section 4.5), which are sent home by overseas migrants, e.g. South Asian migrants in the UAE • gains increased employment, because the remittance payments are used to build homes or invest in local businesses • suffers an imbalanced population, because many young people migrate, leaving a dependent population (children and the elderly) who may face managing farms and businesses on their own.

Figure 6 *The impacts of international migration on host and source locations*

However, migration is very much a political as well as an economic and social issue. In 2015, there was a massive increase in the number of illegal migrants entering the EU (mostly from the Middle East and Africa). Several European countries erected steel fencing along their borders to try to control the flow. Most – especially those from Syria – were attempting to reach Germany and other countries in Northern Europe.

Key word

Source location – Places from which migrants move.

Host location – Places to which migrants move.

Over to you

1 a Describe and explain the international flows of people shown in Figure 3.

  b Using Figure 3, research specific examples of international migrations, then copy and complete the table below.

Migrations from ...	Two examples of migrations of people
South to South	
South to North	
North to North	1) Knowledge economy employees moving from Sydney to London
North to South	

2 Explain why rural-urban migration can **(a)** increase a nation's economic growth, **(b)** be a problem nonetheless.

3 a In pairs, draw a mind map to explain **(i)** the social, and **(ii)** the environmental challenges facing cities such as New Delhi.

  b Which do you regard as the more serious – the social or environmental challenges? Explain your answer.

On your own

4 Distinguish between the following pairs of terms: **(a)** million city, mega city and world city (or hub), **(b)** push and pull factors in migration, **(c)** source and host locations.

5 a Use Figure 2 on page 302 (Section 7.4) to construct flow lines to show migration flows to the UK.

  b Describe the pattern shown on the map, and give possible reasons for it.

6 a Draw a table to show the advantages and disadvantages of **(i)** elite migrants, **(ii)** low-wage migrants to host locations.

  b Which group seems most beneficial to the host location? Suggest reasons why.

Exam-style questions

 1 Using examples, explain the impacts of international migration on host locations. *(8 marks)*

2 Assess the role of international migration in the globalised economy. *(12 marks)*

In this section, you'll learn that one of the outcomes of globalisation has been the emergence of a global culture.

Cuba opens its doors

21 March 2016 – a historic date. On that day, Barack Obama became the first US President to visit Cuba in 57 years!

Cuba's head of government, Fidel Castro, declared Cuba a communist state in 1959, when he took over the country in a revolution that resulted in its isolation from the Western capitalist world for 50 years. Subsidies from the communist government in the USSR supported its economy until 1991, when the Soviet Union collapsed. After 1991, and the ending of subsidies from its main backer, Cuba began a period of development that included accepting some foreign capital (particularly from tourism). Tourism brought many Cubans into contact with foreign travellers and their cultures (e.g. their clothes and their diets) for the first time.

Since 2008, when Fidel Castro resigned due to ill health, his brother Raul has been in power in Cuba. President Obama's visit was a result of the relaxation of strict communist controls by Raul Castro, which has allowed **free enterprise** businesses to be set up for the first time in decades (see Figure 1). This relaxation is similar, but smaller in scale, to that in communist China (see Section 4.6). Since 2012, Cubans have been able to buy and sell houses and cars, take out loans, and set up private businesses. However, on the other hand, guarantees of State employment have gone, and State-owned farmland has also been sold off to private companies.

Cuba's new spirit of openness has led to improved relations with the United States, which began to isolate Cuba from 1959 onwards. An initial sign of thawing relations between the two countries was the reopening of the US embassy in Havana in August 2015 (after its closure in 1961) – followed by President Obama's official visit in March 2016.

However, despite the welcomed political relaxation in Cuba, the accompanying social and economic changes are leading to growing inequality, as Figure 2 shows.

The river where Jonas Echevarria fishes runs through a neighbourhood of fine restaurants, spas and boutiques – symbols of Cuba's private enterprise. New luxury apartments also reflect the wealth of some Cubans. However, whilst private restaurants serve fillet steak to Cuban entrepreneurs, most Cubans make do with plantains, bread and eggs.

Jonas Echevarria lives in a shantytown in a home built from scraps of corrugated tin and timber. His diet involves subsidised monthly rations from the State. The gap between the haves and the have-nots in Cuba is increasing.

Figure 2 *Growing inequality in Cuba – adapted from an article in the New York Times, February 2014*

Key word

Capitalism – An economic system based on private ownership of investment capital and wealth production, whereby goods and services are sold for profit. Based on the concept of 'economic man', a theory developed by Adam Smith (1723-1790), in which people look after their own individual interests first and foremost. Often referred to as 'free enterprise', and the 'market economy', in which market forces determine the price of any service or item of produce.

Communism – A socio-economic system or organisation, where the land and all property is owned by the community or the State, so that every person contributes and receives according to their abilities and needs. Based on the ideas of Karl Marx (1818-1883) and a classless society. Under most communist systems to date, e.g. the USSR (Soviet Union), China, and North Korea, the State owns or controls most resources, banks, and the media.

Figure 1 *Private enterprise in twenty-first century Cuba*

Twenty-first century Cuba

Cuba is changing rapidly. The influx of international tourists, and the spread of satellite TV and the Internet, is broadening Cubans' knowledge of the rest of the world and challenging Cuba's traditions and values – globalisation is diluting Cuban culture. Netflix is now available to Cubans, American tourists are allowed to visit Cuba again, and the cap on remittances sent by Cuban Americans to relatives back home has been lifted. In some locations, **cultural erosion** (e.g. the loss of language and traditional food) has also resulted in changes to the environment. Beach resorts have changed Cuba's coastline, and its coral reefs are now threatened by increased tourist activity. Until recently, Cuba's forced isolation protected it against such coastal developments.

The processes by which Western attitudes and values have moved to Cuba is known as **cultural diffusion**, and is a symbol of globalisation. Closer and faster connections enable the transfer and spread of attitudes and values around the world, and traditions fall to new fashions and trends. The food, music, clothing and architecture of Cuba are changing – and it now takes a big effort to maintain a strong Cuban identity.

Cultural landscapes and diversity

Culture changes all the time. Economic development changes landscapes and creates new ways of living. Peoples' attitudes and values also shift over time, as different influences play a part. Current global changes are affecting the ways in which people see the world. **Glocal** cultures develop where global processes exist at a local level (for example, see Figure 3). This shows how urban environments in British cities (like Leicester's Golden Mile) have been transformed by decades of inward migration. **Ethnic enclaves** (see Section 6.3) like this gain their own identity, where street furniture, road names and cuisine add to the city's multicultural character and strengthen cultural diversity.

Cultural diffusion and the media

Traditionally, cultural differences are often expressed through language, because phrases don't easily transfer from one region to another. The real feel and soul of a culture can be lost in translation. However, as the ownership of the global broadcasting, film and music industries becomes ever more concentrated into the hands of large media TNCs (see Figure 4), the use of an increasingly common vocabulary is starting to erode cultural diversity. This is known as the **global homogenisation** of culture – with everywhere becoming the same.

Figure 3 *Diwali celebrations light up the Golden Mile in Belgrave Ward, Leicester*

Newspapers and magazines

◆ Australia: 101 newspapers, from national to suburban.
◆ UK: Four daily and Sunday newspapers (including the *Times* and *Sun*)
◆ USA: *The New York Post*, plus financial papers such as *The Wall Street Journal*. Eight daily and 15 weekly regional newspapers, plus New York suburban papers.
◆ Russia: A 33% share in *Vedomosti* (Russia's leading financial newspaper).

Television

◆ Studios: Fox is the main brand name, e.g. Twentieth Century Fox, Fox Television Studios in France, India, Australia and New Zealand.
◆ Networks: Fox TV (USA and satellite), My Network TV, plus channels in Eastern Europe, Uruguay, Israel, Indonesia and New Zealand.
◆ Satellite: BSkyB (UK), Foxtel (Australia), SKY (New Zealand, Italy, Germany), STAR TV (Asia including parts of China and India).

Figure 4 *The spread of global culture – some of the companies worldwide owned by Rupert Murdoch's News Corp, which can impact on people's political as well as their cultural thinking. Fox TV in the USA is openly supportive of the Republican Party, and every winning political party in UK General Elections since 1979 has been promoted by the Sun newspaper.*

Cultural exchanges

Information technology and digital communications spread ideas and products faster than ever, which influences how people consume goods and services. Music, TV programmes and films made by large media TNCs now combine with migrations of people to create cultural mixes (or **hybrids**).

What we see on television, read in newspapers, or listen to on music systems, is increasingly provided by a small group of huge companies. Five companies (EMI, Universal, AOL, Time Warner and SonyBMG) now own 90% of the global music market. In recent years, these companies have been cutting their ranges of recording artists, in order to increase their profits. It makes better economic sense for them to sell 10 million CDs or downloads by one artist (e.g. Taylor Swift, Ed Sheeran or Adele), rather than spending money, time and effort marketing ten different artists – each one selling one million.

In some ways, globalisation is the twenty-first century term for 'cultural imperialism'. Previous generations called it 'Americanisation', 'Westernisation' or 'Modernisation'. As the global economy draws people closer together, big brand names – like Coca-Cola, McDonald's, Disney and Nike – become globally famous. Every year, Forbes (a US company) calculates the brand value of 500 top companies. Its top ten for 2015 are shown in Figure 5.

Ranking	Brand	Type of company
1	Apple	Technology
2	Microsoft	Technology
3	Google	Technology
4	Coca-Cola	Beverages
5	IBM	Technology
6	McDonald's	Restaurants
7	Samsung	Technology
8	Toyota	Automotive
9	General Electric	Diversified
10	Facebook	Technology

▲ **Figure 5** *The Forbes top ten brands in 2015*

Changing values

Globalisation can also challenge values – sometimes in a transformative and beneficial way. For example, official data published by Chinese state media in 2011 stated that only 25% of disabled people in China find employment. A year later, author James Palmer wrote: 'Disabled people in modern China are still stigmatised, marginalised and abused'. Yet that same year, China came top of the medals table in London's Paralympic Games. The chance to train and compete on the global stage has helped marginalised groups such as those with disabilities to gain support and training, and gain a more equal status, although there is a very long way to go before any kind of equality is achieved.

The same thing has happened in the UK, where Australian TV comedian and presenter Adam Hills (himself disabled) stated at the end of the 2012 Games that 'Sydney was the first Paralympics to treat Paralympians as equals. London was the first to treat them as heroes.' Increasingly, Western countries are adopting more tolerant policies on other ethical and moral issues, such as gay rights. However, there is still some way to go in places such as Russia, some other European countries, the Middle East and Africa.

▲ **Figure 6** *Paralympian Ellie Simmonds winning gold for Britain at London's 2012 Paralympic Games*

Growing resistance

The spread of global culture has been counter-balanced by concerns about its impacts – and about its perceived exploitation of both environments and people. This has led to opposition by anti-globalisation groups and environmental pressure groups, which reject globalised cultures and the practices of many TNCs in avoiding tax payments. The Occupy movements from 2011 onwards have demonstrated in world cities such as London and New York.

Two examples of recent national resistance to a 'global culture' are Iran and France:

◆ In **Iran** in the early 2000s, the Islamic government led a backlash against Mattel's Barbie dolls, which were confiscated from toy stores when the government denounced its un-Islamic image. Since then, the government has liberalised its position, partly because of the need for international cooperation against radicalism, but also because Iranian youths were increasingly accessing banned social media such as Twitter and Facebook anyway.

◆ Until the early 2000s, **France** paved a way in rejecting globalisation. To protect French culture, the government excluded culture from its agreements on trade. It limited how much foreign culture (music, films, TV) could be broadcast, and 40% of broadcasts had to be French with no more than 55% American film imports. However, Internet downloading of music and films has placed this in dispute. Since 2007, the French government has been more accepting of globalisation, because of successful French TNCs such as EDF Energy.

Figure 7 *The Occupy London encampment outside St Paul's Cathedral in October 2011*

Over to you

1 a In pairs, research 'Images of Cuba' and devise a six-slide PowerPoint presentation to show Cuba's traditional culture, e.g. architecture and people.
 b Compare your initial research with three further images under the search term 'holidays in Cuba'. Explain how and why these differ.
 c Using these images, explain why globalisation sometimes leads to tensions within countries.

2 In pairs, outline the benefits and problems created by cultural diffusion for **(a)** Cuba, **(b)** you.

3 a Search for the phrase 'How Forbes calculates brand value' online. Explain how brands are valued.
 b Use the Forbes website to update the data in Figure 5.
 c Using the dataset, compare the value of Western brands with those from elsewhere.

4 In pairs, discuss the reasons why increased globalisation can lead to greater acceptance of inequalities such as those linked to disability and gay rights.

On your own

5 Distinguish between the following pairs of terms: **(a)** capitalism and communism, **(b)** free enterprise and market economy, **(c)** cultural diffusion and cultural homogenisation, **(d)** ethnic enclaves and cultural diversity.

6 a Select one major film, music or media company and produce a fact file for it (using Figure 4 as a guide).
 b Explain how far this company impacts on your lives as a student group.

7 Write a 500-word report to assess the reasons why some groups object to cultural globalisation.

Exam-style questions

 1 Explain the process of cultural diffusion. *(4 marks)*

 2 Assess the contribution of globalisation to cultural diffusion. *(12 marks)*

In this section, you'll investigate how far globalisation has increased the overall level of development worldwide, whilst simultaneously widening the development gap (plus increasing disparities in environmental quality).

Richer or poorer?

Globalisation has revolutionised the world – creating an explosion in global trade (see Section 4.1), and encouraging countries to trade together in blocs (such as the EU). But this process has led to greater inequalities:

◆ The GDP of most Asian countries has accelerated rapidly, with China experiencing an economic revolution (see Figure 1). Economic growth there has outpaced population growth, so China's GDP per capita increased by 14 times between 1990 and 2015.

Year	USA GNI (PPP) per capita (US$)	China GNI (PPP) per capita (US$)	Ratio USA : China	China's % share of world GDP (PPP) per capita
1990	23 954	980	24.4 : 1	2.2
2000	36 450	2915	12.5 : 1	7.1
2015	55 860	13 217	4.2 : 1	18.0

🔺 **Figure 1** *China closes the gap with the USA*

◆ A new global elite of 'super-rich' has emerged, with China's richest 1% of people owning one third of China's property and industrial wealth.

◆ As Figure 2 shows, every global region increased its GDP per capita from 1980 to 2012. However, the regional growth rates differ markedly, with a big disparity between the developed countries ('Advanced economies' in Figure 2) plus the 'Newly industrialised Asian economies' (also known as 'Emerging countries'), versus the developing economies of Latin America, the Middle East and Africa (particularly sub-Saharan Africa). Between 1980 and 2012, the **development gap** (i.e. the difference between the wealthiest and poorest nations) actually widened.

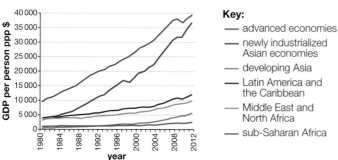

Key:
— advanced economies
— newly industrialized Asian economies
— developing Asia
— Latin America and the Caribbean
— Middle East and North Africa
— sub-Saharan Africa

🔺 **Figure 2** *The changing GDP per person (in PPP$) in different economic world regions between 1980 and 2012*

Economic data

Measuring economic progress is difficult, and data can be misleading. For example, China's growth data since 1980 are based on different data collection methods to those of other countries (and their values are expressed in a currency – the yuan – that is not readily exchanged on global markets). For global comparison, economic values are usually given in US$ and include **single indicators**, such as:

◆ **GNI (Gross National Income)** – The value of goods and services earned by a country (including overseas earnings), formerly known as Gross National Product (GNP). Both are good indicators of wealth (especially of HICs, which earn a great deal of income from overseas investments).

◆ **GDP (Gross Domestic Product)** – The same as GNI, but excluding foreign earnings.

◆ **Per capita** data are averages per person, e.g. total GNI divided by population size.

◆ **Purchasing Power Parity (PPP)** relates average earnings to local prices and what they will buy. This is the spending power within a country, and reflects the local cost of living.

Other indicators exist in **composite** form, i.e. using several sets of data. The most common is **economic sector balance** (the percentage contribution of primary, secondary and tertiary sectors to GNI). As countries develop manufacturing industries, the value of their primary sector output falls in relative terms – and the value of their secondary sector rises. For instance, the value produced by Vietnam's primary sector fell from 50% of GNI in 1990 to 22% in 2014 (TNCs now manufacture a lot of clothes in Vietnam). The primary sector value in Malawi, a Low Income Country (LIC), is still high (30%), while the UK's is low (0.6%).

Using social indicators

The **Human Development Index** (**HDI**) was devised by the United Nations (UN) to provide a measure of life expectancy, education and GDP for every country. By linking GDP to education and health, the HDI shows how far people are benefiting from economic growth. The HDI uses four indicators:

◆ Life expectancy

◆ Education (using two indicators: literacy, and the average number of years of education)

◆ GDP per capita (using PPP$)

Each indicator is converted into a value ranging from 0 (low) to 1 (high), and the four values are then combined into a single Index – the HDI. By backtracking data over nearly 150 years, the UN has been able to estimate trends in the HDI since 1870 (see Figure 3). However, HDI health and education data tend to lag behind economic data, since many countries do not have sophisticated agencies for accurate calculation. Therefore, the most recent data in Figure 3 (published in 2014) are for 2007.

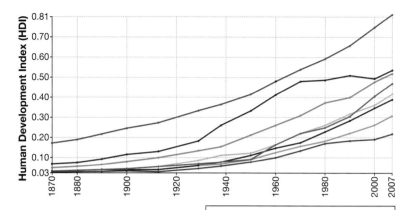

Key:
● OECD
● Central and Eastern Europe (incl. Russia)
● Latin America
● China
● India
● rest of Asia (excl. Japan)
● North Africa
● Sub-Saharan Africa

▲ *Figure 3* *UN estimates of trends in the HDI from 1870 to 2007*

Women and development

In no country – no matter what its level of development – are women equal to men. For example, in many LICs, fathers may prevent their daughters from receiving education beyond puberty, because school is expensive – why support them financially when they will leave home on marrying, and sons won't? This view is hard to shift in many rural societies. Even in HICs, women earn less than men and are less likely to be promoted. So, the UN has developed the Gender Inequality Index (GII; see Figure 4), which uses indicators relating to:

◆ **reproductive health** – as gender inequalities decline, fertility rates and maternal mortality rates fall, and the age of having a first child rises

◆ **empowerment** – women enter politics as they become more empowered

◆ **education and employment** – staying on at school and university opens up more opportunities for women.

Figure 4 shows exemplar data from a much larger database.

▼ *Figure 4* *Ranked countries based on the Gender Inequality Index (GII), 2015 – the higher the GII, the greater the inequality*

Rank, 2015	Country	Gender Inequality Index value, 2014	Maternal mortality ratio (deaths per 100 000 live births), 2013	Births per 1000 women aged 15–19, 2015	% of seats in Parliament held by women, 2014
VERY HIGH HUMAN DEVELOPMENT					
1	Norway	0.067	4	7.8	39.6
2	Australia	0.110	6	12.1	30.5
3	Switzerland	0.028	6	1.9	28.5
14	UK	0.177	8	25.8	23.5
HIGH HUMAN DEVELOPMENT					
50	Belarus	0.151	1	20.6	30.1
50	Russian Federation	0.276	24	25.7	14.5
52	Oman	0.275	11	10.6	9.6
LOW HUMAN DEVELOPMENT					
145	Kenya	0.552	400	93.6	20.8
145	Nepal	0.489	190	73.7	29.5
147	Pakistan	0.536	170	27.3	19.7

Environmental quality indices

The link between economic development and environmental quality is well established – air quality deteriorates as economic development increases. There are many causes, all linked to energy production, industrial processes and road transport. Key pollutants include sulphur dioxide (SO_2), nitrogen oxides (NOx), particulates (PM10), and volatile organic compounds (VOCs). All can lead to respiratory problems. Most HICs have improved their air quality by controlling vehicle emissions and by transferring manufacturing overseas (outsourcing the problem). Most countries record air pollutants, and calculate air quality indices, but in slightly different ways, so international comparisons are difficult.

A widening gap?

Just as there is a gap between the economic growth rates in different global regions (see Figure 2), widening income inequalities also develop within countries. The increasing inequality of income distribution within countries can be measured using the **Gini index** – an index with values between 0 and 100% shown using a Lorenz curve (see Figure 5). Section 7.7 explains how it's calculated.

◆ A low index value indicates a more equal income distribution; 0 represents perfect equality (everyone having the same income). This would be shown in Figure 5 with a curve close to the straight line.

◆ A high index value indicates unequal distribution; 100 corresponds to perfect inequality (where one person has all the income). This would be shown in Figure 5 with a curve far from the straight line.

The advantage of the index is that it measures inequality, rather than giving a single indicator for an entire population, e.g. GNI. However, only about one third of countries publish a Gini index.

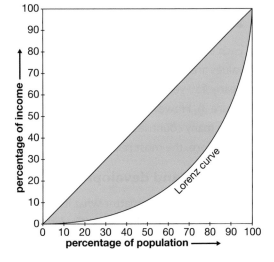

◭ **Figure 5** *The Lorenz curve, used to plot and calculate the Gini index*

China's economic and environmental inequalities

China has a structural problem – most of its economic growth is based on industries exporting low-value products. The Chinese people have been used to provide a cheap labour pool, and – in spite of rising wages – many Chinese workers are not much better off than they were before industrialisation. An east-west divide now exists in China, as incomes decline the further inland (or west) you go (see Figure 6):

◆ Area [A] – the eastern mega-city wealth between Beijing, Shanghai and Guanzhou

◆ Area [B] – areas of intermediate wealth

◆ Area [C] – poorer county level cities in the interior

◆ Area [D] – poor, rural, western interior settlements

Almost all of the major cities and industrial zones are located on the coast, and urban dwellers have higher incomes than rural (see Section 4.6). The 2010 Gini index for China as a whole was 47% – and its inequality is increasing.

Chongqing on the Yangtze River (in Sichuan province) is at the heart of economic developments around the Three Gorges Dam. 30 million people now live there. It's one of China's dirtiest cities, and its filthy air causes thousands of premature deaths. For a quarter of the time, the air quality doesn't reach the Chinese government's own safety standard. China may be doing well economically – but is it at the expense of the environment?

Figure 6 *China's internal inequalities*

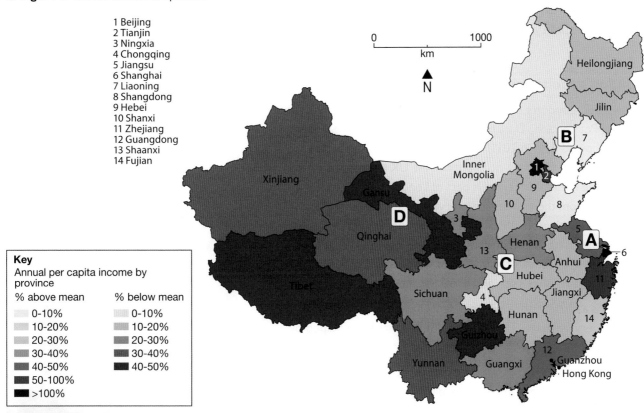

1 Beijing
2 Tianjin
3 Ningxia
4 Chongqing
5 Jiangsu
6 Shanghai
7 Liaoning
8 Shangdong
9 Hebei
10 Shanxi
11 Zhejiang
12 Guangdong
13 Shaanxi
14 Fujian

Key
Annual per capita income by province

% above mean:
0–10%
10–20%
20–30%
30–40%
40–50%
50–100%
>100%

% below mean:
0–10%
10–20%
20–30%
30–40%
40–50%

Over to you

1 a In pairs, research ten economic and social development indicators for ten countries in different global regions that you think represent different levels of development in the period since 2000. Use the World Bank and UN (e.g. HDI) websites.
b Draw up tables and rank your ten countries.
c Plot line graphs to show how your economic indicators have changed over time.
d How far do your selection of countries match the pattern shown in Figure 2?
e Which indicators seem to show levels of development best? Justify your choice.

2 a Using Figure 4, explain the significance of the indicators used to calculate the GII.
b To what extent does a country's level of development contribute to its position in the GII?
c In pairs, discuss and set three targets and actions that would improve the UK's position in the GII ranking.

3 a Use Section 7.7 to plot a Lorenz curve from the data given.
b Calculate the Gini index for these data.

On your own

4 Distinguish between the following pairs of terms: **(a)** GNI and GDP, **(b)** HDI and GII, **(c)** a Lorenz curve and the Gini index.

5 In 300 words, explain why it's difficult to judge the consequences of globalisation.

6 China's Gini index varies each year, but in 2016 had yet to reach 50%. Using the information in this section, explain why China's economic growth may not have been beneficial for all.

Exam-style questions

AS 1 Explain why economic and human indicators of development can sometimes give different impressions of a country's level of development. *(6 marks)*

A 2 Assess the statement: 'Globalisation produces as many losers as it does winners'. *(12 marks)*

London's melting pot

In July 2012, London's *Evening Standard* newspaper ran a story about 'Melting Pot London'. Data from the Office for National Statistics (ONS) showed that, in 2011, the parents of 65% of the babies born in London had been born outside the UK themselves (the UK average is 25%). Diversity is part of London's character. Journalists from the newspaper interviewed staff at North London's Whittington Hospital about the range of different nationalities using their maternity ward (see Figure 1). That diversity is also true of the staff – the maternity ward had a Chilean midwife, Jamaican and Polish nurses, and a Malaysian matron.

London is a melting pot – with residents from every country, speaking almost every language. This growing diversity has been helped by several processes linked to globalisation:

◆ **Open borders**. EU citizens are free to move around the EU as a right. For example, in 2015, there were about 250 000 French people living in London.

◆ **The freedom to invest in businesses or transfer capital.** For example, in the UK, any bank or individual can trade in shares without having to use the London Stock Exchange. Individuals are free to invest, without barriers. There are no restrictions for financial institutions in setting up offices and no government approval is required.

◆ **FDI**. In 2015, the UK attracted over 32 000 jobs (mostly to London) from overseas-owned companies investing in software and financial services. London also attracted 35% of all companies who moved their European headquarters into the UK. The UK also led the rest of Europe in attracting European research and development projects, and was the first choice for European project investments by US companies. It was also a leading recipient of FDI investment from France, Japan, Australia, Canada, India and Ireland.

Processes like those above (and others such as historical links, e.g. the Commonwealth) have led to a cultural mix. For instance, globalisation has helped to establish diverse groups throughout the UK – from Polish communities in Lincolnshire to Latvian groups in Cornwall. In mainland Europe, France has large North and West African communities from its former colonies, and Germany has a large Turkish population of over three million. The dispersal of people throughout the world has led to **diasporas** – that is, distributions of people away from their homelands. For example, the Jamaican diaspora consists of Jamaicans (and those of Jamaican descent) living in the USA, UK, Canada, and Central America, e.g. Honduras.

Matron, Whittington Hospital, North London

> You name it, we've got it. French, Tamil, Turkish, Romanian... the list goes on. The whole United Nations is represented here. There used to be lots of Bengali mothers, and we still have them but it's become more diverse.

▲ *Figure 1* *The Whittington Hospital in North London*

▲ *Figure 2* *Members of the Jamaican diaspora in London, working on their allotments*

Key word

Diaspora – From the Greek word διασττορά, which means 'scattering' or 'dispersion'. This process is the movement or migration of people who share a national and/or ethnic identity away from their perceived homeland.

But there's another view

However, not everyone is happy about globalisation and its effects. Freedoms for some can impact adversely on others. Two big issues in relation to this are immigration (where some people dispute the rights of others to move freely), and the distribution of resources such as water (where developing a resource in one country may have an impact on other countries which regard that resource as shared).

Immigration

International migration is controversial and can cause resentment in some people within host populations. Migrants have become victims of harassment, abuse, violence and exploitation. For example, tensions have been rising both within the EU and also with its immediate neighbours:

◆ **Extreme political parties**, such as Golden Dawn in Greece, The League (or *Lega*) in Italy, and France's National Front, are becoming increasingly popular. Less extreme (though significant) are attitudes in the UK's tabloid press about immigration, particularly uncontrolled immigration.

◆ Since 2014, streams of refugees from Syria – as well as economic migrants from many other countries – have caused tensions between Greece and other Balkan countries (which are the main entry points to Northern Europe) and Turkey.

Trans-border water conflicts

The Mekong is one of South-East Asia's major rivers – flowing for 4200 km from China through Myanmar, Laos, Cambodia and Thailand, to its delta in southern Vietnam (see Figure 3). Since the 1990s a number of dams have been built on the river or its tributaries, which has caused some controversy.

A 1995 treaty, known as the **Mekong River Agreement**, required the governments of Cambodia, Laos, Thailand, and Vietnam to all agree to any proposals for new dams before they go ahead. The treaty aimed to share water allocations within the Mekong river basin.

The Xayaburi Dam in Laos is testing this agreement. Water taken from the main river or its tributaries in an upstream country, will then affect the water flow downstream (in this case in Thailand, Cambodia and Vietnam). Laos benefits from the project in terms of employment and water, while Thailand receives the electricity generated by the dam and has financed it. In 2010, 11 more dams were proposed along the Lower Mekong – nine of them in Laos (see Figure 3).

China
China owns half the length of the Mekong. The upper basin is mountainous, but the southern part is in one of China's poorest regions. Damming the river to generate hydroelectric power would encourage economic development there.

Myanmar (formerly Burma)
Myanmar is the country least affected by the current Mekong proposals. It contains several tributaries of the Mekong, but none of them would be affected by the existing dam proposals. It has no plans to build any dams of its own yet, but any such plans would affect the water flow in downstream countries.

▼ **Figure 3** *The dams built and proposed along the Mekong and its tributaries, plus the likely impacts on the six countries involved*

Key
▨ Mekong basin area
• dam under construction
∘ commissioned dam
• planned dam

Vietnam
The Mekong delta is a fertile area of 50 000 km², which supports 40% of Vietnam's population. Annual floods there allow a large amount of rice to be grown. Continuing to dam the Mekong further upstream will reduce the river flow in Vietnam (10% has already been lost).

Cambodia
Nearly all of Cambodia is within the Mekong basin. It depends on the river for the crucial annual flooding of its rice-growing area. Its lack of reliable energy sources has led to the depletion of its forests for firewood. Hydroelectric power would boost the country's economic development, but it would also displace villagers from fertile land beside the river.

Thailand
Only 36% of Thailand's territory is within the Mekong basin, but Thailand would like the water and electricity generated by the dams for industrial development – and to aid rural investment to stem the flow of migrants to its cities.

Laos
Laos is one of the world's poorest countries, and 90% of its population depends on the Mekong for agriculture. Most water comes from tributaries within Laos, but dams for hydroelectric power or flood control would reduce the flow downstream.

Attempts to control globalisation

The theory of globalisation is based on classic economic freedoms – known as **liberalism**. Its more recent form (a belief in free flows of people, capital, finance and resources) is known as **neo-** (or new) **liberalism**. However, the three examples below show how neo-liberalism may not always work in practice.

1 Censorship

China is a single-party, communist state. As such, globalisation presents a psychological challenge for its leadership – the free flow of information and ideas is perceived as a threat, and the Chinese government also looks nervously at global events like the 'Arab Spring', with its ideas about increasing democracy and less State control. In its own country, the Chinese government enforces the censorship of Internet content, as well as all published material, in order to retain control. There are two types of censorship:

◆ State-controlled (e.g. Chinese news) – where print publishing or broadcasting via TV or radio is run by official State media.

◆ State-monitored – where overseas contacts or media are monitored and censored, including TV, print media, radio, film, theatre, text messaging, video games, literature and the Internet.

2 Limiting immigration

Across Europe, Australia and the USA, there have been debates about migration controls. Right-wing candidates have referred to the topic in a range of countries. The most sensational was a proposal by Republican candidate Donald Trump during the 2016 US Presidential election campaign to control the flow of immigrants from Mexico by building a high wall right along the US-Mexican border.

Debates in the UK have focused on limiting net migration – difficult, with flows from the EU, skills shortages in the knowledge economy, and a booming market in overseas university students. Two arguments used by opponents of immigration are that cheap migrant labour undercuts local wages, and also that the government has not planned adequately for the increased demands on welfare, healthcare, housing and education caused by the influx of migrants.

🔽 *Figure 4* A Chinese steel works in full production

3 Trade protectionism

The free market can be a challenge for national governments. In 2016, cheap Chinese steel (see Figure 4) was being 'dumped' onto global markets at prices heavily subsidised by the Chinese government in order to protect its own manufacturers. The consequences were huge in the UK – the Indian owners of Tata Steel (the UK's largest steel manufacturer, which was losing £1 million a day) put all of its UK steel plants, such as Port Talbot, up for sale and threatened to close them if a deal could not be arranged. A solution would have been to raise tariffs on imported steel to protect domestic producers (as the USA has done), but this is forbidden by WTO rules.

Maintaining cultural identity

Rarely is the impact of globalisation more keenly felt than over the traditional land rights of indigenous peoples. In Canada, such peoples are referred to collectively as the First Nations. They consist of 643 recognised groups, together with Inuit and Métis. These indigenous peoples had much of their land taken away from them during colonial rule in the eighteenth and nineteenth centuries – and it was never returned. However, since the late twentieth century, efforts have been made to acknowledge the rights of the First Nations to compensation payments.

Many cases of resource exploitation in Canada (see Figure 5) have caused conflict with traditional communities, and the Canadian government has been accused of supporting TNCs against indigenous landholders. In 2013, six out of 21 proposed resource projects were close to collapse, because of protests from traditional communities. Their targets included projects such as:

◆ fracking (in New Brunswick)

◆ oil sands and shales mining (e.g. the Bakken project in Alberta)

◆ the Trans Mountain Pipeline between Alberta and Vancouver

◆ the Pacific Trails Pipeline.

Canada's oil industry expects the government to settle claims about ownership and royalties with First Nations representatives. However, conflicts such as these are likely to increase in the future, as increased resource exploitation results in companies exploring more remote regions and indigenous lands.

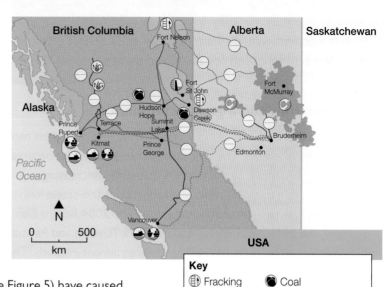

Key
- ⊕ Fracking
- ⊕ Port
- ◉ Oil/tar sands
- ⊜ Tanker
- ⊝ Pipeline
- ◐ Coal
- NG Natural gas
- ◖ Dam
- ⊖ Transmission lines

Figure 5 *The projected pathways of pipelines and electricity transmission lines (called the 'Carbon Corridor) from oil and gas reserves in Alberta and British Columbia to the coast*

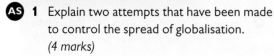

Over to you

1   In pairs, design a diagram that explains how some people view London's melting pot very positively, whereas others take a different view.

2   **a**   Explain how diasporas can arise from globalisation.
    b   In pairs, research one diaspora of people who have come to the UK. Research **(i)** their reasons for coming to the UK, **(ii)** their economic and cultural contribution to the UK. Present your findings.

3   In pairs, explain why government attitudes towards globalisation can never please anti-immigration groups.

4   In pairs, research and present a six-slide PowerPoint presentation about one indigenous people that includes: **(a)** resources being exploited on their traditional lands, **(b)** government attitudes towards them, **(c)** protests that have taken place to establish their rights.

On your own

5   Explain what the following terms have to do with globalisation: open borders, deregulation, diaspora, liberalism and neo-liberalism.

6   Explain how and why each of the following are against the spirit of globalisation: **(a)** censorship, **(b)** limiting immigration, **(c)** trade protectionism.

7   **a**   Identify and explain the key players involved in **(i)** water conflicts in the Mekong basin, **(ii)** conflicts involving First Nations peoples in Canada.
    b   Explain any similarities and differences between the conflicts.

Exam-style questions

AS 1   Explain two attempts that have been made to control the spread of globalisation. *(4 marks)*

A 2   Assess the nature of social, political, and environmental tensions that have resulted from change caused by globalisation. *(12 marks)*

Globalisation and a global conscience

Global broadcasting and 24-hour news coverage is making us more aware of global events. The time delay between an event occurring and it being broadcast to the world is constantly decreasing – in some cases, due to rolling news and reporter-at-the-scene live broadcasts, the two occur simultaneously. The 2004 Boxing Day tsunami, the 2011 'Arab Spring' and the 2015 Tunisian and Parisian ISIS murders all reached global TV audiences almost instantly. These images bring consequences. A greater awareness of global events has led to calls for increased rights and other changes across the world:

◆ The 'Arab Spring' uprisings brought protestors onto the streets, with – for the first time in Arab countries – women challenging their governments.

◆ Organisations (such as Amnesty) campaign for more human rights as global awareness increases.

◆ Environmental pressure groups also challenge the world to find new and better ways of meeting people's needs sustainably.

🔺 **Figure 1** *The 'veins' of the global economy*

Sustaining globalisation

Many people in HICs get used to having what they want, wherever and whenever they want it, e.g. fresh strawberries in Northern Europe in December! Global supply chains fulfil these demands, and seasonal produce is no longer a concern. The UK's supply chain, for example, extends across geographical and seasonal boundaries, so now Kenya (which depends on food exports to support its economy) grows cash crops – often sacrificing its own needs. The 'veins' in Figure 1 show global resources flowing into the UK.

Ecological footprints

Like most HICs, the UK is living well beyond its 'environmental means'. To supply resources for every country at the UK's current level of consumption, would take 3.1 Earths. At current rates of growth in global consumption, UK environmental researchers believe that the world as a whole will need two planets' worth of resources by 2050.

Figure 2 shows the ten countries with the world's largest ecological footprints per capita (i.e. each country's total **ecological footprint**, divided by its population). To live within the planet's means, the global ecological footprint per capita should equal biocapacity per capita – which is 1.7 global hectares. So Luxembourg, in number 1 position in the table, is currently demanding nine times more resources and waste disposal facilities than the planet can either regenerate or absorb into the atmosphere. HICs are living beyond their ecological means.

Rank	Country	Ecological footprint per capita (in global hectares)
1	Luxembourg	15.8
2	Australia	9.3
3	USA	8.2
4	Canada	8.2
5	Singapore	8.0
6	Trinidad and Tobago	7.9
7	Oman	7.5
8	Belgium	7.4
9	Sweden	7.3
10	Estonia	6.9

🔺 **Figure 2** *The ten countries with the world's largest ecological footprints per capita in 2012*

Sustainable living

Can globalisation and sustainable development exist side by side? The following solutions are intended to make a difference.

1 Responding locally – Transition towns

Some local groups and non-governmental organisations (NGOs) promote local sourcing of goods to increase sustainability (re-localisation). Totnes in Devon (population 8000) was the world's first '**Transition town**'. Now a global network exists, using the Internet and social media to spread the idea of '**Transition**'. By 2016, Transition had become a movement of communities in 50 countries, attempting to reduce their carbon footprints and increase their resilience (the ability to withstand and adapt to shocks). At its heart is a belief that healthy local economies are vital to healthy communities, and that change can be driven by ordinary people. It promotes:

IN TRANSITION
From oil dependence to local resilience

◣ *Figure 3* The cover of 'The Transition Handbook'

- reducing consumption by repairing or reusing items

- reducing waste, pollution and environmental damage

- meeting local needs through local production, where possible (e.g. purchasing food from local growers).

In 2012, Bristol introduced the 'Bristol Pound' (a community currency – shown in Figure 4). It aims to encourage people to spend in local, independent businesses in Bristol – rather than in national chain stores or TNCs. However, strategies like this also threaten global economic growth, because they reduce the demand for new items from overseas. Most developed economies actually rely on a throw away culture for their economic growth.

Transition brings economic and social – as well as environmental – benefits. Every £10 spent in local businesses is actually worth £23 to the local economy – through what economists call the 'Multiplier Effect' (e.g. when local employees and suppliers are paid). In that way, local people gain employment as well as involvement in the local economy. However, that same £10 spent in a supermarket is worth only £13 locally, because the supermarket's profits are returned to its Head Office (which might be in a different country, e.g. German supermarkets Aldi and Lidl).

◤ *Figure 4* The Bristol Pound being used in a Bristol deli

There are also disadvantages to the Transition approach. Some services (e.g. transport) are co-ordinated centrally, so it's hard to influence them. It's also been argued that doing Transition in a big city like London could be difficult. However, there are currently about 40 community-scale Transition initiatives across London, such as the Crystal Palace Food Market, Brixton Pound and Brixton Energy.

2 Fairtrade

The WTO policy of trade liberalisation pitches small businesses against much larger rivals, and can mean that factory workers (or the growers of commodities like coffee) receive small shares of a product's value (see Figure 5). **Fairtrade** aims to return a bigger proportion of the revenue to producers or growers. It's an independent not-for-profit organisation, which certifies products by issuing the FAIRTRADE Mark as a guarantee that they are ethically produced and that a fair price has been paid to the producers.

In 2009, Starbucks served its first Fairtrade coffees. It claims to help farmers improve their coffee quality and environmental sustainability, as well as supporting thousands of families. Starbucks' Fairtrade Certified Espresso Roast is sourced from small farms in Guatemala, Costa Rica and Peru. But, in 2014, only 8.5 percent of their coffee was Fairtrade certified. In 2015, Ethical Consumer (a not-for-profit co-operative that promotes ethical shopping) gave coffee retailers scores out of 20 for using products such as Fairtrade. The marks received were generally not high (see Figure 6).

3 Ethical shopping

The UK's retail sector is increasingly aware of ethical issues in shopping. For example, M&S now sells only Fairtrade teas and coffees, plus naturally dyed clothes and fabrics (made by small businesses overseas) in order to reduce its carbon emissions. All supermarket chains now display **ethical shopping** credentials for those who want to make a difference. Local produce – such as Farm Aware meat and named suppliers on food products – is returning to the shelves, and farmers' markets are commonplace.

But are there any downsides to ethical shopping?

◆ Buying organic destroys more forests – less use of fertilisers and pesticides means that more land is needed to produce the same amount.

◆ Fairtrade does raise farmers' incomes, but it also increases potential overproduction – causing prices to fall, which leaves farmers no better off.

◆ Buying local food minimises 'food miles' and helps the local economy – but most consumers still use cars to go shopping and, therefore, end up using more energy than home deliveries from supermarkets.

◆ Growing cash crops, even under fairtrade conditions, can mean that some farmers end up not growing enough food to feed themselves and their families.

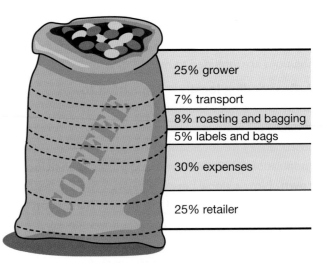

25% grower
7% transport
8% roasting and bagging
5% labels and bags
30% expenses
25% retailer

⚫ **Figure 5** *How the revenue from a bag of coffee beans is distributed*

Brand	Score
Soho Coffee Shops	11.5
Esquires Coffee Houses	10.5
AMT Coffee Shops	8.5
Coffee Republic	7.5
Coffee#1	7.5
Love Coffee	7.5
Muffin Break	7.5
Caffe Ritazza	7.0
Costa – Rainforest Alliance coffee	7.0
Puccino's Coffee Shops	6.0
Caffe Nero	5.5
Pret a Manger – organic coffee	5.5
Starbucks	3.5
Harris & Hoole	2.5

⚫ **Figure 6** *Scores out of 20 given to coffee retailers by Ethical Consumer in 2015*

Key word

Ethical shopping – A deliberate choice of products produced under Fairtrade, organic or cruelty-free terms.

Waste and recycling

Local authorities manage the disposal of most of the UK's waste. NGOs such as 'Keep Britain Tidy' also try to alter people's behaviour, e.g. by campaigning for beach and countryside cleanliness.

In 2012, the UK generated 200 million tonnes of waste – half of which came from the construction industry. Households generated 14%. Since 2004, in line with other European countries, waste management and recycling in the UK has shown a steady improvement (see Figure 7).

◆ In 2013–14, the total amount of waste managed by English local authorities was 25.6 million tonnes. That was 9.1% lower than in 2000–01 – in spite of a 9% increase in population in the same period.

◆ In 2014–15, the total amount of waste recycled (including composting) was 43.7% of the total – compared to 12% in 2000.

◆ In 2000–01, 79% of local authority waste was sent to landfill. By 2013-14, that figure had fallen to 31%.

◆ The amount of local authority waste being incinerated has also more than doubled since 2000–01 to 24%.

However, recycling percentages vary between local authorities. South Oxfordshire District Council achieved 67.3% in 2014-15, and nine other councils achieved over 60%. But the London borough councils of Newham and Lewisham were the lowest, with just 18% in 2013–14. Wales had the highest recycling rate of the four UK constituent countries between 2010-2014 – achieving 55%.

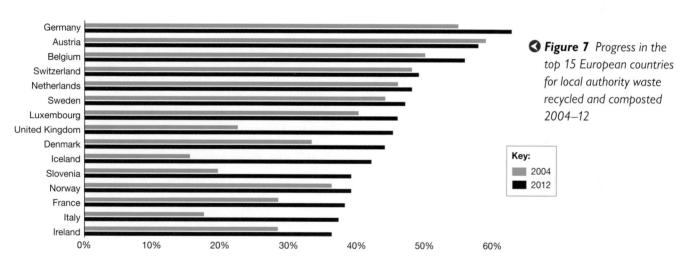

Figure 7 *Progress in the top 15 European countries for local authority waste recycled and composted 2004–12*

Key:
- 2004
- 2012

Over to you

1 a In pairs research the term 'transition network'. Locate the Transition Network website and read its principles.
   b Discuss and explain which principles you support and which you do not.
   c Explain why the transition movement can work in smaller towns like Totnes, but would be a challenge in larger cities.

2 a Devise an investigation into the ethical-shopping and waste-management habits of your local area.
   b Conduct the survey and compare people's **attitudes** (i.e. what they believe in) to their **actions** (i.e. what they do).
   c Write a short report about your findings.

On your own

3 Distinguish between **(a)** transition town and local sourcing, **(b)** Fairtrade and ethical shopping.

4 Draw up a list of advantages and disadvantages of Fairtrade and ethical shopping.

5 Suggest possible reasons for **(a)** the low scores given in Figure 6, **(b)** the variations in Figure 7.

Exam-style questions

 1 Explain how local groups and NGOs can promote local sourcing of food and other goods. *(6 marks)*

2 Assess the actions taken by NGOs and local government in promoting ethical and environmental concerns about unsustainability. *(12 marks)*

Having studied Globalisation, you can now consider the three synoptic themes embedded in this chapter. 'Players' and 'Attitudes and Actions' were introduced on page 147. This page focuses on 'Futures and Uncertainties', as well as revisiting the three Enquiry Questions around which this topic has been framed (see page 147).

3 Futures and Uncertainties

People approach questions about the future in different ways. They include those who favour:

- **business as usual**, i.e. letting things stand. This might involve doing nothing, or only doing what's absolutely necessary when it's unavoidable. This would involve letting globalisation processes continue.

- **more sustainable strategies** towards development, particularly managing landscapes or biomes which are threatened, e.g. rainforests.

- **radical action** which might involve directly challenging globalisation as a means of appropriate development, for example the Transition movement (see Section 4.11).

Working in groups, select two locations (at any scale from continental to local) from this chapter and then discuss the following questions in relation to each:

1 How far has globalisation created unintended consequences – economically, socially, environmentally?

2 Is there evidence to suggest that anti-globalisation movements (e.g. 38 Degrees, UK Uncut) are threatening globalisation processes?

Revisiting the Enquiry Questions

These are the key questions that drive the whole topic:

1 What are the causes of globalisation and why has it accelerated in recent decades?

2 What are the impacts of globalisation for countries, different groups of people and cultures and the physical environment?

3 What are the consequences of globalisation for global development and the physical environment and how should different players respond to its challenges?

Having studied this topic, you can now consider answers to these questions.

Discuss the following questions in a group:

3 Consider Sections 4.1-4.4. What have been the main drivers and players leading to the expansion of the globalised world?

4 Consider Sections 4.5-4.8. To what extent do you think the benefits of globalisation outweigh the problems created?

5 Consider Sections 4.9-4.10. How far is globalisation being challenged as a model of economic development? Do you think that it's likely to be more seriously challenged in future?

Books, music, and films on this topic

Books to read

1. *No Logo* by Naomi Klein (1999)

   A book written about the 'alter-globalisation' movement, assessing negative aspects of globalisation (e.g. worker exploitation and corporate censorship). It explores the benefits of bottom-up developments involving local stakeholders, rather than top-down from TNCs.

2. *Fugitive Denim* by Rachel Louise Snyder (2008)

   A book assessing the stages in the clothing industry – a classic example of how globalisation affects people all around the world. Throughout each stage, the author questions the morals and issues of a globalised world.

3. *Globalization and the Environment: Capitalism, Ecology and Power* by Peter Newell (2012)

   This book analyses the link between key aspects of globalisation and their impact on the environment. It takes into account the management of globalisation and whether this is sustainable.

Music to listen to

1. 'Price Tag' by Jessie J ft. B.o.B (2011)

   This song's lyrics examine a major aspect of globalisation – consumerism and how its obsessive nature feeds parts of globalisation.

Films to see

1. *The Way Home* (2002)

   This film looks at how a young boy becomes accustomed to life in an interconnected city. Then when he goes to live in a rural village with his grandmother, he's in for a shock. It also shows the disparity in places that are turned on and off by globalisation.

2. *Capitalism: A Love Story* (2009)

   A documentary film that investigates corporate greed and the economic crisis that hit in 2008, which affected economies all around the world. It assesses the price that Americans pay for their country's obsession with capitalism.

3. *Bombay Calling* (2006)

   A film about a younger worker in an Indian call centre. This is a brilliant illustration of the economic benefits to host countries of globalisation, but the film also shows the loss of traditional culture as people, especially younger people in the developing world, become more westernised.

Chapter overview – introducing the topic

This chapter studies regeneration – its characteristics, processes and impacts. In particular, it explores why, for many, regeneration is so contested.

In the Specification, this topic has been framed around four Enquiry Questions:

1 How and why do places vary?

2 Why might regeneration be needed?

3 How is regeneration managed?

4 How successful is regeneration?

The sections in this chapter provide the content and concepts needed to help you answer these questions.

Synoptic themes

Underlying the content of every topic are three synoptic themes that 'bind' or glue the whole Specification together:

1 Players

2 Attitudes and Actions

3 Futures and Uncertainties

Both 'Players' and 'Attitudes and Actions' are discussed below. You can find further information about 'Futures and Uncertainties' on the chapter summary page (page 244).

1 Players

Players are individuals, groups and organisations involved in making decisions that affect rural or urban regeneration. They can be individuals, national / international organisations (e.g. IGOs), national and local governments, businesses (from small operations to the largest TNCs), pressure groups and non-governmental organisations.

Players that you'll study in this topic include:

- Sections 5.3, 5.4 and 5.8 – How the **UK government** planned economic change in both London and the UK as a whole, by embracing globalisation. You'll also see how **large TNCs**

(particularly those in the knowledge economy and property development) supported market-led regeneration.

- Section 5.4 – How **government** policies towards migration have brought cultural changes.

- Sections 5.6, 5.9, 5.10, 5.12 and 5.13 – How **local communities and government** drive policies to encourage regeneration. In some cases, partnerships with **IGOs** (e.g. the EU) influence regeneration. Section 5.12 shows the balance between the **private and public sectors** in this process.

2 Attitudes and Actions

Actions are the means by which players try to achieve what they want. For example, those promoting regeneration see it for its business potential. However, others see inequalities resulting from property regeneration and gentrification.

Actions largely concern the following issues:

1 **Economic versus environmental.** Key to regeneration and change are the economic benefits that can result from a change of environmental quality (see Section 5.8). Sections 5.9 and 5.10 explore the actions of local and national governments in achieving regeneration goals.

2 **Economic versus social issues.** Regeneration involves social change, as different groups migrate to newly regenerated and gentrified areas. However, gentrification and regeneration do not always sit well with host populations, some of whom may be excluded (see Sections 5.4 and 5.12).

3 **Community engagement.** Section 5.6 explores the actions taken by local communities in bringing about change.

In groups, consider and discuss the following questions:

1 In the balance of economic, social and environmental changes resulting from regeneration – how far do benefits outweigh problems?

2 How far should governments play a role in the regeneration process?

A good day for Yorkshire!

It's 5 July 2014. Rick, a salesman from Leeds, is up early. He and some friends have spent the night in the small Yorkshire Dales town of Hawes. They're up early to get a good viewing point. In a few hours the first cyclists will appear – travelling through the Yorkshire Dales on Day 1 of the Tour de France (one of the world's biggest sporting events). The race is starting outside France this year, as it often does; beginning in Leeds, it will head north and circle its way around the Yorkshire Dales over some of northern England's most challenging hills. Figure 1 shows the full route of the three UK stages – Day 2 sees it ending in Sheffield. It's a great time for Yorkshire!

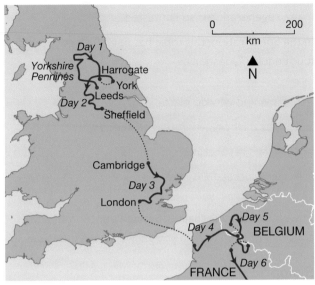

▲ **Figure 1** *The start of the Tour de France in 2014*

A sense of belonging

Where Rick comes from is important to him. Born in Leeds, where he still lives, he proudly declares himself to be a Yorkshireman. July 2014 was a great time to come from Yorkshire, and people there lost no opportunity in announcing their heritage! Those who had moved away also declared their Yorkshire roots with pride. Looking at its landscapes in Figure 2, it's easy to see why.

What we call 'Yorkshire' is actually four counties: West, South, and North Yorkshire, plus the East Riding. Over the years, the Yorkshire region has varied in size – depending on boundary changes by different governments – but at its core is a sense of identity; if you think you're from Yorkshire, then you are!

▼ **Figure 2** *Tour de France cyclists in Yorkshire's Pennine landscape*

Identifying with places

How do we become attached to a place and develop a strong sense of identity with it? Some factors are personal, such as our families, friends, and things we've done there growing up. But what makes Leeds different from London or Cornwall? Much of what makes a place distinctive is geographical in nature, and includes its:

◆ physical landscapes, which result from geology and landscape processes (such as erosion).

◆ human landscapes, which are often linked to physical factors. Local geology has produced Yorkshire's gritstone buildings, e.g. Leeds Town Hall (see Figure 3).

◆ economic past. Leeds Town Hall dates from the city's wealthy Victorian industrial past. Many of West and South Yorkshire's towns and cities have grand civic buildings like this.

Other things which make a place distinctive include:

◆ its religious past. Places of worship – churches, mosques, temples or synagogues – form the heart of many places.

◆ its food and drink, with regional specialities such as Yorkshire pudding.

◆ how it's portrayed in the media. Books, film and TV often characterise places – e.g. *Coronation Street* (Manchester) or *Doc Martin* (Port Isaac in Cornwall).

⬆ **Figure 3** *Leeds Town Hall – typical of the imposing civic buildings of many northern towns and cities, built during their Victorian industrial heyday*

Rebranding places

Place image is also used to encourage economic growth. The 2014 Tour de France was worth a lot to Yorkshire's tourism industry. The Yorkshire region has been through hard times since the 1970s – with the closure of most of its coalmines and textile mills, as well as many of Sheffield's famous steel mills. What's left is a region that sometimes struggles to find an economic future – hence the importance of **regeneration** (see the key word box). Part of that process involves changing people's image of a place from somewhere that's run down to somewhere that has economic potential and excitement. This change of image is called **rebranding**. Regeneration, and the ways in which places change and rebrand, are what this topic is all about.

Regeneration – Redeveloping former industrial areas or outdated housing to bring about economic and social change. Regeneration plans focus on the fabric of a place: new buildings and spaces with new purposes.

Rebranding – Ways in which a place is deliberately reinvented for economic reasons, and then marketed using its new identity to attract new investors.

Over to you

1 Draw a spider diagram to explain how different places in Yorkshire might have benefited from the visit of the Tour de France in 2014.

2 Compare Rick's image of Yorkshire with the image of your own locality. In pairs, devise a six-slide presentation, using both text and photos, to describe your own locality and its characteristics as follows: **(a)** physical and human landscape, **(b)** population and religions, **(c)** buildings, **(d)** distinctive local food and drink options, **(e)** people's image of it, and **(f)** any other characteristics of your choice.

On your own

3 Explain the difference between 'regeneration' and 'rebranding'.

4 Explain the extent to which the place where you live has strong characteristics, and how these may have evolved.

5 **a** Use this section and further Internet research to describe Yorkshire's physical and human features.

b Explain how these could be used to rebrand Yorkshire in people's minds.

c Identify and explain ways in which your local area could be rebranded to attract investment.

In this section, you'll learn how economies can be classified in different ways and how economic activity, social factors and quality of life vary.

I love the north, but …

Roisin works in Sheffield. She's a recent graduate from Sheffield University and loves the city. But like many graduates in northern cities, she's finding it difficult to get a well-paid job. Many of her friends have moved to London, and are earning over £35 000 a year. Should she join them, or stay in Sheffield? It's a question that many in the north have faced before.

Understanding economic change

The drift to London is typical for many young graduates in the twenty-first century. It's arisen because of two long-term economic changes in the UK:

◆ The decline of the primary and secondary sectors.

◆ The growth of the tertiary and quaternary sectors.

These changes are illustrated in Figure 1.

BACKGROUND

Classifying employment

Classifying 31 million jobs in the UK is complex! The most common method is by economic sector:

◆ **Primary sector** – producing food crops and raw materials (e.g. farming, mining, forestry)

◆ **Secondary sector** – manufacturing finished products

◆ **Tertiary sector** – providing services, either in the private sector (e.g. retail, tourism) or public and voluntary sectors (e.g. health care, education)

◆ **Quaternary sector** – providing specialist services in finance and law, or industries such as IT and biotechnology.

But these terms do not fully describe a person's job, position, or their hours. To overcome this, there are other classifications, depending on whether jobs are:

◆ **full-time** (35 hours per week) or **part-time** (under 35 hours)

◆ **temporary** or **permanent**

◆ **employed** or **self-employed**.

Date	Primary	Secondary	Tertiary	Quaternary
1980	0.89	8.9	14.8	2.8
2015	0.48	5.1	22.1	5.98
Trend	Down by 45%	Down by 43%	Up by 49%	Up by 113%

◀ **Figure 1** *Changes to primary, secondary, tertiary and quaternary employment in the UK between 1980 and 2015 (in millions of workers)*

Primary and secondary sectors decline …

In the 1980s, the Conservative government planned changes to the UK economy – often called the **old economy**.

Goods produced by primary (e.g. coal mining) and secondary (manufacturing) industries in the UK were often more expensive than the equivalent goods produced overseas. Reasons for this included that:

◆ British coal was located deeper underground and was more expensive to mine than in many other countries.

◆ UK wages were often higher than those overseas – making British products more expensive. The growth of manufacturing in Asia, with its cheaper labour costs, led to cheap imported goods.

Large numbers of UK mines and manufacturing plants closed during the 1980s, creating derelict land (like the example in Figure 2). Employers like these had provided mainly full-time, well-paid jobs. The closures particularly affected northern England, the Midlands, Wales and Scotland, where unemployment soared.

▲ **Figure 2** *When their industries closed in the 1980s, many former industrial areas ended up like this*

... while tertiary and quaternary sectors grow

To replace the lost jobs in the primary and secondary sectors, the government encouraged the growth of a new **post-industrial economy** – sometimes called the **new economy**. This growth took place in the:

◆ tertiary sector, particularly in tourism and retail. These areas grew because of higher incomes, cheaper air travel and increased car ownership. Some parts of the UK also sought to **rebrand** their past to create a new image (like Liverpool in Figure 3). But, unlike the former industrial jobs, many of these new jobs are seasonal (tourism in summer and retail at Christmas), and are often low paid and part-time.

◆ quaternary sector, which has shown the fastest growth (see Figure 1). This is usually called the **knowledge economy**, and it provides highly specialised jobs that use expertise in fields such as finance, law and IT. The biggest of these fields is banking and finance; international banks in London generate huge wealth.

Quaternary industries can locate anywhere, so they're described as **footloose**. Their locations are often chosen according to financial incentives (e.g. low tax rates) and **connectivity** (good transport links and superfast broadband are vital). Growth in the quaternary sector has concentrated in London's Docklands (see Section 5.3). Quaternary salaries are much higher than the average, so divisions in wealth between the north and south of the UK have widened.

▶ **Figure 3** *Liverpool docks – an example of the regeneration of former wharves and warehouses as luxury apartments, shops, cafes/restaurants, and a marina – a different 'brand' altogether*

Employment, incomes and inequality

Variations in employment lead to significant economic and social inequalities. Figure 4 shows different job types, the percentage of UK employees in each type, and their mean weekly wage. Some data are not surprising – you'd expect a company director to earn more than a worker in an elementary occupation (such as a cleaner), for example. However, the mean weekly wage of 'managers, directors and senior officials' is about four times higher than that of workers in 'sales and customer services', which is more surprising. Based on this are five inequalities, discussed on the next two pages.

Job type	% of all UK employees	Mean weekly wage
Managers, directors and senior officials	9.8%	£907.90
Professional	21.0%	£700.80
Associate professional and technical	14.3%	£610.00
Skilled trades	7.5%	£486.50
Process, plant and machine operatives	5.8%	£440.70
Administrative and secretarial	12.4%	£355.80
Caring, leisure and other services	9.6%	£266.50
Elementary occupations (no training required, e.g. a cleaner)	8.0%	£241.70
Sales and customer services	11.6%	£237.10
	UK average	**£502.20**

◀ **Figure 4** *Job types in the UK, ranked by weekly wage in 2013. The highest income categories are in green, and the lowest in blue.*

Socio-economic inequalities

Job type is the cause of many other inequalities. London's position as the UK capital means that incomes there are far higher than elsewhere. This leads to other inequalities in areas such as health and education.

1 Regional inequalities

Incomes vary regionally. Figure 5 shows the distribution of average incomes by UK economic region. Incomes in London are the UK's highest because:

◆ it's the capital, so incomes are higher in senior positions in government, the civil service and in major company headquarters

◆ those who work in the Docklands-based knowledge economy have higher incomes than the average.

58% of jobs in London occur in the three highest-income categories from Figure 4 on page 197, and only 22% of jobs in the lowest three categories. In areas where the knowledge economy is weaker, such as Yorkshire and Humber, the equivalent percentages are 41% and 31% respectively.

2 Variations in quality of life

Do higher incomes increase happiness? Workers in London and South East England have higher average incomes, but housing and many other costs are also higher there. In 2015, Hampton's estate agents produced a 'happiness map' (Figure 6). The map uses Geographical Information Systems (GIS) to show several combined indicators. It shows where people feel happiest (according to the government's 'Life Satisfaction Index') and where houses are cheapest/most affordable. 'Life satisfaction' means how people assess their quality of life (e.g. their work-life balance). Light-shaded areas in Figure 6 show the UK's happiest and most affordable places – and it looks as if housing affordability is more important to happiness!

3 Occupation and life expectancy

A person's occupation has social consequences. By recording parental occupation at birth and a person's occupation when they die, there is a national database which helps to assess relationships between life expectancy and job type. It uses a different job classification from the one in Figure 4, because it classifies people by role (e.g. managers, skilled worker, etc.). But its results are quite startling, as Figure 7 shows. For example, there is a difference of 5-6 years (depending on gender) between the life expectancy in the highest and lowest occupational groups.

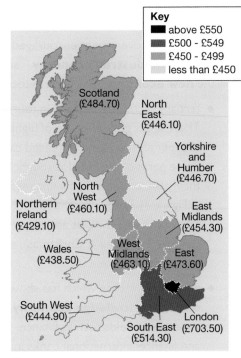

▲ **Figure 5** *Regional variations in mean weekly incomes in 2013*

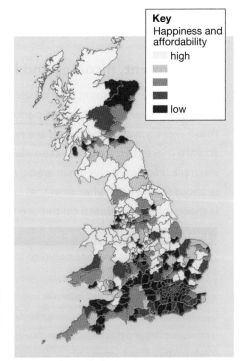

▲ **Figure 6** *A GIS map of happiness in Great Britain*

◀ **Figure 7** *Occupational groups and life expectancy at birth, 2007–2011*

Class	Occupational group	Life expectancy	
		Men	Women
1	Higher managerial and professional	82.5	85.2
2	Lower managerial and professional	80.8	84.5
3	Intermediate	80.4	83.9
4	Small employers	80.0	83.5
5	Lower supervisory and technical	78.9	81.9
6	Semi-routine workers	77.9	81.7
7	Routine workers	76.6	80.8
	Unclassified	74.0	78.5

4 Income and health

The same relationship occurs between income inequality and health. Those on the lowest incomes are said to be the most **deprived** (that is, lacking a reasonable standard of living). In the 2011 Census, people were asked to record the state of their health. Figure 8 plots the percentage of those who reported that their health was 'not good', against their level of deprivation (calculated from other parts of the Census). The population has been divided into 'deciles' (or tenths, based on deprivation). The top 10% are the most deprived ('Decile 1' on the graph), while the least deprived are shown as 'Decile 10'. There is therefore a clear link between income, deprivation, and health.

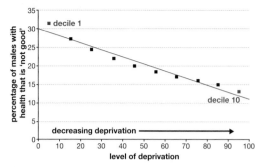

▲ **Figure 8** *Comparing deprivation in males in England and their health*

5 Variations in educational achievement

Figure 9 shows regional variations in the level of good GCSE passes in England, as well as the percentage of those qualified up to university degree level. Again, London has the highest in each column, with North East England the lowest. This does not mean that there is something about London that makes students there more able to pass GCSEs, nor does it mean that schools are poorer-quality elsewhere. The relationship is linked to employment – those with the highest qualifications are more likely to work in London and to move there. Their children are more likely to pass GCSEs with high grades – the result of factors such as paying for extra tuition, or a home culture of doing homework.

	Percentage of students obtaining A/A* grades	Percentage of adults with university degrees
London	25.3	40.5
South East	24.1	32.3
East	21.3	28.1
South West	20.9	29.5
West Midlands	18.4	25.5
North West	19.7	26.6
East Midlands	18.3	25.8
Yorkshire and Humber	17.6	25.4
North East	17.6	24.3

▲ **Figure 9** *Variations in educational achievement by economic region in 2013 (obtained from ONS regional surveys)*

Over to you

1 a In pairs, list the advantages and disadvantages of reducing primary and secondary employment and expanding the tertiary and quaternary sectors.

   b Give examples of jobs in different occupational categories in Figure 7.

 2 Describe and suggest reasons for the regional variation in weekly income in the UK (Figure 5).

3 Describe and suggest reasons for the relationship between:

   a income (Figure 5) and 'happiness' (Figure 6)

   b occupational groups and life expectancy (Figure 7)

   c deprivation and health (Figure 8)

   d income (Figure 5) and educational attainment (Figure 9).

On your own

4 Define the following terms: old economy, new economy, knowledge economy, deprivation.

5 Using the data categories in this section, research deprivation data for your local area. Describe and explain what makes your area similar to or different from the rest of the UK.

6 How healthy is it for one part of a country to have much higher socio-economic characteristics than the rest?

Exam-style questions

 1 Explain how employment changes have affected your local place. *(8 marks)*

A 2 Assess the extent to which economic activity, social factors and quality of life have affected your local place. *(12 marks)*

Global changes, local places

Every day, huge container ships arrive at Britain's newest container port – London Gateway in Essex (see Figure 1). This new port is 30 km east of Central London, and it can cope with the world's largest container ships. In the 1970s, the development of container ships signed the death warrant for the original Port of London (to the east of Tower Bridge), see Figure 2. The Thames simply wasn't deep enough that close to Central London to accommodate them. From being Europe's largest port in 1900, London's dock facilities have had to shift further and further downstream (see Figure 2 inset). The new facilities are still one of Europe's largest ports, but they bear no resemblance to the old port. It's one of many ways in which global changes (in this case, containerisation) affect local places.

London's changing East End

The last of London's original East End docks closed in 1981. Until the early 1970s, they were the UK's largest docks. Living close by were dockworkers and their families. They were poorly paid, and much of their housing was social housing rented from local councils.

But as container ships became larger, and huge cranes replaced the traditional workers to unload them, the docks fell into disuse. Their closure was devastating!

◆ Between 1978 and 1983, over 12 000 jobs were lost. In the 1981 Census, over 60% of adult men were unemployed in some parts of East London.

◆ The riverside downstream from Tower Bridge consisted of abandoned docks and derelict wharves (see Figure 3) – not a good image for a major city.

◆ Nearby, industries in East London's Lea Valley (see Figure 2) also closed, because they needed the port to import raw materials and export finished products.

The population of the area declined, as people left to find work. Between 1971 and 1981, the population of the East End fell by 100 000.

▲ **Figure 1** *A container ship arriving at the new London Gateway container port in Essex*

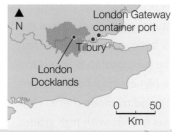

▼ **Figure 2** *London's port facilities, showing the original docks to the east of Tower Bridge (main map, in orange) and the new docks further downstream at Tilbury and London Gateway*

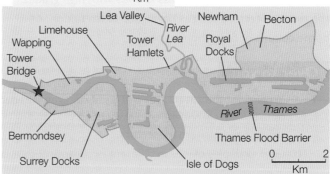

▼ **Figure 3** *Dereliction in London's Docklands (Canary Wharf on the Isle of Dogs) in the early 1980s*

Re-imaging inner cities

Similarly, high unemployment in cities such as Manchester, Leeds and Liverpool (see Section 5.2), gave inner-city areas a poor image with little economic potential. The resultant lack of investment in these communities led to falls in the quality of the living environment, whilst crime rose considerably between 1975 and 1985 (see Figure 4). High levels of deprivation, combined with ethnic and community tensions, led to riots in Liverpool (Toxteth), Leeds (Chapeltown) and London (Brixton) in 1981.

	Burglary	Theft & handling stolen goods	Violent crime	Total crimes
1975	515 429	1 267 674	71 002	2 105 631
1985	866 697	1 884 069	121 731	3 611 883
Increase %	68.2%	48.6%	71.4%	71.5%

⏶ *Figure 4* *Increasing crime in the UK during the era of high unemployment, 1975–85*

The Conservative government reacted by attempting to rebrand inner cities. Starting in 1984, Garden Festivals were held to develop a 'greener' image for inner cities – a process known as **re-imaging**. Later, European Capitals of Culture focused on cultural regeneration in cities (Liverpool won the honour in 2008), and the UK Government now awards City of Culture status – to Derry/Londonderry in 2013 and Hull in 2017.

Regenerating London Docklands

Imagine the potential of 21 km² of available building land – so close to Central London! An area of that size and importance needed a plan as part of a local and national strategy for dealing with dereliction and unemployment. The job went to a government agency, the LDDC (London Docklands Development Corporation). Formed in 1981, its focus was to encourage growth. It brought together key **players**, such as:

♦ property owners keen to purchase land (the former Port of London was government-owned)

♦ architects

♦ construction companies

♦ investors.

The process was known as **market-led regeneration** – leaving the private sector (i.e. the free market) to make decisions about the future of Docklands. The LDDC was given planning powers that by-passed local councils in Newham, Tower Hamlets and Greenwich. As long as planning permission was granted by 1991, companies could obtain tax breaks on new buildings. These tax incentives were designed to attract investors – and they still apply.

The LDDC focused on three things: economic growth, infrastructure and housing (see the next two pages).

⏷ *Figure 5* *New housing in Millwall, which was used to help re-image London Docklands when it was built in the 1980s and early 1990s*

201

1 Economic growth

The LDDC's flagship project was Canary Wharf, now London's second **Central Business District** (CBD) – see Figure 3 (before) and Figure 6 (after). A huge transformation took place in land use and employment, with high-rise office buildings (designed to stimulate quaternary employment) replacing docks and industry. The drive behind the regeneration was to create high-earning jobs. High earners, it was argued, would generate other jobs in a 'trickle down' effect to poorer communities.

Now, companies in Canary Wharf include investment banks (e.g. Barclays, HSBC) and companies in the knowledge economy (see Section 5.2). Every day, 100 000 commuters travel there, as well as 325 000 who work in the City. Employment has grown, and the East End is no longer one of the UK's most deprived areas – but poverty is still present there. In 2012, 27% of Newham's working population earned less than £7 per hour – the highest percentage of any London borough.

2 Infrastructure

Accessibility and connectedness have both been key to the success of the Docklands regeneration. New transport developments (or **infrastructure**) have included:

◆ extending the Jubilee Line on the London Underground

◆ developing the Docklands Light Railway (a surface light rail network, covering most of Docklands)

◆ building new roads, such as the Limehouse Road Link

◆ creating London City Airport (5 km from Canary Wharf), to provide access to the City and Canary Wharf for business travellers.

3 Population and housing

The Docklands population has also been transformed:

◆ Many older people have moved out – often retiring to the Essex coast, e.g. to Southend-on-Sea.

◆ The older residents have been replaced by a much younger generation; the average age of Newham's population was 31 in 2011, compared to a UK average of 40. Newham's age-sex structure is shown in Figure 7.

◆ The ethnic composition of the East End has always been diverse, but large-scale immigration since 2000 has increased its mix. Newham is now London's most ethnically diverse borough.

Figure 6 *Canary Wharf after redevelopment*

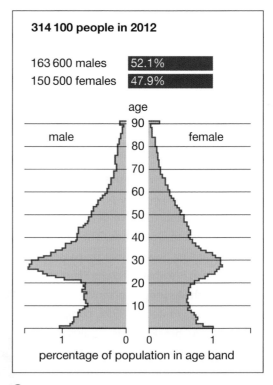

314 100 people in 2012

163 600 males   52.1%
150 500 females   47.9%

Figure 7 *The age-sex structure of Newham in 2012*

Before regeneration, most housing in Docklands was rented from local councils at low-cost. Two big changes then occurred:

◆ In the 1980s, the government introduced the Right to Buy scheme, which gave those living in council housing the right to buy it at a reduced price. A lot of East End housing then transferred from the public to the private sector – greatly reducing the amount of social housing available. Nearly half of the housing in East London boroughs is now in the rented private sector. Lower-income people living in social housing have been forced out.

◆ One of the aims of the Docklands regeneration was to increase housing supply – not of social housing, but of housing in riverside locations where the docks used to be. This led to **gentrification** – what has been called 'the march of the middle classes into East London'. Riverside property in old warehouses became desirable and very expensive.

As a result, traditional communities have been broken up and there has been a major change in how the East End 'feels', as the reactions in Figure 8 show.

Some problems remain

Regeneration has not removed deprivation from Docklands; high-income earners are in the minority.

◆ Those in poor health are often unable to work, and are concentrated in what remains of low-cost social housing. This has led to high deprivation in Tower Hamlets and Newham (see Figure 9).

◆ The borough of Tower Hamlets had the lowest average life expectancy in London in 2012 (77 years).

Over to you

1 In pairs, draw a table to show the economic, social and environmental achievements and problems created by the regeneration of Docklands.

2 Assess how significant the following have been in changing East London: physical factors, accessibility, the role of planners, and population change.

On your own

3 Define the following terms: containerisation, dereliction, regeneration, re-imaging, deprivation, infrastructure.

4 Type 'Index of Multiple Deprivation for Newham' into Google. Identify **(a)** the main type of deprivation in East London, **(b)** how deprivation levels can vary.

5 Research an example of regeneration in your local area. Write a 500-word case study of its purposes and its impacts, and include illustrations.

Key word

Gentrification – A change in social status, whereby former working-class inner-city areas become occupied and renewed by the middle classes.

> *Essex is now the home of the East End, as all the original East End people have moved out – Romford Market is where you will find true East End people. Not Newham or Tower Hamlets!*

> *My great-grandmother came from Bethnal Green, and moved from working in London to Tilbury Docks. I think it's a good thing their East End has gone – it was unremitting poverty.*

⬆ **Figure 8** *How local people see the East End now*

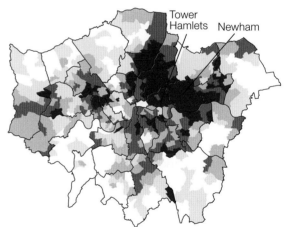

⬆ **Figure 9** *Deprivation across London's boroughs and wards*

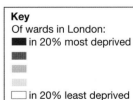

Key
Of wards in London:
■ in 20% most deprived
▨
▦
░
☐ in 20% least deprived

Exam-style questions

AS 1 Explain how increasing connectedness has shaped the economic and social characteristics of a place you have studied. *(8 marks)*

A 2 Assess how far past and present connections have shaped the economic and social characteristics of a place you have studied. *(12 marks)*

A sense of belonging

Do you feel that you 'belong' in the place where you live or study? How many people do you know there? The following passage comes from a landmark study of an East London community in the 1950s by Michael Young and Peter Willmott. It was based in Bethnal Green (see Figure 1) – about 2 km east of the City of London. One of the people in their research was a Mrs Landon. In the following passage, she talks about shopping locally and the number of people she sees that she knows. How much does this sound like where you live?

Figure 1 *Chilton Street in Bethnal Green in the 1950s, near where Young and Willmott carried out their research*

> 'Some days', says Mrs Landon, 'you see so many you don't know which to talk to.' She kept a record over a week of all the people she saw in the street and whom she considered herself to 'know'. There were 63 people in all, some seen many times and 38 of them relatives of at least one other person out of the 63'.

From Young and Willmott (1957) 'Family and Kinship in East London'

Geographers and sociologists argue that **centripetal forces** were at work here – in other words, forces that drew people together. The Bethnal Green area had a strong sense of community. One of the most significant factors was that of family ties. Sons worked in the same firms as their fathers, daughters as their mothers. Many had a large family network; in the table in Figure 2, the 45 couples taking part in the research had 1691 relatives living in Bethnal Green between them (excluding their parents)!

	Number of relatives, excluding parents, living in Bethnal Green					
	No relatives	1–4	5–9	10–19	20–29	30 and over
Number of couples (45)	4	8	12	9	7	5

Figure 2 *Number of relatives (other than parents) living in Bethnal Green*

It's all change

If Bethnal Green in the 1950s sounds different from where you live, you're not alone! The East End of old has almost disappeared. Change has affected everywhere from Cornwall to Cumbria. Few places are as they were in the 1950s – with close family and friendship ties, local work and a strong sense of community. Although changes are felt most strongly locally, they often have causes well beyond the locality.

The causes of change

The example of London's East End shows that what happens in one place can often result from decisions made elsewhere. Three changes have profoundly affected places in the UK: globalisation, employment change and inward migration. Each has largely forced people apart – what geographers and sociologists call **centrifugal forces**.

1 Globalisation

Globalisation resulted in manufactured goods being produced more cheaply overseas. This affected manufacturing in London and elsewhere (see Section 5.2). The container revolution in shipping also badly affected East London. The closure of the original Port of London led to internal migration, as people left their close family networks to find work elsewhere. Population characteristics changed, as one population left and another replaced them. This led to the break-up of communities and families. Bethnal Green's population now consists of many people who work in London's global knowledge economy. They're out at work all day, and are more likely to eat in restaurants than shop in local markets, so – unlike Mrs Landon opposite – they know few people where they live.

⬆ **Figure 3** A three-bedroom flat for sale in Chilton Street in Bethnal Green in 2015 – price £1 million! Compare this with the same street in Figure 1.

2 Employment change

Employment change from the old economy to the new has also had profound impacts. More people are now in higher-income jobs; in the 1951 Census, 18% of the UK population had professional or managerial occupations, whereas in 2011 it was 31%. More people buy their own property now – and also invest in it – so many urban areas have **re-urbanised**. Inward migration, gentrification and regeneration have revitalised many places. But newcomers have also displaced existing residents, because house prices and rents have risen so much that they can no longer afford to live there (see Figure 3). Many have therefore been forced away from places where they grew up.

> **Key word**
>
> **Re-urbanisation** – A flow of people back into cities to live (reversing decades of population decline)

3 Inward migration

A growing economy and ageing population has led to a need for overseas migrants to provide workers. Inward migration changes the character of places. Former residents may be less likely to identify with their local area once they leave. However, inward migration also creates new identities. Brick Lane in East London (Figure 4) has seen waves of migrants escaping persecution, e.g. Protestants from France in the seventeenth century and Jewish people in the nineteenth and twentieth centuries. Now it's home to a large Bangladeshi community.

⬆ **Figure 4** Brick Lane in East London

Investigating change

Investigating change in the place where you live or study is fertile territory for geographical fieldwork! There are many ways of collecting data – both quantitative and qualitative – to observe, measure and record changes happening locally.

Quantitative approaches to studying places

Quantitative data are sometimes called 'hard data' – collected on scientific principles (e.g. objectivity). Most quantitative data consist of tables of figures. They can be **primary** (collected yourself, unprocessed) or **secondary** (collected and processed by another party).

1 Surveys

Quality of life broadly means people's wellbeing in a social or environmental sense, rather than economic.

◆ Socially, it includes health, safety, quality of housing, and the sense of community.

◆ Environmentally, it includes the quality of air, buildings, or noise.

Each can be assessed in the field.

Social data can be collected using a Quality of Life survey (see Figure 5). It can either be done by you (e.g. by surveying different places in a locality), or you can use it as a questionnaire to ask residents.

Environmental factors can be measured in several ways, such as:

◆ a decibel meter. You can download free apps onto your smartphone and record noise levels.

◆ an environmental quality survey (EQS).

Each gives you a numerical score that you can use to judge quality of life.

2 Profiling

Profiling uses Census data to identify the population characteristics of an area. An example is shown in Figure 6 for Canning Town in East London. To obtain Census data, visit the Office for National Statistics (ONS) website, or use a search engine e.g. 'Census data Canning Town'. Many local councils publish Census data and GIS maps to show several indicators, e.g. multiple deprivation.

Ask recipients to score their local area for the following qualities. Each statement should be scored out of 10 (where 10 is best).
Is the area:

1 safe for people to walk about in terms of traffic? ☐

2 safe for people to walk about in terms of the threat of crime? ☐

3 clean and free from graffiti and litter? ☐

4 close to local shops? ☐

5 close to local transport for getting to work? ☐

6 safe for local children (e.g. away from busy roads) to walk to primary school? ☐

7 easy and safe (e.g. good pavement quality) for elderly people to walk to places? ☐

8 well provided for with leisure amenities (swimming pool, pub, community hall)? ☐

9 provided with quality housing (well kept, maintained, with gardens)? ☐

10 provided with a mix of housing for different ages (families, couples, elderly)? ☐

Finally: Add up a total score out of 100 = ☐

▲ **Figure 5** *An example Quality of Life survey*

Ethnicity (%)	
White	31.6
Southern Asian (India, Pakistan, Bangladesh, Sri Lanka, Afghanistan)	10.0
Black or Black British	11.4
General health (%)	
General health: not good or with a limiting long-term illness	13.8
Housing (%)	
Living in social rented housing	36.0
Owner-occupied housing	23.0
Living in a flat	61.1
Employment status: aged 16–74 (%)	
Employed full-time	37.7
Employed part-time	9.7
Permanently sick / disabled	4.0
Unemployed	6.9
Highest qualification level reached (%)	
No education qualifications	23.7
Qualified with university degree	31.0

▲ **Figure 6** *A 2011 Census profile of Canning Town in East London*

Qualitative approaches for investigating places

1 Photos

Photos of places can be a source of fascination, and mean that you can study your own place using images that people are likely to have, or tell you about. An older generation can show photos of changes that have taken place, as well as giving interviews (see below).

2 Interviews

Interviews help to record people's lived experiences. They can be **structured** (where everyone is asked identical questions) or **semi-structured** (with core questions for everyone, but also the freedom to go 'off script' where appropriate). Interviews can be carried out with individuals, or with focus groups. Figure 8 shows a semi-structured interview used by Queen Mary University of London (QMUL) students in Canning Town, and the results that they gathered. You could use this in your own locality.

1 Is there a strong sense of community here?

Yes	33%
No	64%
Not sure	3%

2 How many of your neighbours do you know by name?

< 4	44%
4-9	33%
10 +	19%
Did not want to answer	4%

3 What do you like about living here? *[Partial results]*

Nothing	20%
Family / friends / community	17%
It is a quiet area	12%

4 What do you dislike about living here? *[Partial results]*

Crime	32%
Dirty streets	10%
Lack of facilities	9%
Everything	9%
Noise	8%

5 If you could do one thing to improve the area, what would it be? *[Partial results]*

More facilities, including those for children, and community schemes	30%
Reduce crime / more police	18%
Cleaner streets	16%
More green areas / parks	6%

6 How would you describe living in Canning Town?

> Before, people used to stay and you got to know them … the trouble is now I hardly know anybody any more down this street.

> In cases where the properties are let and the turnover of residents is quite high, you lack that close community – because people tend not to stay in the same place for very long.

> Front gardens are full of rubbish and they don't care where it goes or anything.

▲ **Figure 7** *The Canning Town survey from QMUL*

Over to you

1 In pairs, design two diagrams to show **(a)** the centripetal forces that made Bethnal Green such a close community in the 1950s, **(b)** the centrifugal forces that have occurred since then.

2 Identify how the **population** of your local area is being affected by **(a)** globalisation, **(b)** employment change, **(c)** inward migration.

3 In groups, trial the use of **(a)** a quality of life survey (Figure 5), **(b)** a Census profile (Figure 6), **(c)** an interview survey (Figure 7) to investigate changes in your place. Present your results.

On your own

4 Find Chilton Street in Google Maps. Annotate a copy of the map to explain how its location has meant it has been affected by all the socio-economic changes affecting East London since the 1950s.

5 Identify the evidence in Figures 6 and 7 that suggest that Canning Town has been affected by globalisation, employment change, and inward migration.

6 Explain how far the **character** of your locality is being affected by globalisation, employment change, and inward migration.

Exam-style questions

 1 Explain how the identity of one place you have studied has been affected by change. *(6 marks)*

 2 Evaluate the extent to which economic and social factors have affected the identity of one place you have studied. *(16 marks)*

In this section, you'll learn how and why some places are more economically successful than others.

Sydney, global city

Each weekday, Lizzy waits for the commuter ferry from McMahon's Point on Sydney Harbour's north shore. As commutes go, hers is amongst the world's finest – with the view in Figure 1 to appreciate during the trip. It's ten minutes' journey to Sydney's CBD. Lizzy is in her mid-40s, British-born, and married with two children. The family decided to migrate when her IT employer offered her a position in Sydney as project manager. After a brief Internet search on Sydney and its climate … decision made!

Lizzy and her family are part of a trend. In 2015, Sydney's population reached a record 4.8 million – an increase of 400 000 in four years, due mostly to international migration. Over 1.2 million British-born people now live in Australia (see Figure 2), and 1.5 million (over 30%!) of Sydney's residents were born overseas. That makes Sydney one of the world's most multicultural cities; at least 250 different languages are spoken there.

△ **Figure 1** A classic view of Sydney harbour

Country of birth	Number of residents in Australia	Percentage of the total number of overseas-born Australian residents
UK	1 221 300	5.2
New Zealand	617 000	2.6
China	447 400	1.9
India	397 200	1.7
Philippines	225 100	1.0

◁ **Figure 2** The countries of birth of some of the overseas-born people living in Australia in 2014

'Successful places'

Sydney is part of an economically successful region along Australia's south-east coast, stretching 2000 km from Brisbane to Melbourne. What's made it so?

The Australian Bureau of Statistics identifies Australia's most advantaged areas, using its Index of Relative Socio-economic Advantage and Disadvantage (like the UK's IMD). It found that this region's cities had a large proportion of high-income jobs in the 'knowledge economy' (see Section 5.2). In the 2015 rankings produced by Loughborough University's Globalization and World Cities Research Network, Sydney was one of the world's 'Alpha' cities (just below the top two of London and New York).

Sydney's economy is like that of other World Cities (see page 172) with strengths in the quaternary sector. Its Gross Regional Product (like GDP, but calculated regionally) was US$337 billion in 2013 – Australia's largest.

- With overseas-owned banks and TNCs, it's the leading financial centre for the Asia-Pacific region (see Figure 3).

- In 2011, there were over 450 000 businesses based in Sydney. These included half of Australia's top 500 companies (e.g. Qantas and Westfield) and two-thirds of regional headquarters of global TNCs (e.g. IBM and Vodafone). It makes for a crowded CBD as Figure 3 shows!

- Sydney also has a young economically active workforce, with a median age of 36 (compared to the UK's 41). But it comes at a price; some of Australia's remote rural areas ('the outback') are declining and losing young people and their skills.

- It has low levels of multiple deprivation. Sydney has areas of deprivation, especially in the western suburbs, but employment levels are generally high with above average incomes.

Most knowledge-economy employers are 'footloose' – they are not tied to raw materials, so they can locate anywhere. Sydney attracts businesses partly because of its beaches, harbour environment and climate. Its time zone also allows business trading in the USA and Europe – essential for investment banks. Since 1985, Australia's national governments have embraced globalisation by:

◆ de-regulating banking and finance (allowing any overseas bank to operate there)

◆ focusing the country's inward migration policy on well-qualified professionals. Skills are in short supply and must be supported by inward migration.

But it costs!

Australia's average income is higher than in the UK (see Section 5.2); adult salaries in 2015 averaged AU$82 000 a year (about £40 000, or £770 a week). Household salaries – with more than one earner – averaged AU$145 000 (£70 000) a year.

The average income in Sydney is the world's seventh highest of any city, but it's also very expensive to live there. Because of demand, property in Sydney is extremely expensive (see Figure 4); it ranks between the world's fifteenth and fifth most expensive city (depending on the survey), and is definitely Australia's most expensive. But it also ranks tenth in the world for quality of life, and its residents have the world's second-highest purchasing power (after Zürich in Switzerland).

Figure 3 *The Sydney offices of many TNCs (located along the Darling Harbour waterfront)*

40 of the 43 foreign banks operating in Australia have regional headquarters in Sydney, including:

◆ Banco Santander	◆ Credit Suisse
◆ Bank of America	◆ Deutsche Bank
◆ Bank of China	◆ HSBC
◆ Barclays	◆ Royal Bank of Scotland
◆ BNP Paribas	
◆ Citi	◆ Société Générale

Figure 4 *Houses with a view of Sydney Harbour – but at a price. These harbourside houses cost several million Australian dollars each!*

BACKGROUND

Living in the sun-belt

Sydney's climate makes it one of the world's 'sun-belt' cities – cities with a sunny and warm climate that have experienced rapid growth. Many are in the USA (e.g. Miami), where companies have moved away from the colder north-east. Plenty of workers are willing to join them. This has led to the growth of 'gated communities' for high-income earners – select housing on small estates, serviced by security staff. These estates attract high-income earners, paying high annual service fees (as well as the high price of the property itself).

Meanwhile … in the rust-belt

If you've seen the 1978 American film *The Deer Hunter*, you will also have seen Clairton, Pennsylvania, where it's based. It's a small town of about 6000 people near Pittsburgh. It also has the USA's largest coke-manufacturing facility (coke is processed coal for the steel industry). Its population peaked at about 20 000 in 1950, when the USA was supplying war-damaged Europe with steel to help it rebuild.

And there's the problem. Like many American mining and steel-manufacturing towns and cities, Clairton is part of a region known as the **rust-belt** (see Figure 5). The term refers to the decline in metal manufacturing. It was once the world's biggest heavy industrial region, with coal (mined locally in the Appalachian Mountains), steel (using iron ore from Michigan), and engineering. However, its decline has been continuous since the 1950s – leading to **de-industrialisation**.

There are several reasons for the decline of the rust-belt:

◆ Overseas companies (e.g. in China) produce cheaper coal and steel.

◆ Mining companies have mechanised to cut costs (resulting in job losses).

◆ Lower wage costs in the south-eastern USA have led to the relocation of the steel and car industries (80 000 jobs were lost in car manufacturing in Michigan during the period 1993-2008, while 90 000 were gained in states such as Alabama, Tennessee, Virginia and Texas).

Now the US coal industry only survives because of government subsidies that keep prices low – costing US$2.9bn (£1.9bn) in 2014-15. Without subsidies, companies cannot compete globally and would be forced to close – creating a **negative multiplier** (see Figure 6).

Key word

Negative multiplier – A downward spiral or cycle, where economic conditions produce less spending and less incentive for businesses to invest (therefore, reducing opportunities).

The rust-belt region has undergone economic restructuring. High-income jobs in the primary and secondary sectors have been replaced with low-wage tertiary jobs in retail and local government.

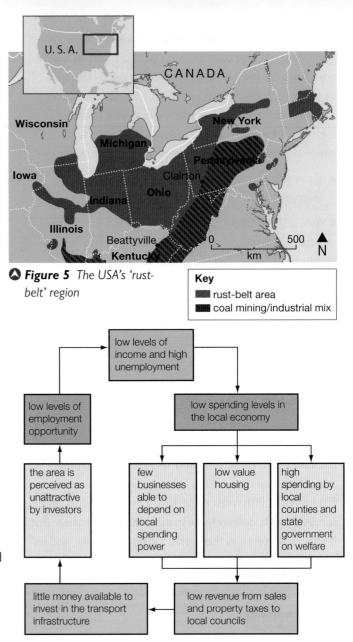

🔺 **Figure 5** The USA's 'rust-belt' region

Key
- 🔲 rust-belt area
- 🔳 coal mining/industrial mix

🔺 **Figure 6** The cycle of decline (negative multiplier) which threatens the American rust-belt

City	Percentage change in population (2000–2012)	Population (2012)
Detroit, Michigan	−26.6	698 582
Gary, Indiana	−23.3	79 170
Youngstown, Ohio	−20.7	65 405
Flint, Michigan	−19.6	100 412
Cleveland, Ohio	−18.6	390 928
Saginaw, Michigan	−17.9	50 763
Dayton, Ohio	−14.8	141 539
Buffalo, New York	−10.9	260 568

🔺 **Figure 7** Population decline in rust-belt cities, 2000–2012

The decline has caused problems such as:

- population decline (see Figure 7) and a brain drain, as people leave to seek work elsewhere

- high unemployment and crime, including drugs

- reduced revenue for councils as consumer spending falls (the USA has a local sales tax paid at the till), but also increased spending commitments caused by those claiming welfare.

Beattyville – a declining rural settlement

Beattyville, Kentucky, is in 'coal country' – the coalfield stretching through the Appalachian Mountains (see Figure 5). The coal industry's decline has caused many social problems. In the 2012 Census, Beattyville was one of the USA's poorest towns. Its population of 1270 (in 2013) lived mainly in trailer homes or log cabins (see Figure 8) – like British 'sink estates', with reputations for poverty and crime.

- Its median annual household income was $12 000 (£8000) – the national median was $54 000 (£27 000).

- Half of its families lived below the poverty line.

- One third of teenagers left high school without graduating. Only 5% of residents had college degrees.

- Homelessness forced families to live together with three or more generations under one roof.

- Drug crime was rife, based on re-selling prescription drugs/opioid painkillers. In 2013, drug overdoses accounted for 56% of all accidental deaths in Kentucky.

- Men's life expectancy was eight years below the US average, at 68.3 years.

It's a priority for regeneration – but how?

❷ *Figure 8 Life hidden away in a forest side road for some of the poorest US citizens in 'coal country', Kentucky*

Exam-style questions

(AS) 1 Explain the reasons for the economic success of one place you have studied. *(8 marks)*

(A) 2 Evaluate the reasons why some places are 'economically successful', while others are not. *(16 marks)*

Over to you

1 Draw a spider diagram to show why Sydney is 'economically successful'. On it, classify the reasons into **(a)** economic, **(b)** social, **(c)** environmental, and **(d)** political. Which reasons seem the most important?

2 In pairs, list factors that could threaten Sydney's future economic success.

3 Summarise Sydney's economic success and its possible future using a SWOT analysis (**S**trengths, **W**eaknesses, **O**pportunities, and **T**hreats) of Sydney as a place in which to live and work.

4 In pairs, research one sun-belt location. Research **(a)** similarities, **(b)** differences between it and Sydney as 'successful' places.

On your own

5 Consider the rust-belt industries (coal, steel, engineering, cars). Draw a mind map to show ways in which these are connected.

6 Research **two** towns/cities from Figure 7. Find out **(a)** which other industries have declined, **(b)** where the population has moved to and why.

7 Copy Figure 6. Annotate and develop it further to show the particular problems facing Beattyville.

8 What are the arguments for and against subsidising the US coal industry?

9 Write a report entitled 'Priorities for regeneration' about problems facing the rust-belt. In it, identify two lessons from sun-belt cities that could be used to regenerate the region.

Opening day at the community shop

On a wet Friday in November 2014, a sizeable crowd gathered in Grampound, a village of 800 people in mid-Cornwall. They were there for the opening of the new community shop (see Figure 1). For most people outside Grampound, the event passed by unnoticed. But this opening was of real significance – and shows how people can engage with a local project they believe in.

Grampound faces many similar issues to other villages. Located in mid-Cornwall, between St Austell and Truro (see Figure 2), it struggles to keep basic services. Villagers who work in Truro or St Austell shop in big supermarkets on their way home from work, rather than in the village. The previous general store closed in 2013, when the owners retired and couldn't find a buyer. Grampound (see Figure 2) still has a small primary school, a pub, a furniture shop and a smoked foods shop, which serves local hotels.

So why start a community shop? Data from the 2011 Census revealed that 25% of Grampound's population was aged over 65, and a third of households consisted of single people. Cornwall also has the lowest mean weekly income in the UK and the highest percentage ownership of old, fuel-inefficient cars. Social isolation (particularly for the elderly), and high travel costs, are a reality for many people – and shopping locally for day-to-day necessities helps to relieve both. A community shop and coffee shop – owned and run by the village – seemed to be the perfect solution to these issues.

Community action

Despite the challenges, Grampound remains a thriving community. In 2008, it won a competition, run by energy company, Calor, to find the UK's best community. Grampound won in the South West region, and came second in the UK finals. In their feedback, Calor commented on the number of clubs and societies for all ages and backgrounds, and a strong sense of belonging and engagement in the village.

It's that sense of engagement that gave birth to the community shop – organised by a group of volunteers.

◆ Out of 280 households in Grampound, 257 became shareholders in the new shop, raising £20 500.

◆ The Prince's Countryside Fund, a national charity with Prince Charles as patron, also awarded £19 000.

◆ Grants from the Parish Council, charities keen on rural development, and from a company which built two wind turbines within the parish, added another £10 000.

In total, the village raised over £50 000!

▲ *Figure 1* *The grand opening of Grampound's community shop*

▲ *Figure 2* *Grampound and its location in Cornwall*

The wider significance

What makes Grampound unusual? Raising money takes place everywhere, but engagement in communities varies. Grampound is a working (as opposed to a tourist) village; similar involvement is harder in villages with second homeowners. The size of the settlement also counts, because people generally engage less in urban areas.

Grampound shows that certain factors determine how engaged people become with the places in which they live:

◆ **Key people** who are willing to stand for elections, raise money, or simply organise activities.

◆ **A range of activities**. In spite of its small population, Grampound has a thriving carnival every September, and its website lists 14 clubs and organisations.

◆ **Politicians**. Grampound's county councillor lives in the village. He organises a monthly local-produce market (see Figure 3), which in 2015 had about 20 stalls. His newsletters give information on everything local, as well as updates on key issues. Residents feel informed and supported.

Grampound's residents also engage politically:

◆ In the 2013 Parish Council Election, the turnout (the percentage of people who voted) was 63%. Nationally, the turnout in Parish Council Elections is below 30%.

◆ The turnout in Grampound for the County Council Election in the same year was still as high as 43%, against a county average of 33%.

Wider engagement

Historically, General Elections have higher turnouts than Local Elections.

◆ That said, the national turnout has fallen from 82.6% in the 1951 General Election, to 76% (in 1979), 71.4% (in 1997), to just 66.1% in 2015.

◆ Turnout also varies geographically; it tends to be higher in rural than in urban areas (see Figure 4).

◆ Particular issues, such as devolution, can also reverse voter apathy. In September 2014, the turnout in the Scottish independence referendum was 84.6%.

Traditionally, the national turnout in the UK for EU elections has been low. In 2014 it was 36%, and has never exceeded 38.5%. In continental Europe, the turnout varied between 90% in Belgium (home of many of the EU's major institutions) and 18% in the Czech Republic. Political engagement reduces as people become more removed from the centre of power – producing **voter apathy**.

Figure 3 *The monthly village produce market in Grampound*

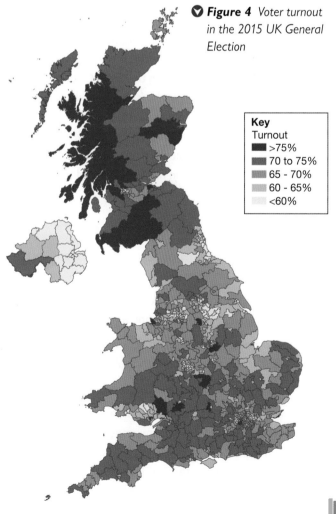

Figure 4 *Voter turnout in the 2015 UK General Election*

Key
Turnout
- >75%
- 70 to 75%
- 65 - 70%
- 60 - 65%
- <60%

Why does engagement vary?

Grampound works as a community because residents engage in its activities. Generally, community engagement varies according to:

◆ **age**. 18.7% of Cornwall's population is aged 65-84, compared to 14.2% nationally, so there are more people with time to devote to community activities. Those aged over 60 are also more likely to vote in elections.

◆ **gender**. More women engage in community work then men. In Grampound, in 2015, women were in the majority on many working groups and committees.

◆ **ethnicity** and **length of residence**. In the 2011 Census, Cornwall had 400 'short-term residents' (i.e. living in the UK for 3-12 months). That's 8 per 10 000 usual residents (the units used in Figure 5). Figure 5 shows the proportion of non-UK-born short-term residents in England and Wales. Half of those were in the UK as full-time students; a quarter were working short-term; and the remainder were either visiting relatives or on extended holidays/gap years. Engagement with communities is unlikely to develop in so short a period.

Deprivation also influences voting and engagement – to quote news website East London Lines, '*the poor don't vote*'. However, when combined with **ethnicity** it *increases* the likelihood of voting! Ed Fieldhouse, Professor at Manchester University, found that voting in the 2001 General Election was greater within East London's Bangladeshi community than amongst Londoners as a whole – even though the area was poorer. He believes that people vote where they have faced prejudice, or exploitation at work, and that traditions of community organisation have transferred there from Bangladesh.

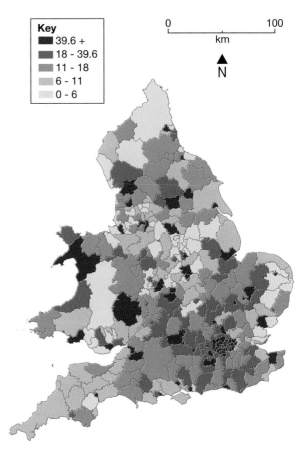

Key
■ 39.6 +
■ 18 - 39.6
■ 11 - 18
▨ 6 - 11
▨ 0 - 6

0 ———— 100
km

▲ N

🔺 **Figure 5** *The proportion of non-UK-born short-term residents per 10 000 usual residents in England and Wales*

Regeneration and engagement

Regeneration brings change to communities, and can therefore cause conflict. Conflicts occur because

◆ the process is top-down – designed by planners, developers, and local or national government, and imposed from above.

◆ most schemes are based on economic motives. Even housing projects are usually based on sales and land values, rather than social housing for people.

◆ groups disagree about what regeneration is about, and who it's for.

Inequality is often a powerful factor, and breeds resentment where local people feel that only those with high incomes will benefit from regeneration. This is especially true in university towns and cities, where property regeneration attracts new residents (e.g. students, or different socio-economic groups) whose lifestyles are different from those of long-term residents.

Regeneration schemes in Cornwall

Support for regeneration schemes in Cornwall has varied, depending on whether they are part of local, regional, or national policies.

- **Locally**, in 2008, Grampound's Parish Council sought residents' views about a proposal to build 69 new houses in the village – which was overwhelmingly supported. A third of the housing was classed as 'affordable', and residents felt that the extra population would help to maintain the village's services. The developers worked with the Parish Council on the design, and also held public meetings to gauge support.

- **Regionally**, plans for a waste incinerator as part of local economic expansion at St Dennis in central Cornwall caused protests – yet Cornwall Council still decided to build it. It was built in Cornwall's poorest area but created only seven jobs, and its 100-metre-high chimney (see Figure 6) is regarded as an eyesore. People fear its toxic emissions.

- In line with **national** policies, Cornwall Council has supported renewable energy. Wind turbines (see Figure 7) often cause protests from those who claim that they spoil the countryside. However, Grampound's residents were persuaded by an offer of £15 000 annually from the energy company for community projects.

▼ **Figure 6** *The waste incinerator at St Dennis, Cornwall*

> Building the incinerator in the first place was the wrong decision for St Dennis and for the whole of Cornwall. Now, on top of the environmental concerns and towering over the community, it's posing a serious health and safety risk.

Steve Gilbert, former MP for the area

Over to you

1. **a** Design two methods to investigate engagement and political participation in a community.
   b Use your methods to investigate the following in your community: clubs and activities; fund-raising for projects; voting behaviour in Local/Parish, County, General and EU Elections.
   c Present your findings and compare them with the data for Grampound.

2. In pairs, discuss and draw a spider diagram to show how and why attachment to, and involvement in, places can vary according to age, ethnicity, gender, length of residence (e.g. new migrants, students), and levels of deprivation.

On your own

3. Explain the factors that resulted in the following:
   a Grampound's community shop and local housing scheme receiving residents' support.
   b The waste incinerator at St Dennis and local wind turbines enjoying mixed support.

4. Research one community project in your own community. Find out its aims; who planned it; how the finance was arranged; and how much support it received.

5. Research and compare this with a regeneration project in your locality or region. Find out its aims; who planned it; how it was financed; and how much support it received.

6. Using your own evidence – and this study of Cornwall – write 500 words discussing whether 'participation in local projects only ever works in small communities'.

▲ **Figure 7** *Wind turbines, like these near Grampound, often cause great objections – but they also have their supporters*

Exam-style questions

AS 1 Using examples, assess the reasons why the degree to which people engage with places varies. *(12 marks)*

A 2 Evaluate the reasons why people's lived experience of places and engagement with them varies. *(16 marks)*

In this section, you'll learn about a range of ways available to investigate and evaluate the need for regeneration.

How to spot regeneration

Regeneration is easy to identify; just look for cranes, earthmovers, fencing, and concrete – lots of it! Figure 1 shows construction work in 2015 to upgrade the Custom House railway station in London's Docklands. It serves Canning Town, which is one of London's most-deprived areas. There was already a station at Custom House, because it's on the Docklands Light Railway. This new construction work is the latest addition – Crossrail (a new link to connect East and West London). So regeneration is like geological layers – imprinting new structures for the future on those of the past. But what kind of regeneration will it be?

Transport is critical to London's infrastructure, and to the regeneration process. When the Jubilee Line on the London Underground was extended in the 1990s, it not only brought commuters to Canary Wharf, but also provided development opportunities for places along its route. With that came a change of image. Bermondsey was one of London's most-derelict landscapes after the old docks closed (see Section 5.3). However, when the Jubilee Line extension arrived, the area became a property hotspot as investors bought up old warehouses and terraced houses ripe for gentrification.

Now, two decades later, the latest property hotspots are along the Crossrail link (see Figure 2). Bringing Crossrail to Custom House has brought opportunities for regeneration to an area that many considered needed a boost.

⬆ **Figure 1** *Work on the Crossrail station in Custom House, East London*

⬇ **Figure 2** *The new Crossrail link to connect East and West London*

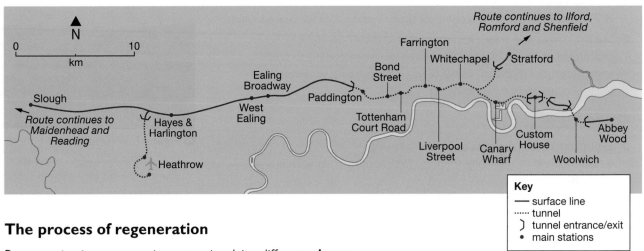

The process of regeneration

Regeneration is an economic process involving different **players**. In the UK, it usually involves partnerships between central government (which is funding Crossrail) and the private sector (which funds property re-development). But it goes more widely – like 1980s Docklands, it transforms an area socially (with new housing and schools) and environmentally (through the creation of public spaces). There is therefore considerable physical change – demolishing buildings, decontaminating polluted sites, and then reconstructing housing, retail and leisure facilities.

216

Custom House – the need for regeneration

Custom House is a ward near Canning Town, in the London borough of Newham. In the 2001 Census (when regeneration was planned) it was one of London's most-deprived areas:

◆ **Economically** – the area needed employment. In 2001, only 37.6% of adults were in full-time work (compared to the London average of 51.6% and the Canary Wharf area of 60.3%).

◆ **Socially** – the area needed improved housing, health facilities and education. In 2001, 71.6% of the housing stock was rented, and much of it was of low quality (see Figure 4). Also, 43.1% of adults had no educational qualifications.

◆ **Environmentally** – the closure in the 1980s of the docks and their associated industries resulted in environmental decay (with a legacy of derelict land and deindustrialisation).

Analysing the need for regeneration

Custom House has been undergoing regeneration since the mid-2000s. It is part of a project known as CATCH (**CA**nning **T**own and **C**ustom **H**ouse), which until 2010 was mostly funded by central government. Its focus has been on **community-led regeneration**, i.e. identifying the community's needs as a way of driving change forward. Its aim has been to create suburbs with a mixture of owned and social housing.

The programme has cost £3.7 billion so far, so it only moves forward when a case can be made to justify spending. Preparing a case for regeneration involves statistical evidence, such as Census data. The 2011 Census included data about deprivation – household incomes; those living in social housing; and those claiming out-of-work benefits. These three indicators are shown for the whole of Newham in Figure 3a. A simple analysis of the data shows interesting patterns:

◆ The wards of Canning Town North and South have some of the highest deprivation.

◆ At a glance, it looks as though relationships exist between different sets of data, e.g. that the percentage of people living in social housing might be related to the percentage claiming out-of-work benefits.

These data can be analysed using two simple techniques – scatter graphs (see Figure 3b) and correlation techniques, such as Spearman Rank (see page 307).

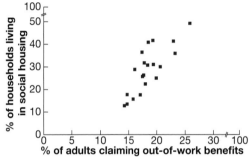

▲ *Figure 3b* *The relationship between the % of households living in social housing and the % of adults claiming out-of-work benefits in Newham in 2011*

Individual wards in Newham	Percentage of households living in social housing (2011)	Percentage of adults claiming out-of-work benefits (2011)	Median household income est. (2012/13)
Beckton	36.5	17.4	34 100
Boleyn	25.8	17.5	31 630
Canning Town North	49.3	25.8	28 910
Canning Town South	36.0	23.2	32 870
Custom House	41.4	22.9	31 840
East Ham Central	15.7	15.8	32 380
East Ham North	17.6	17.1	32 340
East Ham South	31.1	19.4	31 430
Forest Gate North	28.9	16.1	34 270
Forest Gate South	26.4	17.7	34 550
Green Street East	12.8	14.2	31 570
Green Street West	13.5	14.7	31 080
Little Ilford	30.1	20.6	29 730
Manor Park	22.5	18.0	31 580
Plaistow North	41.7	19.3	30 740
Plaistow South	25.1	19.9	31 750
Royal Docks	30.8	18.4	38 580
Stratford and New Town	31.8	17.8	35 840
Wall End	17.7	14.7	32 330
West Ham	40.9	18.5	32 310

▲ *Figure 3a* *Three indicators of deprivation for individual wards in Newham.*

Regenerating Custom House

The CATCH programme in Custom House initially focused on four areas to benefit the community:

1 Housing:

◆ 10 000 affordable new homes, particularly family-sized houses, were planned for construction by 2020.

◆ Much of the existing social housing (like that in Figure 4) was poorly built and needed renovation.

2 Employment:

◆ There was job creation and training for local people, plus offices and workspaces were made available for small businesses.

◆ New local shops and a supermarket were opened.

◆ Public transport was improved.

3 Education:

Replacement buildings for local primary and secondary schools were built. In 2015, 59% of Newham's students achieved five or more GCSEs at A* to C – more than double the figure in 1996 (27.9%), and better than improvements nationally.

4 Health:

◆ A new health centre, library, community centre and children's play areas were opened.

◆ The streets were made safer by redesigning them using traffic calming and open spaces.

However, since 2010, government funding has been cut. The focus is now on private-sector investment, using the stimulus of Crossrail to attract investors. This means that property development – not community need – is now driving regeneration. The differences can be seen in the recent development known as Hallsville Quarter in Canning Town (see Figure 5).

Hallsville Quarter

The Hallsville Quarter development began in 2014. It's a partnership between property company Bouygues Development, Newham Council, and One Housing Group (a not-for-profit house-building company). It aims to regenerate Canning Town (both North and South wards) by developing a supermarket, shops, open spaces, bars and restaurants, new homes, small-business premises, a cinema and a hotel.

◆ The first phase in 2015, near Canning Town Station, created 179 private and affordable homes, a supermarket and car parking

◆ The second phase has 349 homes (almost all private sector), a hotel, retail and restaurant units, and landscaping. The homes include 160 apartments for private sale, 134 private rented sector homes, and 55 shared-ownership properties.

⬣ **Figure 4** *Housing in Canning Town before regeneration*

⬤ **Figure 5** *An artist's impression of the Hallsville Quarter development, as used by its developers on their website*

Media portrayals of regeneration

The Media portrays regeneration in different ways – and often shows conflicting opinions, as Figure 6 shows:

◆ Figure 6a is from London's *Evening Standard* property section in 2015.

◆ Figure 6b is a community viewpoint, from the London Tenants Federation.

Creating images of places

Figure 5 – an artist's impression of the new Hallsville Quarter development, seen within the local landscape – is designed to attract people to the area. Images like this are powerful in representing place and using images to attract people. Other media (music, photography, film, art and literature) can also help to represent places. These include:

◆ music (e.g. The Specials' 'Ghost Town', about the depressed condition of Britain's inner cities in 1981)

◆ photography (e.g. Figure 1 of Bethnal Green in Section 5.4 – an image which almost romanticises a past life, which contained much poverty)

◆ film (e.g. *Get Carter*, portraying a violent Newcastle in 1970)

◆ art (e.g. the impressionists' views of London in the late 19th century)

◆ literature (e.g. Ian McEwan's novel *Saturday*, about one day in London in 2005).

Figure 6a The regeneration of Custom House – as reported in the Evening Standard *newspaper*

Crossrail is a huge boost for Custom House. It is currently out on a limb, served only by the Docklands Light Railway, but Crossrail to the West End is only 15 minutes. The vision is for tens of thousands of new homes and waterside living.

Royal Wharf is one of the new neighbourhoods, with 3385 homes designed around a high street, parks, shops, restaurants and school. Family houses are part of the mix and half of this development will be open space. Flats start at £250 000. Four-bedroom townhouses cost from £1 million.

Figure 6b The regeneration of Custom House and Canning Town – the community's view

Betty O'Connell:
I've lived here for 30 years. Residents were initially in favour of regeneration, as our homes were below standard. We worked with the council to put together a 'residents' charter' – but that's been thrown out. What resulted wasn't the council working with us, but them saying 'this is how we are going to do it'.
We thought the new homes would be council homes, but they are housing association homes, and rents are higher. Residents can't get their furniture into the new flats. People's health has deteriorated because of stress. There are a lot of older people here who need help from social services to adapt their homes. Now the council won't even talk to us.

▲ *Figure 6* Two views of the regeneration process from 2015

Over to you

1 a Using Figure 3a, explain how each indicator signals a need for regeneration.
   b Research similar data for your area, using your county or borough name plus 'ward profiles'.

2 a Using Figure 3b, describe the relationship in the scatter graph.
   b Using data in Figure 3a and guidance on page 307, carry out a Spearman's Rank correlation on the two variables in Figure 3b.
   c Test your correlation for significance, and explain its strength.
   d Repeat questions 2b and 2c for another pair of variables in Figure 3a.

3 In pairs, research and prepare a brief presentation on how your area, city or county is portrayed in music, photography, film, art, and literature.

On your own

4 Using Figures 1 and 2, explain the importance of Crossrail to London.

5 Research one regeneration project in your own area. Find examples of:
   a images to portray how the future will look, compared with the present
   b data to support why regeneration is needed
   c a brief outline of the plans
   d how different media report the regeneration in different ways.

6 Read the different views about regeneration process in Figure 6. Explain these conflicting views about regeneration.

Exam-style questions

1 Explain two ways in which the need for regeneration can be identified. *(6 marks)*

A 2 Using examples, evaluate the need for regeneration in different places. *(16 marks)*

It's only a tree, surely!

The tree pictured in Figure 1 is not just any old tree. In a 2015 poll of over 10 000 people (conducted by The Woodland Trust), it gained over a third of the votes – earning it the title of 'England's favourite tree'. It's a 250-year-old pear tree near Cubbington, Warwickshire. However, this tree's future is now in doubt, because it lies on the proposed route of HS2 – the planned high-speed rail link between London and Birmingham. For some, this tree and its future symbolises the opposition to HS2, because of the environmental damage it will cause. Others, however, think that HS2 will begin to bring Britain's rail network into the twenty-first century, and will help to regenerate cities like Birmingham. The debate is a familiar one – economic need versus environmental cost.

▲ **Figure 1** 'England's favourite tree' – a pear tree in Warwickshire

Why build HS2 now?

The planned construction of HS2 reverses all governments' transport policies since 1945, which have consistently been in favour of expanding the road network (rather than the railways). In 1958, when the first motorways opened, only 4.5 million road vehicles were registered in the UK. By 2013, there were 35 million! However, the road-building programme has simply not kept pace with growing vehicle ownership and use. As a result, the UK now has the most congested roads in Europe.

◆ The worst congestion is in the economic core of London and South East England – extending to Birmingham and Manchester (via the M6), and Leeds (via the M1). A government report in 2008 estimated that by 2025 road congestion would cost the UK £22 billion each year in lost time.

◆ Rail travel does offer an alternative to the car, and also allows people to work on laptops while they travel, but rail passenger traffic is also at its highest ever level. From 2002-3, when just under 1 billion rail passenger journeys were made, by 2014 the total had risen to 1.65 billion. Some rail routes are already close to capacity.

With a rapidly growing population, the UK needs new infrastructure just to catch up – let alone plan for future economic growth.

◆ **Route**: Two phases. Phase 1 will be a high-speed link (travelling at up to 400 km per hour) between London Euston and Birmingham Curzon Street (via north-west London and the Chiltern Hills). Phase 2 will then lead north-west to Manchester and north-east to Leeds (via the East Midlands and Sheffield).

◆ **Benefits**: Improved journey times between major cities (e.g. the London to Birmingham journey time will be cut from 80 to 49 minutes). An estimated 60 000 construction jobs will be created.

◆ **Problems**: The planned route will pass right through the Chilterns AONB. Like the TGV high-speed network in France, there will be no intermediate stations, so communities along the route will not gain from it.

▲ **Figure 2** The plan for HS2

◆ **Time**: If approved, construction will begin in 2017 (with Phase 1 high-speed services to Birmingham in 2025), and will extend to Scotland via the existing rail network when Phase 2 is complete by 2030.

◀ *Figure 3* *An artist's impression of a section of the HS2 rail line passing through the Chilterns*

The role of central government

Some projects, like HS2, are just too expensive for private companies to invest in (HS2's estimated cost in 2015 was £50 billion). As an engineering project, HS2 is huge (as Figures 2 and 3 imply). Also, as a service, transport rarely makes a profit, so although UK rail services are privatised (i.e. run by private companies), central government subsidised them by £3.8 billion in 2015. Without those subsidies, the existing rail companies would operate at a loss.

So, if private companies can't afford to build HS2, central government is left as the only organisation able to provide sufficient capital. This expense is viewed as an investment, because the government will gain:

◆ franchising fees from train companies to run services

◆ an economic multiplier – i.e. growth – which should result from the improved transport links generating higher company profits, and also jobs, from which the government will receive taxation revenue.

Another infrastructure project requiring central government funding will be the essential expansion of airport capacity in the UK. Each project is subject to a **cost-benefit analysis**, where costs are weighed against economic growth. One objection to HS2 is that it might only benefit London. However, improved accessibility between northern cities should also produce economic growth there.

Meanwhile, in Cornwall …

Cornwall's population is growing rapidly; traditionally it's attracted retired people, but now families are increasing in number. Without new housing, where will they all live?

South West England needs housing. A 2014 report showed that it needed 27 000 new homes that year, just to meet the extra demand, but only built 16 100. Housing is essential to regeneration – otherwise how can people live where there may be work?

◆ Nationally, housing shortages have driven up house prices in relation to earnings. Cornwall is a low-wage county, and houses are expensive in relation to earnings (costing between 12 and 16 times the average income).

◆ The demand for low-cost social housing in Cornwall rose 40% between 2010 and 2011.

The government has two challenges:

◆ Can the loss of greenfield landscapes be balanced against housing needs?

◆ Should planning restrictions be lifted in Green Belts or Areas of Outstanding Natural Beauty (AONB)?

Key word

Cost-benefit analysis – A process by which the financial, social and environmental costs are weighed up against the benefits of a proposal.

Meeting the demand for housing

The need for extra housing is urgent, because of:

◆ a rapidly **rising population** – due in equal part to increasing immigration and birth rates

◆ an increase in the **number of households** – high divorce rates mean that more housing is needed as households divide up and become smaller

◆ **overseas investors** buying up property as a safe investment, especially in London and major cities, and then leaving it unoccupied

◆ the need for more **affordable housing** – since the 1980s, a lot of social rented housing stock has been lost through the Right to Buy scheme (see Section 5.3), leaving a shortage of suitable properties for those on low incomes.

The fracking debate

As well as in housing, the UK government also has the power to make decisions affecting development via the planning laws (normally overseen by local councils). In recent years, central government has been attracted by the perceived benefits of **fracking**, which could add to the UK's home-produced natural gas supplies and reduce the need for imported gas.

The decision to allow fracking is about improving the UK's **energy security**. However, as with housing developments, this decision also conflicts with landscapes of value, such as National Parks (see Figure 4). In December 2015, MPs voted to allow fracking companies to drill underneath these protected areas from their boundaries.

Local opposition occurs in every location where test drilling to find shale gas occurs (see Figure 5). The size of the UK's total shale gas reserves is unknown; the British Geological Survey estimated in 2012 that known reserves would not even provide a single year's supply! Although drilling is increasing, the public are more opposed to shale gas than any other energy source.

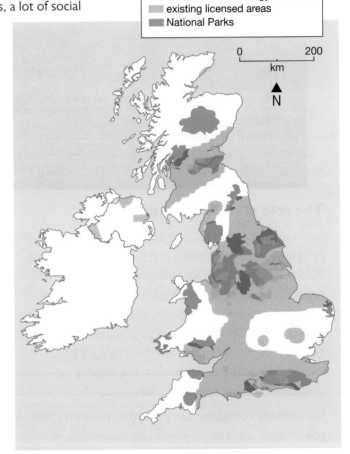

Key
- areas available for new licenses
- identified shale energy reserves
- existing licensed areas
- National Parks

0   200
km

N

🔺 **Figure 4** *Areas in the UK where permission to search for shale gas by fracking has been granted*

Key word

Fracking – The process of drilling down into horizontal layers of shale deep underground and then injecting a mixture of water, sand and chemicals at high pressure into the shale to fracture it and release gas trapped in the rock, which can then be brought to the surface.

◀ **Figure 5** *Protesting against fracking for shale gas in the UK*

Deregulation – what's that?

A minor – but significant – news story was broadcast early in 1986. It was that the UK's financial sector was to be deregulated. Margaret Thatcher's Conservative government made this decision, which resulted in a new era of prosperity for the UK's financial sector – and transformed London into a major World city (see Section 5.3).

Deregulation involved the following changes:

♦ Instead of the London Stock Exchange having a monopoly on all share dealings, any bank, financial adviser – or even individuals – could trade in shares. It opened up the freedom of individuals to invest.

♦ Barriers stopping overseas banks and other financial institutions from setting up offices in London were also removed. Until then, only UK banks could trade there.

Deregulation allowed foreign investors to invest in the UK without seeking UK government approval. The results transformed the UK's economy to the extent that banking, finance and business services now account for 30% of the UK's GDP (compared to just 15.5% in 1986). The London Docklands regeneration (see Section 5.3) also created space for these expanding financial institutions to set up large offices in Canary Wharf (see Figure 6).

Similar government decision-making occurred when the UK joined the European single market in 1992 (allowing free movement of labour within the EU). The UK's membership allowed people seeking work from other member countries to enter the UK (and vice versa). This movement of people helped to balance the UK's ageing population through increased taxation revenue. Although immigration is controversial for some, there is no doubt that economic growth can only come from expansion and an available labour market.

Before

After

⬆ **Figure 6** *Canary Wharf before and after 1986, showing the expansion of the financial sector following deregulation*

Over to you

1 **a** Use Figure 2 to identify where HS2 might be expensive to construct, because of the environmentally sensitive nature of the area.

   b What solutions could be adopted to deal with the environmental issue, which might affect the cost of construction?

2 **a** Research one group which is against and one group that is in favour of HS2. What are their basic points of disagreement?

   b In pairs, draw up a list of advantages and disadvantages of HS2.

   c Do you support HS2? Explain in a 300-word statement.

3 In pairs, research arguments for and against **(a)** expanding Heathrow and Gatwick airports, **(b)** increasing the number of licenses to frack for shale gas.

On your own

4 Explain why central government is likely to get more involved in decisions about issues such as where large areas of new housing should be built, or where fracking should be allowed, rather than local planning officers.

5 Explain the possible implications for the UK of a population that's likely to reach 70 million by 2035.

6 Explain the arguments, decades later, for and against **(a)** the deregulation of the financial sector, **(b)** the free movement of labour within the EU.

Exam-style questions

 1 Using examples, assess the role of central government in regeneration. *(12 marks)*

 2 Evaluate the role played by central government in regenerating places. *(16 marks)*

In this section, you'll learn how rural areas brand themselves, and how local government policies aim to attract inward investment.

Welcome to Newquay, Cornwall!

On summer Saturdays, the departures board at Newquay Airport looks impressive, with international flights to Dublin and Dusseldorf, national flights to Aberdeen, Belfast, Birmingham, Doncaster, Edinburgh, London and Newcastle, and local services to the Scilly Isles. It's the summer holiday season, of course, and on summer Sundays the departures board also includes Alicante and Frankfurt. That's quite impressive for one of the UK's most rural airports (see Figure 1).

But that's on a Saturday in the middle of the tourist season. Most of those flights only depart on summer Saturdays – in the winter, the airport is quiet and under-used (apart from 2-3 local flights each day to the Scilly Isles, there are three daily flights to Gatwick and one to Manchester). Newquay Airport reflects Cornwall's biggest economic problem – it lacks a year-round economy.

◇ **Figure 1** *Newquay (Cornwall) Airport blending in to the rural scenery*

- Cornwall's 'old economy' (see Section 5.2), which consisted mainly of primary sector jobs, has declined (see Figure 2). Apart from the autumn harvest season, these old industries provided year-round, permanent jobs.

- Cornwall's 'new economy' varies. Its quaternary 'knowledge economy' is small, whilst its biggest industry is tourism, a tertiary activity. However, jobs in tourism are mainly low-wage, part-time and seasonal.

Therefore, Cornwall's rural areas are less productive than before – so geographers refer to a **post-production countryside**. Cornwall's big problem is how to develop a high-income economy that will provide well-paid jobs all year round, to replace the primary sector jobs now lost.

◇ **Figure 2** *Reasons for the decline of Cornwall's primary sector jobs*

Industry	Reasons for its decline
Farming	Falling farm revenues, as supermarkets seek to pay their suppliers the lowest possible price (especially for milk).
	Cheaper imported food from countries where wages and other costs are much lower. For example, EU member states can produce milk for 16p a litre (about half the cost incurred by UK farms).
	A reduction in EU subsidies and government grants led to a rapid decline in dairy farming.
Fishing	EU quotas allocated some Cornish fish stocks to fishing vessels from other European countries.
	Stocks of many types of fish, such as cod, have declined due to over-fishing of young fish in previous years (before they could breed).
Tin and copper mining	The rich veins of tin and copper worked by Cornish miners in the nineteenth and early twentieth centuries have now mostly been mined out.
	Tin prices have collapsed, due to cheaper overseas competition.
	The strength of the pound has made the UK's tin more expensive overseas, which finally led to the closure of Cornwall's last tin mine in 1998.
Quarrying	The St Austell area of mid-Cornwall has some of the world's largest and best china clay reserves. In the 1960s, 10 000 people worked in the Cornish china clay industry. But, by 2015, the French TNC owner (Imerys) had cut the labour force down to 800 and moved much of its operation to Brazil.
	Fewer – larger – quarries now use technology, rather than people, to extract the china clay.

Cornwall's isolation

Cornwall is remote from the rest of the UK (see Figure 3), and is therefore not ideal for operating national or international businesses. It lies far from the UK's core economic area of London, the Midlands (e.g. Birmingham) and the north (e.g. Leeds) – where most economic activity takes place. Journey times by road or rail are long and expensive, and business travellers often face overnight stays away from home on top of high travel costs. To create jobs locally, Cornwall needs investment.

Rebranding the countryside

Cornwall is never short of tourists. Many overseas visitors come to trace their Cornish ancestry. Due to its mining heritage in the eighteenth and nineteenth centuries – brought to life in the BBC's dramatisation of *Poldark* – many thousands of Cornish miners emigrated to countries such as Australia or Canada to work. One song by West Country folk group Show of Hands, 'Cousin Jack', is about miners who left Cornwall in the late nineteenth century when new mineral discoveries were made overseas. Cornwall's heritage and literary 'branding' are among its assets – as is its scenery, which is often used in film productions.

Other rebranding strategies for Cornwall's post-production countryside include:

- **farm diversification** and the sale of specialised products – many farm shops now sell 'Cornish' food products, such as authentic pasties, Cornish cheeses, hand-made ice cream, and award-winning beers and wines

- **'foodie'** restaurants at Padstow (started by Rick Stein, TV chef), at Watergate Bay near Newquay (Jamie Oliver's 'Fifteen'), and Rock – all of which have helped to rebrand the Cornish coast as a **destination tourism** location

- **spectacular gardens**, which are the result of Cornwall's mild climate and Victorian ancestors, who explored overseas and brought back sub-tropical plants that have thrived in gardens such as the Lost Gardens of Heligan

- **outdoor pursuits** and adventure activities, such as the Extreme Academy at Watergate Bay near Newquay, which offers lessons in rock-climbing, surfing and para-surfing (see Figure 4).

Each of these strategies attracts both domestic and international tourists. Whilst Cornish foods tend to attract a domestic market, many German, French and Dutch tourists are attracted by Cornwall's gardens. Watergate Bay and Newquay's beaches regularly host the international Northern Hemisphere surfing championships – attracting participants from Australia, New Zealand and the USA.

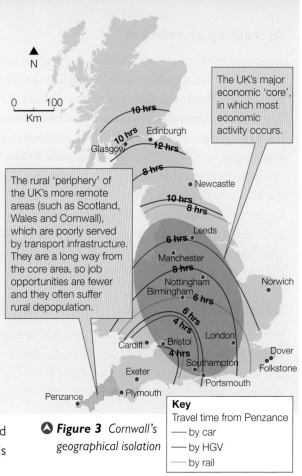

The UK's major economic 'core', in which most economic activity occurs.

The rural 'periphery' of the UK's more remote areas (such as Scotland, Wales and Cornwall), which are poorly served by transport infrastructure. They are a long way from the core area, so job opportunities are fewer and they often suffer rural depopulation.

Figure 3 *Cornwall's geographical isolation*

Key
Travel time from Penzance
— by car
— by HGV
— by rail

Key word

Destination tourism – The decision to visit an area for a short period, based on a single attraction (e.g. a wish to visit a particular restaurant or attraction). Other nearby places then receive visitors on the back of this.

Figure 4 *The Extreme Academy at Watergate Bay near Newquay in Cornwall, which specialises in rock-climbing, surfing and para-surfing*

Attracting investment

To attract investment, Cornwall has to compete with other areas that, together with Cornwall, all qualify for government **Regional Aid** (given to companies that wish to invest there); see Figure 5. All of these areas were recognised by the EU as being less economically advantaged, and would therefore qualify for government assistance and investor incentives through funding from the EU towards moving or set-up costs. The areas concerned include parts of the UK's 'periphery' (i.e. outside the 'core' shown on Figure 3), such as Scotland and Wales, as well as regions affected by the decline of the 'old economy', such as the North East.

Local Enterprise Zones

Within the general umbrella of 'Regional Aid' are specific **Enterprise Zones** which attract particular forms of aid. In 2015, there were 44 such locations in the UK (all within the areas shown in Figure 5).

Rather than simply trying to attract any business anywhere, Enterprise Zone incentives are focused into small areas, which can then be 'branded' to attract particular companies and organisations. Investor incentives include:

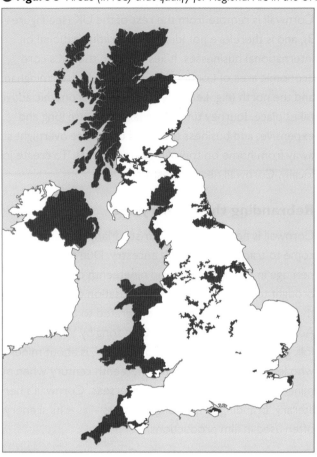

▼ **Figure 5** *Areas (in red) that qualify for Regional Aid in the UK*

◆ council business tax discounts of up to 100% for every business (up to a limit of £160 000 per year) for up to five years. A tax saving of £160 000 a year could help to pay for the creation of five new jobs.

◆ a planning-free environment, in which no planning permission is needed for building (beyond the normal building regulations required for safety).

◆ the provision of superfast broadband.

A few urban Enterprise Zones offer tax reductions against the cost of new buildings and equipment, as well as the cost of employing or training new employees to encourage development of the 'new economy'.

Newquay Aerohub

In 2014, Cornwall Council obtained Enterprise Zone status for Newquay Aerohub Business Park (right next door to Newquay Airport); see Figure 7. The Aerohub is a partnership between Cornwall Council and private-sector investors, who are aiming to begin the process of **diversifying** (broadening) Cornwall's economy away from its dependence on tourism. The new Business Park's 'brand' is its location, which aims to attract investment for an aviation and aerospace 'hub' (or focus). It was hoped that 700 high-value, skilled permanent jobs would be generated there in the first year, and Figure 6 lists those companies which located there by 2015 (mostly, as expected, from the aviation sector).

Aircraft-related industries

- Agustawestland – a Flight Training Centre for helicopter pilots, engineers
- Apple Aviation – aircraft maintenance and repair
- Bristow Helicopters – operates Search and Rescue operations off the coast of South West England (formerly run by the Ministry of Defence)
- British International Helicopters – operates offshore operations for the Ministry of Defence and Royal Navy
- Cornwall Air Ambulance Trust – two helicopters for emergency ambulance care
- Cornwall Aviation Heritage Centre – an aviation museum

- Flynqy Pilot Training – trains pilots for private and commercial licences
- Gateguards – manufactures replica aircraft for museums, collectors and film companies
- Skybus – operates passenger flights between Cornwall and the Isles of Scilly

Others

- Ainscough Wind Energy – maintains wind turbines
- Bloodhound SSC – a Research and Development centre for the Bloodhound Super Sonic Car, which will attempt a land speed record of up to 1000mph
- CIS (UK) Ltd – modular buildings for building sites

The above list of established companies (achieved in just a year) makes Newquay Aerohub look impressive, as does its location (see Figure 7). But by the end of 2015, only 450 jobs had been created and few of these were 'new' jobs. For example, Bristow Helicopters took over Coastguard Rescue from an RAF Squadron based in Cornwall, as part of a government privatisation. Similarly, British International Helicopters took on jobs formerly done within Cornwall by the Ministry of Defence. Each of these companies has created jobs which have been 'displaced' from the public sector into the private, and are therefore not 'new' jobs as such. In the same way, the Air Ambulance was previously based at Royal Cornwall Hospital in Truro, and has simply transferred its base. By 2015, little 'new' investment had come into Cornwall's new Aerohub.

Opinion is divided about the success of Enterprise Zones. Those in urban areas tend to attract companies quickly. For example:

- In south Lancashire, an established science and innovation centre near Warrington attracted 14 new businesses employing 160 people in its first year.

- Liverpool city centre also gained projects worth £165 million (generating 2000 new jobs). But its disused docks proved harder to 'sell' to developers (who face high demolition and clear-up costs).

⚫ **Figure 6** *Companies established at Newquay Aerohub Business Park by 2015*

⚫ **Figure 7** *The proximity of Newquay's Aerohub Business Park to the main runway of Newquay Airport*

Over to you

1 a Study Figure 1 and explain why Newquay Airport is so important for Cornwall's economy.

b How sensible would it be to increase flights to and from Newquay in winter?

2 In pairs, judge the success of rebranding Cornwall's economy to diversify away from tourism.

3 a In pairs, research a PowerPoint presentation to show Cornwall's 'brand' – location, environment, quality of life.

b Using this, create a poster to attract business to Cornwall.

4 List the advantages and disadvantages of Newquay Aerohub Business Park as a location.

On your own

5 Distinguish between 'Regional Aid' and 'Enterprise Zones'.

6 Based on the information provided, devise a SWOT chart (Strengths, Weaknesses, Opportunities and Threats) to show the progress of Cornwall's regeneration so far.

Exam-style questions

 1 Assess the attempts by governments to regenerate rural areas. *(12 marks)*

 2 Evaluate the success of local government policies which aim to attract inward investment to rural areas. *(16 marks)*

In this section, you'll learn how urban areas rebrand themselves, and also how local government policies aim to attract inward investment.

Living with the past

Each day, people lucky enough to live in a property with a waterfront view in the East London suburb of Silvertown next to the old Royal Docks, wake up to the sight of old dockyard cranes (see Figure 1). These cranes have no specific function any more (they once helped to load and unload merchant ships moored at the dockside). However, they do serve as a reminder of the industrial heritage of London Docklands – over three decades since its decline – so they serve as part of the Docklands 'brand'.

Figure 1 *Silvertown, East London*

Although London Docklands' identity is now very saleable, it was far from that in the early 1980s. The government, which owned the closed docks, was keen to sell off the land and encourage new post-industrial developments. However, it was hard to find a positive identity for the area at that time. In the early 1980s, the East End had a reputation for political unrest, unemployment and dereliction, arising from the docks' closure. It also had a very vociferous Labour-voting, working-class population, which repeatedly protested against any regeneration proposals by Margaret Thatcher's Tory government. Local people wanted community regeneration – with better housing and employment. The government wanted large-scale change, and to bring in outsiders, so it avoided the local Labour-run councils as much as possible.

Turning the past to advantage

East London now has a very different image in people's minds. It's ethnically diverse, its economy is growing fast, and it also successfully hosted the 2012 Olympic and Paralympic Games. This new positive image is partly to do with branding and gentrification (see Section 5.3), but it's also to do with embracing the area's past character. London now celebrates its industrial heritage, rather than demolishing it, and many other UK cities have also focused on regenerating old industrial buildings and their canal- or river-side locations (see Figure 2). Many tourists are attracted to these places, and tourism has now given former industrial cities a new market – particularly at weekends. More importantly, regeneration has helped to position inner-city environments as desirable places in which to live or to work.

Relic feature	Rebranded as	Examples in the UK
Industrial past	• Industrial heritage environments	• Beamish, County Durham • Ironbridge in Shropshire
Derelict land and docks	• Brownfield sites for rebuilding and expansion of office space • Waterside locations for housing	• London Docklands, Birmingham's canal-side and Jewellery Quarter • Bristol's and Liverpool's docks
Old factories and warehouses	• Loft and new-build apartments, bars, restaurants • Shopping centres and office space	• Leeds' and Manchester's canal-sides, London Docklands, Newcastle's and Gateshead's quaysides • Fort Dunlop, West Midlands

Figure 2 *Different ways of rebranding former industrial sites*

Regeneration in Glasgow

Most of the shipyards now lie empty along the River Clyde in Glasgow. London's dockside was all about trade; Glasgow's was about shipbuilding. Just as the cities of the American rust-belt were once founded on manufacturing steel (see Section 5.5), in Glasgow it was the shipbuilding industry which supported a whole industrial region. Engineering, steel and coal were key local industries – all linked closely to shipbuilding. So, when cheaper overseas competition in the late twentieth century led to the collapse of Glasgow's shipbuilding industry, the other (related) industries fell with it – a collapse referred to as 'the **domino effect**'. In 2015, only three shipyards still survived along the Clyde – two owned by BAE Systems Surface Ships, which builds advanced warships for the Royal Navy and other navies globally (shown in Figure 3), and further downstream another shipyard building car ferries.

⌄ **Figure 3** *Cranes at the BAE shipyard on the River Clyde in Glasgow*

Like East London, the regeneration of the Clyde has not been an easy sell. It's been managed by Glasgow City Council, in partnership with the Scottish government, and (like Cornwall in Section 5.9) a programme of diversification has been undertaken. Since 2000, tertiary and quaternary industries have grown in the following areas:

- **Arts, culture, sport and tourism**. Glasgow was European Capital of Culture in 1990, the UK's City of Architecture and Design in 1999, and also hosted the successful Commonwealth Games in 2014. The city has a wealth of architecture from famous architects, such as Charles Rennie Mackintosh. The UK and Scottish governments have invested in the new Burrell Collection to create an internationally famous art museum to attract more tourists, as well as the Scottish Exhibition and Conference Centre, the Glasgow Science Centre and the Riverside Museum of Travel and Transport – all located along one particular stretch of the Clyde. Clydebank's heritage-listed Titan Crane (a former crane at John Brown's shipyard in Clydebank, see Figure 4) has also been refurbished as a visitor attraction and local landmark. The aim has been to establish Glasgow as a leading destination for tourists – through a brand known as 'Glasgow: Scotland with Style'. With increasing tourism comes associated employment opportunities in hotels, bars, restaurants and retail, a multiplier effect.

- **Residential development**. As with London Docklands, investment from private property developers has been encouraged to build homes along the Clyde, together with shops and restaurants, to regenerate the former industrial areas.

- **Media**. The BBC Headquarters for Scotland's TV and radio broadcasting opened in 2007, and the commercial broadcaster STV is located nearby – both in former shipyard areas.

⌄ **Figure 4** *The Titan Crane on Clydebank. Formerly used in shipbuilding, it's now a visitor attraction.*

Kick-starting regeneration in Plymouth

Regeneration is a constant process. In the 1960s, the historic naval city of Plymouth, Devon, used to attract people from all over the world to see its rebuilt city centre (see Figure 5). After suffering severe German bombing during the Second World War, its newly designed open boulevards and pedestrianised streets were seen as innovative – and marked a new era of civic design in urban Britain. The tallest building in Figure 5 – the Civic Centre – is now heritage listed as an example of modernist architecture.

Unfortunately, Plymouth's economy has declined since the 1960s, and its city centre now looks dated. Plymouth's naval shipyards have been reduced in size, and the city now competes with Portsmouth to keep ship repair and servicing going. The Royal Navy still provides 10% of the city's GDP, but (like Cornwall) Plymouth's remoteness (see the inset map) makes investment hard to attract.

A 'Vision for Plymouth'

Like Cornwall Council (see Section 5.9), Plymouth City Council is playing a key role in the city's regeneration. Central government spending cuts have forced it to sell off some of its buildings to the private sector (including the listed Civic Centre building in Figure 5). A redevelopment project – Vision for Plymouth – involves the Council and Chamber of Commerce (a committee representing local businesses). So far, regeneration projects (shown in Figure 6) include:

◆ a new shopping complex in the city centre (Drake Circus) – a form of retail-led regeneration. Leisure shopping attracts people to city centres and creates jobs, particularly when visitors combine shopping with a meal out or a cinema trip. But opinions differ about its success – some feel that Drake Circus has taken business away from other parts of Plymouth.

◆ a cruise terminal to attract international tourists. Plymouth's history encourages tourism, particularly through its links with the USA (the Pilgrim Fathers sailed to America from Plymouth in 1620). But cruise ships are few in number – only 26 in 2014.

◆ a rebranded Plymouth Science Park, which has 70 businesses employing 800 people. The Park's links with Plymouth's two universities and its teaching hospital have attracted companies related to marine engineering, medicine, and renewable energies. This has helped to expand Plymouth's growing knowledge economy.

◆ a proposed sport and leisure partnership with Plymouth Argyle football club (for a new stadium complex to include a cinema, hotel and ice rink).

▲ **Figure 5** *Plymouth's newly rebuilt city centre in the early 1960s – the concrete Civic Centre (in the background) is now heritage listed as an example of period architecture*

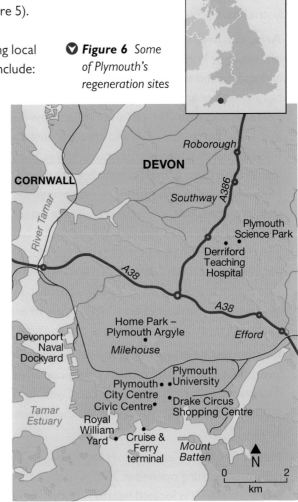

▼ **Figure 6** *Some of Plymouth's regeneration sites*

Royal William Yard, Plymouth

While most regeneration is about demolition and renewal, some is about restoration. Local preservation societies have long campaigned for one of Plymouth's most significant buildings – the Royal William Yard (see Figure 7). This building is a former Royal Navy supply store, which dates from the Napoleonic Wars and is Grade 1 listed (for both its historic exterior and its interior). The restoration has been carried out by Urban Splash – a private-sector regeneration company – which has also been involved in projects in Bristol (Lakeshore) and Sheffield (Park Hill).

▲ *Figure 7* *The Royal William Yard in Plymouth*

As a historic Grade 1 listed building, the Yard has been expensive to restore, because specific materials and techniques have to be used. The process has been completed in phases over 20 years, and includes shops and restaurants and over 200 apartments. It's located close to the new cruise liner terminal (see Figure 6), but 2 km from the city centre (which some feel is a disadvantage for businesses).

Unfortunately, not everyone agrees about regeneration, and there are tensions between groups wanting to preserve urban environments and those seeking change. One example is the Civic Centre in Figure 5. It's Plymouth's tallest building, but it needs considerable amounts of money spent on it in order to maintain it. Plans for it include new apartments and a hotel. However, some people simply want to see it demolished!

Exam-style questions

AS 1 Explain the role of local councils in urban regeneration. *(6 marks)*

A 2 Evaluate the success of named urban regeneration projects. *(20 marks)*

Over to you

 1 Select East London, Glasgow, or Plymouth. In pairs, use phrases like 'regeneration in Plymouth' to research three media sources (e.g. YouTube extracts from radio or TV programmes, or advertising) to show how 'place identity' is used as part of rebranding that location. Prepare a five-minute presentation.

2 **a** In pairs, draw a Venn diagram with three circles – one for each location. On it, write down similarities and differences between the regeneration programmes in each place.

   b Now use three colours to highlight the economic, social and environmental aspects of regeneration in each location.

   c Identify and explain common factors in the three places.

3 Research how regeneration projects in your nearest urban area have **(a)** maintained its character, **(b)** changed its character, **(c)** been most successful.

On your own

4 Explain why historic buildings can **(a)** be an important part of regeneration, **(b)** present regeneration with problems.

5 Explain why residential development is essential to successful regeneration.

6 Identify and assess the roles played by local councils in regeneration in Glasgow and Plymouth.

7 **a** Using your Venn diagram from question 2, score each regeneration project from 1 (poor) to 5 (excellent) in each location. Justify your scores.

   b Write a 500-word report on 'The key elements of successful regeneration'.

Changes in Barking and Dagenham

The East London borough of Barking and Dagenham has a place in social history. It was here that, in 1966, women machinists at Ford's Dagenham car assembly plant (the largest in Europe – see Figure 1) went on strike for equal pay with men. At the time, this London borough offered a good quality of life. In the 1960s, Ford employed 40 000 people at its Dagenham plant – and council re-housing schemes (such as at Becontree) offered residents far better housing than the old East End slums from which many had moved.

However, the borough's more recent history is less happy. In 2002, Ford ended car assembly at Dagenham – after 71 years. It still has a factory there (producing Ford's diesel engines), but most of the work is now done by robots – so, by 2015, Ford employed just 3200 human workers in Dagenham. To make matters worse, Sanofi (a giant pharmaceutical company) also ended production in Dagenham and closed its operation there in 2013 – after 77 years.

Figure 1 *Ford's car assembly plant in Dagenham, photographed in 2013*

Deprivation in Barking and Dagenham

Ford's decline, in particular, has left a legacy of deprivation: economic (unemployment), social (poor health), and environmental (derelict land). Between 2004 and 2007, deprivation levels in Barking and Dagenham increased so much that it became the twenty-second most-deprived local authority in the whole of England – and, by 2015, it had moved up to ninth place. By 2015, Barking and Dagenham had:

◆ London's highest adult unemployment rate (9.8% of working-age adults), which was more than double the rate there in 2001 (when it was just 4.5%).

◆ 27% of residents earning below London's living wage (the London average was 21%).

Deprivation in Barking and Dagenham is widespread across the borough. Figure 2 shows the deprivation level in small areas called **Lower Super Output Areas (LSOA)**, see opposite. This borough map shows the percentage of households who are deprived in at least three of the following characteristics in each LSOA:

◆ **Employment** (where any adult in a household is unemployed or long-term sick).

◆ **Education** (where no one in a household has five GCSE passes at grade C, or is a full-time student).

◆ **Health and disability** (where anyone in a household has 'bad or very bad' health, or a lasting health problem).

◆ **Housing** (where accommodation is overcrowded, shared, or without central heating).

Figure 2 *Deprivation levels in Barking and Dagenham in 2011 – where households display at least three deprivation characteristics. The named areas within the boundary lines are council wards (local districts). The areas shaded within them are the LSOA (see the text).*

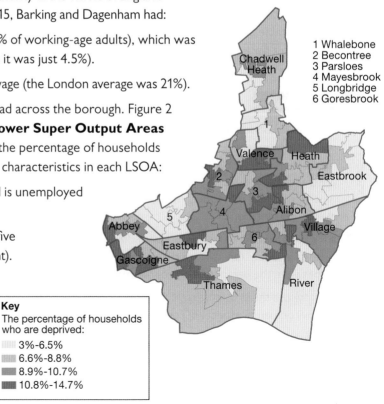

1 Whalebone
2 Becontree
3 Parsloes
4 Mayesbrook
5 Longbridge
6 Goresbrook

Key
The percentage of households who are deprived:
- 3%-6.5%
- 6.6%-8.8%
- 8.9%-10.7%
- 10.8%-14.7%

Measuring deprivation

Deprivation data are obtained from Census data, which are collected and measured in small neighbourhoods known as LSOA (Lower Super Output Areas) – the smallest areas shaded in Figure 2. LSOA cover the whole of England and are areas of similar population size (about 1500 residents, or 650 households). There are 32 844 LSOA in England. Data from each LSOA can then be aggregated into larger areas (such as districts, counties or London boroughs).

The Figure 3 table shows seven deprivation indicators for four London boroughs and the City of London. These seven indicators divide into three broad areas: economic, social and environmental (see below).

Economic

◆ Income deprivation (Column 1). This measures the proportion of people experiencing deprivation linked to low income, e.g. they are out of work or on low earnings.

◆ Employment deprivation (Column 2). This measures the percentage of working-age people who would like to work but cannot (due to unemployment, sickness, disability, or care responsibilities).

- **City of London:** In Central London, where a high proportion of jobs (especially in the knowledge economy) are located.

- **Kingston-upon-Thames:** A mainly residential, high-income borough in South-west London.

- **Newham:** An East London borough that has undergone substantial regeneration, and in which are located the Olympic Park (see Section 5.12) and Westfield Shopping centre.

- **Tower Hamlets:** Adjacent to Newham in East London; it remains London's most deprived borough – in spite of the Docklands regeneration of Canary Wharf being located there (see Section 5.3).

Social

◆ Education, skills and training deprivation (Column 3). This measures the lack of educational attainment and skills in the population.

◆ Health deprivation and disability (Column 4). This measures the risk of premature death and the reduction in quality of life caused by poor physical or mental health.

◆ Crime (Column 5). This measures the risk of personal and material crime.

◆ Barriers to housing and services (Column 6). This measures both the financial affordability of housing and also how close housing is to local services. London boroughs generally score badly on the former, but well on the latter.

Environmental

◆ Living environment deprivation (Column 7). This measures local environmental quality, including housing quality and external features (such as air quality and the level of traffic accidents).

The data for each of the above indicators are collated for all 32 844 LSOA. Each indicator is then ranked and divided into ten equal bands, known as **deciles**. LSOA ranked within decile 1 are amongst the 10% most deprived LSOA, and those ranked within decile 10 are amongst the 10% least deprived. The numbers within the Figure 3 table are deciles (averaged across all LSOA within each borough). The **Index of Multiple Deprivation** (IMD – shown as Column 8) is an overall measure of deprivation (which combines all seven indicators into a single index).

Local Authority name (2013)	1 Income	2 Employment	3 Education, skills and training	4 Health deprivation and disability	5 Crime	6 Barriers to housing and services	7 Living environment	8 Overall IMD
Barking and Dagenham	2.5	3.2	3.6	3.9	2.1	1.7	3.9	2.6
City of London	8.2	8.2	8.8	7.3	9.3	2.3	1.8	6.7
Kingston-upon-Thames	7.2	8.2	8.2	8.9	6.3	5.0	4.2	7.7
Newham	3.0	4.2	4.7	4.4	2.0	1.2	2.8	2.8
Tower Hamlets	2.8	4.3	5.5	3.4	2.8	1.7	2.3	2.8

⬆ **Figure 3** *Average rankings for deciles of deprivation in four London boroughs and the City of London*

Regeneration in Barking and Dagenham

How can deprivation be reduced? The answer is regeneration. The site of the former Ford assembly plant is large, has its own dock, and is ideal for industry or housing. Several regeneration projects have been established to combat deprivation left by the withdrawal of such a large employer. Figure 4 describes five sites identified for regeneration.

Beam Park (Dagenham South)
Located on land previously occupied by Ford, the western side will create 40 000 m² of workplaces. A hotel (Premier Inn) and pub (Brewers Fayre) already employ local residents.

Barking Town Centre
A 1960s concrete town centre in need of refurbishment for retail, commercial and new residential spaces. By 2014, over 400 homes and 1000 m² of commercial space had been created.

Gascoigne Estate
The most deprived housing estate in the borough (see Figure 2). Its regeneration will provide 1500 new homes by 2024, together with schools, a community centre, retail and office spaces, and outdoor leisure spaces.

Barking Riverside
The site of a former power station, this is London's largest regeneration site. Plans for seven new residential neighbourhoods will create 11 000 homes, parkland, five schools, health centres, places of worship, community facilities, as well as 65 000 m² of commercial, retail and leisure space, which should create up to 6000 jobs by 2020.

Dagenham Dock
An industrial site, which includes 200 fuel and chemical tanks, and derelict land. Now a sustainable business area, e.g. a plastic bottle recycling company (recycling 10% of the UK's plastic bottles), and an anaerobic digestion plant (producing biogas).

Measuring the success of regeneration

Its success can be judged using four criteria:

◆ **Economic** – increased employment and income.

◆ **Social** – lower levels of deprivation.

◆ **Demographic** – improved life expectancy and health.

◆ **Environmental** – e.g. reduced pollution and dereliction.

▲ **Figure 4** *Some of the regeneration sites in Barking and Dagenham*

▼ **Figure 5a** *Comparing job categories in Barking and Dagenham with London and the UK as a whole. High-income jobs are shown in yellow, low-income jobs in blue. All data are for 2013.*

Regeneration in Barking and Dagenham – success or not?

Job category	Mean weekly wage (£)	% of all Barking and Dagenham employees	% of all London employees	% of all UK employees
Managers, directors and senior officials	£907.90	6.8	11.6	9.8
Professional (e.g. teachers, doctors)	£700.80	13.3	22.5	21.0
Associate professional and technical (e.g. IT staff)	£610.00	9.2	16.3	14.3
Skilled trades (e.g. trained electrician)	£486.50	21.1	8.3	7.5
Process, plant and machine operatives	£440.70	9.5	4.7	5.8
Administrative and secretarial	£355.80	13.1	11.7	12.4
Caring, leisure and other services	£266.50	11.2	7.9	9.6
Elementary occupations (i.e. no training required)	£241.70	15.2	9.6	8.0
Sales and customer services	£237.10	9.6	7.5	11.6
UK average	**£502.20**			

Name	No of jobs (000)	Median income, 2013 (£)	Annual % change	Income percentiles				
				10	25	40	60	80
Barking and Dagenham	41	£517.40	6.0	£150.40	£304.90	£446.10	£604.60	£806.60
Tower Hamlets (contains Canary Wharf)	217	£804.90	5.0	£333.50	£535.50	£690.40	£951.10	£1,449.60
Newham (contains Olympic Park)	63	£475.70	-5.8	£122.40	£277.70	£395.60	£550.70	£736.20
UK	**24 385**	**£403.90**	**0.0**	**£118.60**	**£244.40**	**£339.90**	**£479.30**	**£689.90**

Notes for Figure 5b:

- The number of jobs refers to the numbers of jobs available in that area.
- Annual % change is the increase or decrease in the number of jobs, 2012–13
- The '10' percentile are the lowest-earning 10%; the '80' percentile are the top 20% earners, and so on.

 Figure 5b *Income data in Barking and Dagenham compared to the UK as a whole and two other East London boroughs that have undergone substantial regeneration.*

Indicator	Barking and Dagenham	London
Male life expectancy (years)	76.3	78.2
Female life expectancy (years)	81.2	83.4
Infant mortality rate per 1000 live births	5.3	4.4
Coronary heart disease (% of the population)	2.5	2.2
Diabetes (% of the population)	5.9	5.0

 Figure 5c *Health data in Barking and Dagenham compared to London as a whole*

Exam-style questions

AS 1 Assess the success of regeneration in one area you have studied. *(12 marks)*

A 2 Evaluate the contribution of economic regeneration to reducing deprivation in one area you have studied. *(20 marks)*

Over to you

1 Using Figure 2, describe the pattern of deprivation within Barking and Dagenham.

2 Use the additional information provided as part of Figure 3 to explain why deprivation varies between the five localities.

3 **a** Draw an annotated map to show the five regeneration schemes in Barking and Dagenham. Use Google Maps to supplement information about each site. Annotate economic regeneration in one colour, social regeneration in a second, and environmental in a third.

  b Which type of regeneration seems most significant in Barking and Dagenham?

  c How far should these five regeneration projects leave Barking and Dagenham better off?

4 Using this section, and Google Maps/Earth, explain the economic benefits of locating in what you regard as the **best** of the five sites shown in Figure 4 to companies that wish to move there.

5 Use Google Maps and other online maps and photographs to show how regeneration is changing **(a)** the area in Figure 1, and **(b)** the regeneration map in Figure 4, both from 2013.

On your own

6 Use the data provided in Figures 5a–c to show **(a)** why regeneration has been necessary in Barking and Dagenham, **(b)** evidence that regeneration brings economic improvement, and **(c)** how regeneration could reduce health deprivation.

7 Use the data provided in Figures 5a–c to write a 750-word report titled 'Regeneration in Barking and Dagenham – success or work in progress?' Use the data to show **(a)** any progress already made in reducing deprivation, **(b)** the successes of economic, social, demographic and environmental regeneration, and **(c)** any further improvements that are needed.

In this section, you'll learn how different players have different criteria for judging the success of urban regeneration.

A scene of change

Large cranes have now been part of East London's skyline for over three decades – from the Docklands regeneration of the 1980s and 1990s (see Section 5.3), to London's 2012 Olympics (see Figure 1), and more recently with the construction of Crossrail. This period has been a great time for construction companies and their employees; those who learnt their skills as apprentices in the Docklands regeneration of the 1980s, are now approaching their fiftieth birthdays. But who have these projects actually benefitted; who has carried them out; who has paid for them; and how successful have they been?

Figure 1 *The final construction stages of the Olympic Athletes' Village, now known as Stratford East, as it neared completion in 2011*

Different sources of investment

Not all regeneration projects are the same. For a start, they are often paid for in different ways. Some investment comes solely from the private sector, some solely from the public sector, and some from partnerships between the two.

1 Private-sector investment

A recent example of private-sector investment is Stratford's Westfield Shopping Centre in East London – now Europe's largest shopping centre (see Figure 2). This is an example of **retail-led regeneration**, which has created over 10 000 new jobs.

◆ Westfield, an Australian property company with a 50% stake in the Centre, borrowed £700 million to build it.

◆ It is recovering its costs by leasing space in the Centre to retail companies. In its first four years, the Centre's annual turnover was £1 billion, which added hugely to the local economy.

2 Public-sector investment

A recent example of public-sector investment is London's 2012 Olympic and Paralympic Games.

◆ The UK government bid for the 2012 Games, supported by the London Assembly and its Mayor.

◆ The Games cost £9.3 billion to host, and these costs were recovered through, for example, ticketing, TV sponsorship and the post-Games sale of apartments and houses in the Athletes' Village (see Figure 1).

◆ Against many expectations, the 2012 Games were delivered on time, under budget and at a profit.

3 Public-private partnerships

A recent example of public-private partnership investment is the London Docklands regeneration of the 1980s and 1990s.

◆ Although portrayed as '**market-led regeneration**' (see Section 5.3), it actually involved a partnership between the government (which handed over land and financial grants) and property developers (who ensured that the regeneration would create economic growth, jobs, and housing).

◆ The government regarded the costs as an economic investment (growth would create tax revenue) as well as being socially advantageous (by reducing local unemployment and social problems).

Figure 2 *Westfield Shopping Centre in Stratford, East London*

Public- or private-sector investment?

Figure 3 assesses the pros and cons of receiving investment from the public and private sectors (as well as both of them working together in partnership).

▼ *Figure 3* *An analysis of public- and private-sector investment*

	Public sector — Policy led		Mixed public and private sector	Private sector — Investment and profit led	
	National	**Local**		**Multiple partners**	**Single company**
Players	MPs, Government officials, regional (e.g. London Assembly) or local councils		Both sets of players	Company directors, shareholders, employees	
When?	Used for any infrastructure or public-service projects that the private sector views as being too costly or risky, or where the expected income will be lower than the costs (projects like this can be used to kick-start a depressed economy)		Where the cost is high, but it can still be shared with the private sector	Where the project will lead to benefits for more than one company or partner	New investment by a single company, to expand its product range and increase its profits.
What?	• **Infrastructure** national transport (e.g. HS2, Crossrail), water or energy projects, all based on public need • **Major national events** e.g. the 2012 Olympic and Paralympic Games • **Health services** where there are clear public benefits		**Housing**, e.g. private housing developments, as part of which the government subsidises the construction of lower-cost (affordable) housing	• **Technology** e.g. car companies sharing research into new engine technology • **Supply chains** e.g. Intel chips made for a variety of computers	Expansion of individual companies and their range of products, e.g. Apple, Microsoft
Benefits	It provides services and infrastructure where they are needed – not where they will make profit		Often works well in the UK, e.g. NHS	Energetic, often creative, and competitive (which can bring costs down)	
Costs	Sometimes seen as slow, with inaccurate budgeting and delivery times		The different sectors need to understand each other	It serves the company's interests, which may not be those of the public (e.g. the VW scandal in 2015 about the falsification of fuel consumption data for its cars)	

A case study of regeneration: London's 2012 Olympics

The cheering crowds; Team GB's third place (overall) in both the Olympic and Paralympic medal tables; 28 days of celebration and positive promotion of London as a city and global destination – what's not to love about the 2012 Games? As a direct legacy of London 2012, the city now has its first new park in over a century. The Queen Elizabeth Olympic Park (see Figure 4) still contains almost all of the original Olympic venues. Unlike any Olympic Games except Sydney, the Olympic sporting venues are still in use as a long-term legacy of 2012. Yet no regeneration project is problem-free, and the decision to host the Games was contested from the start.

▲ *Figure 4* *The Queen Elizabeth Olympic Park, along the Lea Valley in East London*

Major features of the Olympic Park, post-2012

- 560 acres of parkland, with trails, play areas, walking and cycling routes
- The Olympic Stadium – now home to West Ham football club, as well as major athletics events (e.g. The Invictus Games, the 2017 World Athletics Championships)
- The Aquatics Centre – now a local swimming baths
- The Orbit – a viewing platform offering views of the City and East London
- The Velodrome – still hosting national and international cycling events
- The Copper Box – the former handball arena, now a multi-sport arena for community use and a basketball venue
- Here East – the Media Centre during the 2012 Games and now the home of technology companies including BT Sport
- Lee Valley Hockey and Tennis Centre
- Lee Valley VeloPark – a centre for track cycling, road racing, BMX and mountain biking

The regeneration vision

The London Olympics were not the end of the regeneration story. In the 2001 Census, East London contained some of the UK's poorest areas. The Mayor of London at that time, Ken Livingstone, believed that the Games could help to build a process of **convergence** – in other words, to reduce the gap between London's poorest and wealthiest areas. The Games themselves, and the subsequent Queen Elizabeth Olympic Park, were designed to improve East London's 'brand' as a place to visit, to live in and to work, as well as to attract fresh investment as a result of the area's improved image. The Westfield Shopping Centre is part of this rebranding/re-imaging.

Since 2012, the perimeter of the Queen Elizabeth Olympic Park has been planned as an area of growth (see Figure 5). This includes:

◆ office developments in a district branded as the 'International Quarter', consisting of 400 000 m² of offices (which are likely to be large banks – like Canary Wharf), a 4-star hotel and 330 new homes (creating 25 000 new jobs).

◆ residential areas, e.g. Chobham Manor and Sweetwater, which will provide 9000 new homes by 2025.

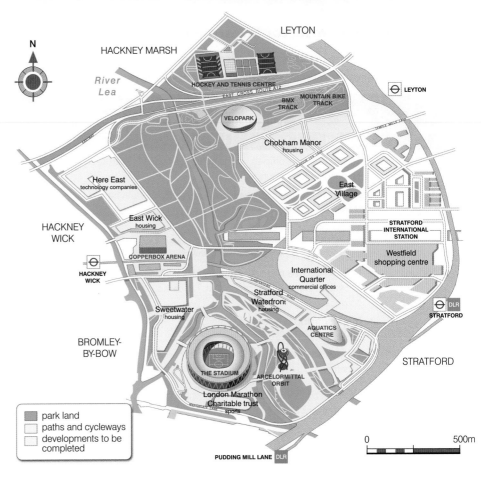

▲ **Figure 5** *The Queen Elizabeth Olympic Park and developments around it*

Key players before and after the 2012 Games

1 UK central government agency
The London Legacy Development Corporation is an appointed (unelected) agency, funded by central government, to oversee the legacy development of the Olympic Park. Their success criteria include the use of the Olympic venues post-2012, increasing employment and more housing.

2 Local government – the elected councils
Four London boroughs shared the hosting of the Games (Tower Hamlets, Newham, Hackney, and Waltham Forest), and they all wanted regeneration to continue post-2012. However, they have no planning control over new developments – this is done by the London Legacy Development Corporation.

3 Regional government – the London Assembly
The Mayor of London and the elected London Assembly were responsible for ensuring that transportation was effective during the Games, as well as supporting the expansion of housing and the East London economy after 2012.

4 Stakeholders in the local economy
The Olympic venues were sited on a former industrial estate, where 207 (mostly locally owned) companies employed 5000 people – all of whom were compensated to move. However, there were still objections. Most companies relocated within the local area, but many workers still faced a longer commute.

5 Environmental stakeholders

The collapse of manufacturing in the Lea Valley following the 1981 closure of the original London Docks, led to widespread dereliction (see Figure 6). The Queen Elizabeth Olympic Park has cleaned up and re-landscaped the whole area. New wetlands now form part of the park, and breeding boxes and nesting sites have ensured rising numbers of species such as newts, fish, bats and birds.

6 Stakeholders in people

◆ The biggest – but most contested – legacy of 2012 has been over housing. Increasing prices in this newly desirable area have made property extremely expensive. **Affordable housing** is needed for those on low incomes to rent.

◆ The Athletes' Village – now known as Stratford East – is standing on the site of a former housing co-operative for low-income residents (known as Clays Lane). Its 450 residents were relocated to social housing throughout London – breaking up the community. A promise to re-house them after 2012 was never honoured.

◆ The initial plan, post-2012, was to re-model the Athletes' Village into 3000 affordable housing units. With government cuts, that intention has now been reduced to 800 – and, instead of just low-income groups, those earning £60 000 a year now qualify!

You can read alternative views about 2012 on the website: 'gamesmonitor.org.uk'.

▲ *Figure 6* *Dereliction along the Lea Valley*

Key

++ Strong agreement

+ Some agreement

— Strong disagreement

– Some disagreement

◀ *Figure 7* *A conflict matrix for analysing attitudes between different players*

Over to you

1 a In pairs, summarize the differences between public- and private-sector investment.

   b Identify examples of each in your local area.

2 Use Figure 3 to help you explain why **(a)** the public, not the private, sector funded London's 2012 Olympics, **(b)** the private, not the public, sector funded the Westfield Centre.

3 a In pairs, list the major players in the 2012 Games and the area's subsequent regeneration.

   b Draw three circles to form a Venn diagram. Label them 'economic', 'social' and 'environmental'. Place each player somewhere inside a circle, or where it overlaps with another.

   c Copy Figure 7. Complete it to show which players **(i)** agree with each other, **(ii)** oppose each other, and explain why.

4 Visit the website 'gamesmonitor.org.uk' and locate its blogs. Investigate the blogs and research other social media to explain why views differ about the success of East London's regeneration.

On your own

5 Using examples, explain why different players would judge the impacts of the 2012 Games differently.

6 Research the phrases 'Olympic Park' and 'Artists' impressions of Olympic Park'. Find images to compare the reality versus the image of the Park. Explain how and why they differ.

Exam-style questions

(AS) 1 Using examples, assess the roles of different players in the urban regeneration process. *(12 marks)*

(A) 2 Assess the reasons why different players have different criteria for judging the success of urban regeneration. *(12 marks)*

Changing times

In spite of its geographical disadvantages (see Section 5.9), Cornwall's economy did well between 2000 and 2010 – growing faster than the UK average (as a result of sustained investment). The investment received by Cornwall increased employment and gave tourism a boost. However, most of this investment came from the public sector (in particular, from the UK government and the EU). Since 2010 – and the political decision to impose austerity and major cuts in the public finances – that funding model has been contested. The private sector is now being relied upon to provide most of the additional investment required by Cornwall.

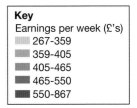

Key
Earnings per week (£'s)
267-359
359-405
405-465
465-550
550-867

Rural disadvantage

Regeneration is harder in rural than in urban areas, and the rural economy faces major challenges (see Section 5.9). Figure 1 shows that low average incomes are typical of large swathes of rural Britain. With a much lower population density in rural areas, the struggle to maintain sufficient customers to make a profit, means that private investors prefer to invest in urban rather than rural areas.

The lack of rural investment naturally leads to a lack of opportunity and high-income employment in Cornwall. As a result of this, Cornwall's young, well-qualified residents are forced to leave and find work elsewhere – causing a 'brain drain'.

◆ In 2011, Cornwall had England's lowest full-time average earnings, which at £25 155 per year were just 77% of the UK average.

◆ However, Cornwall Council believes that the average income figure is even lower when part-time workers are included. It claims that £21 993 is the real average. As Section 5.9 explained, there are a lot of part-time and seasonal workers in Cornwall.

◆ 20% of Cornwall's working-age population earns less than the living wage (£7.45 per hour in 2015).

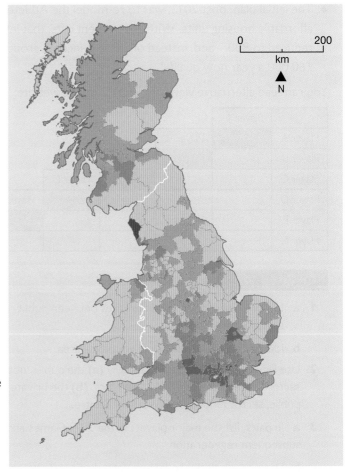

Figure 1 *Average weekly earnings across Great Britain. Notice just how widespread the areas of lowest income are.*

Regeneration and convergence funding

Prior to 2010, a lot of public-sector investment was put into regenerating Cornwall – using grants from both local and national government and the EU. Between 1999 and 2007, this investment worked under a funding model known as Objective One (a type of **convergence funding**, designed to raise the incomes of rural areas).

How Objective One funding worked

One third of new businesses fail in their first year. So, the principle of Objective One was simple: **match funding**. This process aimed to reduce the risks faced by new businesses – through matching the capital raised by individual business people (with approved business plans) to **pump-prime** their new businesses and reduce their initial costs. In that way, a higher proportion of new businesses would hopefully survive and thrive in the longer term.

New potential businesses could approach funding bodies, such as local councils. One investment in a farm shop worked as follows:

◆ The family who owned the farm raised £20 000 to support their business plan (which was expected to cost £320 000 overall).

◆ Their bank lent them a further £20 000 (making £40 000).

◆ Cornwall Council then matched the new amount (making £80 000).

◆ The South West Regional Development Agency (SWRDA) then matched that amount (making £160 000).

◆ Finally, the business was awarded EU Objective One funding to match the SWRDA funding – creating the necessary £320 000 for the business to start.

So, this funding model was able to turn £20 000 into £320 000! That sounds like a free grant of £280 000! However, costs like this need to be judged against the benefits gained from them. Ten years later, the farm shop's annual turnover was £700 000 and it employed 20 people!

◆ By 2007, Objective One had backed 580 projects in Cornwall with £230 million.

◆ Until 2005, the Cornish economy grew faster (at 5.8%) than the UK average (5.4%), and had the fastest growth rate of any EU region.

Figure 2 *Lobb's farm shop in mid-Cornwall, which used a funding model like the one described.*

Key players in Cornwall's regeneration

1 The EU
Convergence funding has been granted to Cornwall since 1999.

2 UK central government agencies
◆ Before it was abolished in 2010, the South West Regional Development Agency made investment grants that came from central government.
◆ Most investment grants have been cut and are now given out directly by central government.

3 Local government
◆ The public sector (the NHS and Cornwall Council) is Cornwall's largest employer.
◆ Since 2010, Cornwall Council has had no start-up funding to offer potential businesses. However, it does offer rebates on business taxes as part of its Enterprise Zone at Newquay (see Section 5.9).
◆ A Local Enterprise Partnership supports business growth, but little funding is available – only 'help with exports, training or start-up advice'.

4 Stakeholders in the local economy
◆ The biggest players are the banks, which have cut back their investment in small businesses since the banking crisis of 2008.
◆ The biggest industries in Cornwall are tourism, food and farming. They want greater economic expansion.

5 Environmental stakeholders
◆ Cornwall's biggest asset is its scenery and environment. The National Trust, Royal Horticultural Society and English Heritage all own large areas of land.
◆ Cornwall has huge potential for wave and wind renewable energies.

6 Stakeholders in people
Education, e.g. Combined Universities in Cornwall (see Figure 4 on page 242), Cornwall FE Colleges.

Regeneration projects

Watergate Bay, Newquay

The Extreme Sports Academy at Watergate Bay targets a young adult age group. It offers courses in surfing, wave skiing, and kite surfing. The owners also run the Watergate Bay Hotel. Both are open all year round and employ 50-60 people.

Next door is 'Fifteen', owned by TV-chef Jamie Oliver, which opened in 2006. This restaurant trains local young people in catering. Fifteen 16-24 year olds from disadvantaged backgrounds are selected for training each year. They work in the restaurant, training at Cornwall College and in the kitchens, supported by professional chefs. Profits from the restaurant fund further training and development.

⌃ *Figure 3* *Watergate Bay*

◀ *Figure 4* *CUC, at Falmouth*

Combined Universities in Cornwall (2005 onwards)

To increase the available range of university courses, and develop a 'knowledge economy' in Cornwall, University College Falmouth and Exeter University joined forces to create the Combined Universities in Cornwall (CUC), together with Truro and Penwith Colleges.

- As well as offering degree courses, CUC also helps its graduates to set up their own businesses, or secure jobs in knowledge-based companies in Cornwall – trying to cut the 'Cornish brain drain'.
- The 'student economy' in Falmouth has resulted in a healthy property rental market, and a thriving evening economy of bars and restaurants.

Wave Hub (2010-present)

Wave Hub is a wave-power research project, situated 16 km off Cornwall's north coast. It was installed on the seabed in 2010, and acts as a 'socket' for wave-energy converters to be plugged into. A cable takes electricity from the Hub to the mainland. Its capacity is 20 MW (the equivalent of 6-8 wind turbines).

It cost £42 million to build – with funding from the SWRDA (£12.5 million), the EU (£20 million) and the UK government (£9.5 million). Four private-sector developers will install different devices into the Hub, to test which one works best. The first was installed in 2014. The project will earn £76 million over 25 years for Cornwall's economy, and create 170 jobs. Its biggest potential is the creation of a new wave-power industry for Cornwall.

Superfast Broadband (2011-16)

By 2016, over 95% of Cornwall had access to fibre broadband (the first county to achieve this), and it also had the greatest take-up as a percentage of people. Cornwall now has the world's largest rural fibre network. It cost £132 million: £53.5 from the EU Regional Development Fund, and £78.5 million from BT. It encourages businesses, particularly knowledge-economy companies and those who work from home. An independant evaluation has shown that around 2000 jobs have been created, with an economic impact of around £200 million per year.

▶ *Figure 5* *The advertising used for Superfast Broadband in Cornwall*

The Eden Project

The Eden Project opened in 2001 (see Figure 6). It consists of two large conservatories (known at Eden as 'biomes'), which exhibit the world's major plant types – as well as an education centre about sustainable living and a hostel for residential groups. It has transformed the landscape from a former china-clay quarry, to a completely re-imagined environment.

Before

After

Benefits

In its first ten years, The Eden Project:
- generated £1.1 billion for the Cornish economy (a seven-fold return on its cost)
- attracted 13 million visitors
- employed 650 people directly (many of whom were previously unemployed), as well as supporting 3000 related jobs (e.g. supplying food)
- used 2700 local suppliers
- raised Cornwall's profile, alongside the contributions of Tate St Ives and Rick Stein's restaurants in Padstow
- increased employment in Cornish tourist-related industries
- developed Cornish tourism as a year-round sector, and gave tourists somewhere to visit on a rainy day (although 79% of its visitors arrive between Easter and October half term, the remainder now visit out of season)
- encouraged wider investment by the South West Regional Development Agency, to help regenerate St Austell town centre.

The costs: funding sources	
National Lottery	£56 million
EU and South West Regional Development Agency	£50 million
Commercial loans	£20 million
Other loans and funds generated by the Eden Project	£14 million
Total	**£140 million**

🔺 *Figure 6* *The Eden Project*

Over to you

1 a Working in groups of four, select one of the five regeneration projects (ensuring that each project has been selected by at least one group). Copy and complete the table below, and assess how successful you think your chosen project has been. You might find it helpful to develop a scoring system for each factor.

How well does it suit Cornwall's people and environment?	
What are its costs (economic, social, environmental)?	
What are its benefits (economic, social, environmental)?	
Does it: • provide year-round employment? • diversify the economy? • overcome Cornwall's geographical isolation? • make good use of Cornwall's environment? • help to develop a 'knowledge economy'? • help to prevent further 'brain drain'?	

b Present your assessment of your chosen project, with your reasons, to the rest of the class.

c Now, as a class, reach an agreed rank order to decide which are the most successful projects.

2 Evaluate Newquay's Aerohub Business Park in Section 5.9. How does its success compare with the regeneration projects in this section?

3 Based on these examples, discuss and decide whether the government was right to withdraw most public-sector funding from 2010 onwards.

On your own

4 a Copy Figure 8 in Section 5.12 (the conflict matrix). Complete the players for the examples in this section from Cornwall.

b Identify which groups **(a)** agree, **(b)** disagree, and decide why.

5 Prepare a debate in class on the motion that 'Rural areas should stand on their own two feet without the burden of public-sector funding'.

Exam-style questions

AS 1 Using examples, assess the success of one or more rural regeneration projects. *(12 marks)*

A 2 Evaluate the role of different players in the success of rural regeneration projects. *(20 marks)*

Having studied this topic, you can now consider the three synoptic themes embedded in this chapter. 'Players' and 'Attitudes and Actions' were introduced on page 193. This page focuses on 'Futures and Uncertainties', as well as revisiting the four Enquiry Questions around which this topic has been framed (see page 193).

3 Futures and Uncertainties

People approach questions about the future in different ways. They include those who favour:

- **business as usual**, i.e. letting things stand. This might involve leaving regeneration to a mix of government and private companies. This is well exemplified in Sections 5.12 and 5.13.

- **more sustainable strategies** towards regeneration, particularly the engagement of local communities in the regeneration process.

- **radical action** which might involve directly challenging processes such as market-led regeneration as a suitable model for change.

Working in groups, select two examples of regeneration from this chapter and then discuss the following questions in relation to each one:

1 How far has regeneration created unintended consequences – economically, socially, environmentally?

2 Why is regeneration such a contested issue?

3 Should regeneration be placed in the hands of communities rather than central government?

Revisiting the Enquiry Questions

These are the key questions that drive the whole topic:

1 How and why do places vary?

2 Why might regeneration be needed?

3 How is regeneration managed?

4 How successful is regeneration?

Having studied this topic, you can now consider answers to these questions.

Discuss the following questions in a group:

4 Consider Sections 5.1-5.4. How and why have urban places changed so radically since the 1980s?

5 Consider Sections 5.5-5.7. How should the need for regeneration be judged, and by whom?

6 Consider Sections 5.8-5.10. Who seems to manage regeneration best – local or national government?

7 Consider Sections 5.11-5.13. How far do the processes of regeneration need to be different in rural compared to urban areas?

Books, music, and films on this topic

Books to read

1. *Remaking London: Decline and Regeneration in Urban Culture* by Ben Campkin (2013)

   The author of this book seeks to address the issues that come with the regeneration of areas in London, and how it can in fact work against the principles of regeneration in the first place.

2. *London* by Edward Rutherford (1997)

   This book looks at London's journey through time, and the countless number of regeneration developments that it's gone through to meet changing needs.

3. *City Rebranding: Theory and Cases* by Keith Dinnie (2010)

   This book looks at some of the reasons why urban areas choose to rebrand and regenerate – and also the strategies used to do so.

Music to listen to

1. 'Ghost Town' by the Specials (1981)

   This song addresses the themes of unemployment and deindustrialisation that were taking place in the UK during this period.

2. 'The Hood Ain't The Same' by Draze (2014)

   This song looks at the effects of gentrification in the US city of Seattle. There are many references to changing urban areas.

3. 'Ill Manors' by Plan B (2012)

   This protest song, written not long after the 2011 London riots, addresses some of the issues being created for the working classes by the regeneration of London.

Films to see

1. *The War to Live in London: Regeneration Game* (2015)

   A short documentary looking at the winners and losers of the regeneration of London's housing (a major problem for the city).

2. *My Brooklyn* (2012)

   A documentary that assesses the impact of gentrification on local people and businesses in Brooklyn, New York.

Chapter overview – introducing the topic

This chapter studies how places vary both demographically and culturally, with changes driven by local, national and global processes.

In the Specification, this topic has been framed around four Enquiry Questions:

> 1 How do population structures vary?
> 2 How do different people view diverse living spaces?
> 3 Why are there demographic and cultural tensions in diverse places?
> 4 How successfully are cultural and demographic issues managed?

The sections in this chapter provide the content and concepts needed to help you answer these questions.

Synoptic themes

Underlying the content of every topic are three synoptic themes that 'bind' or glue the whole Specification together:

> 1 Players
> 2 Attitudes and Actions
> 3 Futures and Uncertainties

Both 'Players' and 'Attitudes and Actions' are discussed below. You can find further information about 'Futures and Uncertainties' on the chapter summary page (page 298).

1 Players

Players are individuals, groups and organisations involved in making decisions that affect population composition. They can be individuals, national / international organisations (e.g. IGOs), national and local governments, businesses (from small operations to the largest TNCs), pressure groups and non-governmental organisations.

Players that you'll study in this topic include:

- Sections 6.3 and 6.4 – How UK population diversity is planned by the **government** as part of economic change and, in Section 6.8, how migration is part of government strategy. **TNCs** support this in order to guarantee a skills supply. In Section 6.3, you'll see how migration policies have brought cultural impacts.

- Section 6.10 – How **planners and developers** make controversial decisions about place regeneration, and the effects that this can have on diversity in cities.

- Sections 6.12 and 6.13 – How **local communities, local companies** and **local government** can drive policies to manage diversity. In some cases, partnerships with **NGOs** can also drive change (e.g. in Slough).

2 Attitudes and Actions

Actions are the means by which players try to achieve what they want. For example, those promoting diversity see it for its business potential, as well as its social impacts on a broader and more tolerant society. However, others see diversity in wealth, as well as conflicts between host populations and new migrants. Success in each case depends on the attitudes of different players.

Actions largely concern the following issues:

1 **Economic versus social – the big picture.** Diverse populations and migration are the result of decisions made at national level to counteract skills shortages and an ageing population (see Sections 6.3, 6.4 and 6.10).

2 **Economic versus social – the local picture.** Urban populations, particularly in London and other large cities with diverse migrant communities, are often more tolerant than those in smaller towns and rural communities (see Sections 6.5 and 6.9).

3 **How attitudes may vary.** For example, Sections 6.6 and 6.13 (concerning Cornwall) show how different people vary in their attitudes towards places, while Section 6.12 (concerning Slough) shows some of the challenges of diversity.

> In groups, consider and discuss the following questions:
>
> 1 Diverse populations – good or bad for the UK?
> 2 Diverse populations – good or bad for local places?

In this section, you'll learn how diversity varies in different parts of the UK.

It's off to Uni!

It's mid-September and (like half a million others at this time of year) Tom is packing ready to start university. He's off to study Geography at Queen Mary University of London (QMUL), shown in Figure 1. It's a big change – not just because Tom is leaving home for the first time, but because he comes from Ottery St Mary (a small town of 5000 inhabitants in East Devon). Tom has lived in Ottery St Mary all his life, but he decided that it was time for a change. QMUL is situated in the borough of Tower Hamlets in London's East End (just 3 km from the edge of the City of London – and a world away from Ottery St Mary).

▲ **Figure 1** *Queen Mary University of London (QMUL) in London's East End*

Big city, big changes

Tom's life is about to change significantly. Most changes will be due to the differences between living in a rural area and living in one of the world's fastest-growing and most densely populated cities. The population of Tower Hamlets was about 270 000 in 2014 (a density of 14 000 people per km2), which is double the number of people living in the whole of East Devon – at 135 000, that's a population density of just 170 per km2!

But it's the differences in the people themselves that Tom will probably notice the most. It would be hard to find two more contrasting populations than those of East Devon and Tower Hamlets!

Even outside his university campus, Tom will be surrounded by a younger population than he's used to. About half of Tower Hamlets' residents (49%) are aged between 20 and 39 – and, at 29, the borough has the UK's lowest median age (see Figure 2). London's median age as a whole is 33, and the UK's average is 39.4. By contrast, the average age in Tom's home area of East Devon is 47.

- ◆ Tom will notice fewer older people than in Devon, because Devon has long been a popular area to retire to. By contrast, Tower Hamlets has the UK's lowest proportion of over-65s (just 6.1% of the population). The London average is 11.1%, England's is 16.3% – and East Devon's is 30%.

- ◆ Tom will also find public transport different. There's a London Underground station right across the road from QMUL (with services 20 hours a day on weekdays and 24 hours a day at weekends). At home in Devon, bus services run half hourly from Ottery St Mary to the nearest city (Exeter) – whereas QMUL is served by over 100 buses an hour!

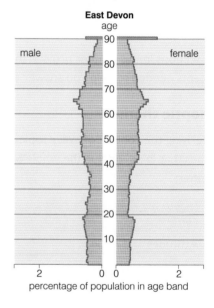

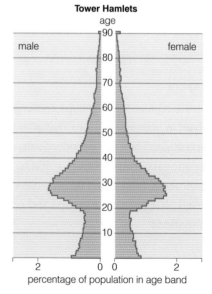

▲ **Figure 2** *Comparing the age-sex diagrams for East Devon and Tower Hamlets*

△ Figure 4 *A street in Shadwell, an inner-city suburb of Tower Hamlets in East London*

Diverse places

Above everything else, Tom will probably notice East London's wide **ethnic mix** (see Figure 3).

◆ Over 93% of the residents of East Devon (and 96.6% of those in Ottery St Mary) classed themselves as White British in the 2011 Census. That compares to just 45% in London, which – with 37% of its population actually born overseas – makes it the world's second most ethnically **diverse** (varied) city, after New York.

◆ Tower Hamlets is one of the UK's most ethnically diverse areas. Just under a third of its population counts itself as White British, but there is also a large Asian or Asian British population and a substantial Black or Black British community.

▽ Figure 3 *Comparing the ethnic diversity of East Devon and Tower Hamlets*

Place	% White, British	% Mixed	% Asian or Asian British	% Black or Black British	% Chinese
England and Wales	83.4	1.80	5.87	2.81	0.82
Tower Hamlets	31.2	2.85	30.58	6.47	1.58
East Devon	93.2	0.90	1.51	0.68	0.30

Diversity is represented in the cityscape too, with shops and services reflecting the population's varied background (see Figure 4).

It's likely that Tom will end up living in London for longer than his three-year degree course. London's economy is growing faster than that of the rest of the UK, so, although it's a very expensive place to live, there are more graduate jobs there – and salaries are higher. London's expanding **knowledge economy** (the economy based on financial services, law, advertising and the media) absorbs a large proportion of UK graduates each year. As a result, London's workforce has become younger, but also more diverse – the number of job vacancies on offer is greater than the number of skilled people available in the UK to fill them, so many employers have to recruit extra skilled workers from overseas.

△ Figure 5 *Broad Street in Ottery St Mary, where Tom comes from*

Over to you

1 a Draw a table to compare the populations of East Devon and Tower Hamlets.

  b Add a column to show how your area compares. Google phrases such as 'White British population in …', adding the name of your county or borough.

2 Compare Tom's experience with that of students who apply for jobs or university places away from your local area. Devise a six-slide presentation to compare your area with a particular university location, by comparing:

- their physical and human landscapes
- their populations and diversity
- other characteristics of your choice.

On your own

3 Explain the difference between '**ethnic mix**' and '**diversity**'.

4 Suggest ways in which the presence of students can change the population and diversity of a university town or city.

5 Suggest possible economic and social reasons why **(a)** many sixth formers from rural areas choose a city location for university, **(b)** many graduates end up staying within 20 km of their old university.

In this section, you'll learn how population structure varies from place to place and over time.

Looking beyond the headlines

In December 2013, the *Daily Express* reported that England was Europe's most overcrowded country – four times more densely populated than France. The *Daily Express* had already become known for its anti-immigration stance, so its message was clear – the newspaper's readers seemed to be encouraged to believe that England was 'full'.

It's important to note that the *Daily Express* was talking about England, not the UK. England certainly has some areas that are very densely populated. London, for example, has 5500 people for every square kilometre. But, if you divide the population of the whole of the UK (65 million in 2015) by its area (243 610 km2), that gives a much lower average population density of 267 people per km2. In world rankings, that makes the UK the forty-second most densely populated country. By comparison, India is twenty-first (with 426 per km2) and Bangladesh is eighth (with 1218 per km2).

There are two important factors to note when thinking about the *Daily Express* report:

♦ England is more densely populated than the rest of the UK – it has 413 people per square kilometre, compared to Wales (149), Northern Ireland (135), and Scotland (68). Perhaps that's why the *Daily Express* deliberately chose to refer to England, and not to the UK as a whole!

♦ Urban areas – that is, settlements over 10 000 people – account for 89% of the country's population, but only 7% of its area. So 93% of the UK is not urban, and is therefore rather less crowded!

The population is growing!

The UK as a whole has a rapidly increasing population. Figure 1 shows that it's grown steadily since 1965, and also that the rate of growth has been faster since 2000 (particularly in major cities like London, Cardiff and Manchester). This presents a challenge for planners who have to decide where the extra people will live and what resources they will need (schools, hospitals, GP surgeries, workplaces, public transport infrastructure).

As well as the above population density differences between the different countries that make up the UK, there are also significant regional differences in population growth rate within England. In 2012, The Office for National Statistics (ONS) projected the population growth rate for the nine English economic regions between 2012 and 2022 (see Figure 2).

Date	Population (million)	% increase in five-year period
1965	54.3	–
1970	55.6	2.4
1975	56.2	1.1
1980	56.3	0.2
1985	56.6	0.5
1990	57.2	1.1
1995	58.0	1.4
2000	58.9	1.6
2005	60.4	2.5
2010	62.8	4.0
2015	65.0	3.5

⊙ Figure 1 *The growing UK population, 1965–2015*

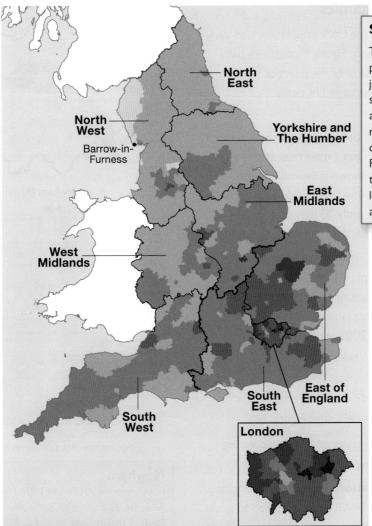

Slowest population growth

The population of the North East is projected to grow the slowest by 2022 (by just 3%). The ten English areas with the slowest projected population growth rates are all in the North East or North West regions. Some areas are even expected to decline – e.g. the population of Barrow-in-Furness (see the map) is actually expected to fall by 2% by 2022. Its naval shipyards no longer provide as much construction work as they did.

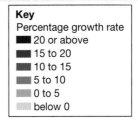

Key
Percentage growth rate
- 20 or above
- 15 to 20
- 10 to 15
- 5 to 10
- 0 to 5
- below 0

◀ **Figure 2** *Population growth rate projections for the nine English economic regions, 2012–2022*

Fastest population growth

London's overall population is predicted to grow by 13% by 2022; the East of England's by 9%; and that of the South East by 8% (compared to England's average population growth rate of 7%). Eight of the ten fastest-growing areas in England are London boroughs (e.g. Tower Hamlets has the highest growth rate of 22%).

The reasons for the regional differences in population growth rate can be divided up into economic, demographic and social:

◆ **Economic**. London's high population growth rate has resulted from its expanding knowledge economy, which has led to an influx of highly qualified workers and their families. Meanwhile – by comparison – in the North East, a slower population growth rate has resulted from the continued collapse of its traditional industries (coal mining, steel manufacture, engineering, shipbuilding), which has led many workers to leave the region with their families to find work elsewhere.

◆ **Demographic**. London's booming economy has led to rapid **internal migration** (within the UK), particularly of young graduates, and **international migration** (from overseas).

◆ **Social**. A longer average life expectancy, caused largely by falling **mortality rates** amongst those aged over 65, means that the UK's population as a whole is now living much longer. Falling mortality rates (e.g. due to increased cancer survival), and improved care of the elderly have resulted in increasing life expectancy – especially in London, the North East, North West and Midlands. In London, male life expectancy rose from 73.3 in 1993 to 80.0 years in 2013.

Population structure and dynamics

As well as differences in growth rate, the UK's population also varies in density and structure (by age and sex). Figure 3 contrasts two extremes within London. Both are typical of the parts of London in which they are located – and of most large UK cities.

Inner London

Figure 3A shows the age-sex structure of Newham (an East London borough). Not every London borough has a male majority like this (see the data panel), but the age structure shown is typical of every Inner London borough – dominated by 21-40 year olds. Newham is easily **accessible** for both work and leisure (only 10 minutes from London's financial centre and 15 minutes from the West End's clubs and theatres).

Planning has resulted in the regeneration of London Docklands and the Lea Valley (the area of the 2012 Olympic Park). Since the collapse of London's original port facilities and traditional industries in Newham, planners have used the resulting derelict land for housing expansion. With so much new housing, between 2001 and 2011, Newham's population was the fastest growing in the UK – by 25%! This population growth came from two main sources:

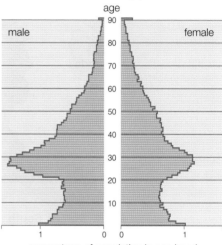

Newham (A)

percentage of population in age band

- ◆ **Natural increase** – Since 2000, Newham has had the UK's highest general fertility rate (see the data panel under Figure 3A).

- ◆ **Internal migration** of mainly university students and young graduates to work in the City's knowledge economy.

Since 2000, the **international migration** rate has also increased. Between 2001 and 2011, Newham's non-UK-born population grew by 72 285. By 2011, the borough had 165 424 foreign-born residents – 55% of its population. Having arrived and settled in the area, many of these migrants then married and began having children – hence Newham's high general fertility rate.

Newham
Population (in 2012): 314 100
Area: 36.22 km²
Density (people per km²): 8672
Average age: 31
General fertility rate (2014): 76.6 per 1000
Gender breakdown: male 52.1%, female 47.9%

Outer London

Figure 3B shows the population details for Kingston-upon-Thames (an outer suburb in South-west London). Kingston is one of London's wealthiest boroughs, with a large number of higher-income couples, who moved there to raise families. It's still very accessible, with 12 trains per hour for the 30-minute journey into Central London. But housing there is expensive and much of the land is protected parkland. At its outer edge, planning for further growth is restricted by London's Green Belt – protected to prevent urban sprawl (see Figure 4).

Unlike Newham, Kingston-upon-Thames is dominated by a UK-born population. Overseas-born migrants constitute only 20% of the population (with the three biggest sources being India, Poland and Ireland).

Kingston-upon-Thames
Population (2012): 163 900
Area: 37.25 km²
Density (people per km²): 4400
Average age: 37
General fertility rate (2014): 57.4 per 1000
Gender breakdown: male 48.9%, female 51.1%

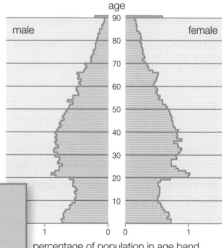

Kingston-upon-Thames (B)

percentage of population in age band

⬤ **Figure 3** *The age-sex structure of the London boroughs of Newham (A) and Kingston-upon-Thames (B)*

Beyond the city

Away from cities, Britain's population becomes increasingly rural. The terms 'rural' and 'urban' are not absolutes; it is hard to define where urban areas begin or end, or when rural areas becomes 'remote rural' (such as the Scottish Highlands; see page 252). Geographer Paul Cloke devised a model in the 1970s to show **the rural-urban continuum** to represent this spectrum (see Figure 4).

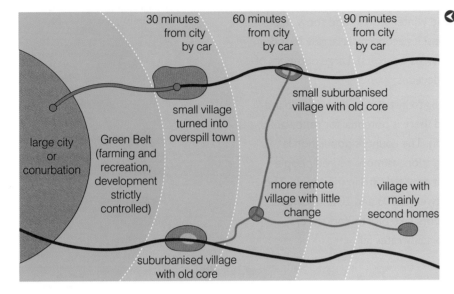

30 minutes from city by car

60 minutes from city by car

90 minutes from city by car

small suburbanised village with old core

small village turned into overspill town

large city or conurbation

Green Belt (farming and recreation, development strictly controlled)

more remote village with little change

village with mainly second homes

suburbanised village with old core

◀ **Figure 4**
The rural-urban continuum

Rural-urban fringe

At the outer edge of most cities is a **rural-urban fringe** (a blurred boundary between countryside and city). Urban areas expand and absorb what was once open countryside – by a process of **urban sprawl**. Historically, railways and roads determined which places would grow – easily accessible villages surrounding cities became **suburbanised**, as planners approved the building of housing estates to form **dormitory suburbs** (homes for commuters). These expanded settlements have gradually adopted urban functions, with shops, schools, and health and leisure facilities. As the travel distance from the city increases, the suburbanised settlements gradually give way to places that are more remote and less accessible to the city – and beyond them are more sparsely populated rural areas.

Mixed rural areas – North Yorkshire

North Yorkshire is a predominantly rural county, although it has all of the components outlined in Figure 4. It contains remote upland areas of the Pennines and North York Moors, both of which are losing population (particularly younger people). The county's population is also older than that of London (see Figure 5). Planners there face the challenge of trying to encourage new employment and housing.

North Yorkshire contains smaller towns and cities, such as Harrogate and York, coastal resorts such as Scarborough, and a mix of small towns and villages between. These areas (unlike the remote upland areas) are increasing in population, due mainly to internal migration from those seeking country living (including newly retired people), especially in accessible rural and coastal locations. International migration is small, with less than 5% of the population born outside the UK. It's also an ageing population, as mortality rates fall.

North Yorkshire
Population (2012): 602 600
Area: 8864 km2
Density (people per km2): 68
Average age: 39.8
General fertility rate (2014): 60.3 per 1000, stable
Gender breakdown: male 49.3%, female 50.7%

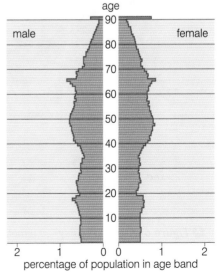

North Yorkshire

age

male

female

percentage of population in age band

⬆ **Figure 5** *The age-sex structure of North Yorkshire*

Remote rural areas – Highland, Scotland

At the UK's extremities lie sparsely populated **remote rural** areas, such as Highland county (see Figure 6). The UK's largest county, it has one small city (Inverness, population 62 000), but the majority of Highland is sparsely populated (see Figure 7) – with villages, remote hamlets and farms far away from any neighbours. Accessibility to the rest of the UK is a problem (apart from Inverness Airport). The county's roads often meander around mountains and lochs, and winter storms and snow make it difficult to commute to other Scottish cities. Its economy is largely based on tourism and farming (with remote, small farms or crofts). Large numbers of agricultural jobs have been lost in recent decades.

However, Highland Council won EU funding to install superfast broadband – thus reducing the region's isolation – and there is potential for renewable energy industries (especially wave and tidal). The county's population is now increasing rapidly, due to internal migration (as more retired people and families seek a rural lifestyle), but most growth is still in the towns. The population of Inverness grew by 17.2% between 2003 and 2013. However, there is relatively little international migration. Like North Yorkshire, Highland's population is also ageing, as mortality rates fall.

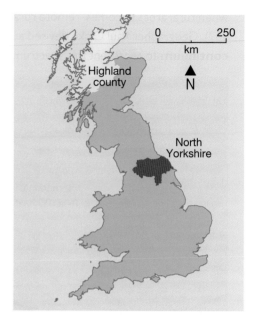

Figure 6 *Highland county – a remote rural area of the UK*

Scale 1: 1 000 000

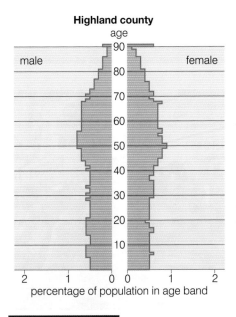

Highland county

 Figure 5 The age-sex structure of Highland county in Scotland

percentage of population in age band

Highland

Population (2013): 232 950
Area: 25 659 km²
Density (people per km²): 9
Average age: 43.2
General fertility rate (2014): 56.1 per 1000, increasing
Gender breakdown: male 48.9%, female 51.1%

Key word

General fertility rate – The number of live births per 1000 women aged 15-44. It is a measure of current fertility levels. This is different from '**fertility rate**', which is the average number of children a women would have during her childbearing lifespan.

Over to you

1 a In pairs, make an A3 copy of, and complete, the following table to show the population characteristics of the four places discussed in this section.

Characteristic	Newham	Kingston	North Yorkshire	Highland County
Population density				
Average age				
Largest age group(s)				
General fertility rate per 1000				
Impact of internal migration				
Impact of international migration				
Accessibility				
How physical factors affect it				
Challenges for planners				

b Identify the most significant differences, and suggest reasons for them.

2 Explain why (a) internal migration is important for all four regions, (b) international migration affects some places more than others, (c) declining mortality rates are affecting places like North Yorkshire and Highland.

On your own

3 Distinguish between (a) internal and international migration, (b) general fertility rate, fertility rate and birth rate, (c) rural-urban fringe and dormitory town.

4 Make a copy of Figure 4 and annotate it to show how the following could affect the rural-urban continuum: (a) an airport like that at Inverness, (b) superfast broadband in rural areas, (c) declining mortality rates and an ageing population.

Exam-style questions

 1 Explain how changing fertility rates are affecting different places. (4 marks)

2 Assess the impact of population change on different places in the UK. (20 marks)

In this section, you'll learn how population characteristics vary from place to place and over time.

Village life – in the city!

Writing in 2010, journalist Caroline McGhie looked for the best places to find a market, a bakery or a coffee shop. Knowing that rural pubs and shops were struggling, she found that only in cities were 'local' shops and services thriving. It's easy to see why – traffic in Britain's cities is congested, so buying what you need day to day in your local area makes sense. Whether it's local shops in London suburbs or gay pubs in Canal Street, Manchester, local communities frequently thrive in cities. The business company Imbiba Partnership summarizes it clearly on its website – 'London remains, quite simply, a collection of villages.'

Welcome to Southall

Once you understand how 'localised' urban communities can be, big cities like London start to make sense. For example, you can see that by looking at the people and shops in Broadway, the main retail street in Southall, West London (see Figure 1). The local connections with the Punjab in North India are clear in the shop signs.

It's nearly seventy years since large-scale Indian immigration began in the UK (after the end of British rule in India in 1947). Many migrants from the state of Punjab in North India ended up settling in Southall. Punjabi soldiers, many of them Sikhs, had served with distinction in the British Army – including during both World Wars. After Indian Independence, many former Indian soldiers and their families decided to move to Britain to live. Many of them came from middle-income farming families in Punjab, and their families still own land there today. Some of these Indian migrants gained employment in London's manufacturing and service industries; many others found jobs in the rapidly expanding Heathrow Airport, 15 minutes away.

In some parts of Southall, half of the residents are Asian or Asian British – with their roots in India (see Figure 2A). Other Indian communities settled elsewhere in West London, and there is also a sizable Indian community in East Ham.

However, Figure 2B emphasises just how localised particular communities can be. It shows that many Asian or Asian British residents with a Pakistani background ended up living in different areas of the capital – with the main concentration in East London. Concentrations of particular communities like those shown in Figure 2 are called **ethnic enclaves**. Together, they form a patchwork of peoples and communities that between them help to create London's **cultural diversity**.

▲ **Figure 1** Shops in Southall Broadway in West London

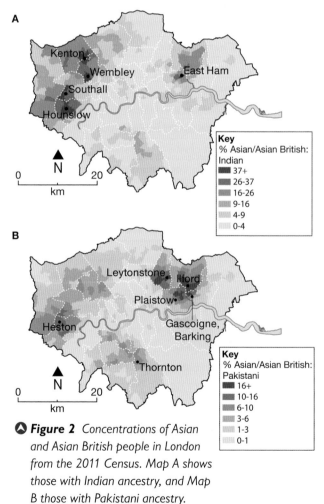

A

Key
% Asian/Asian British: Indian
- 37+
- 26-37
- 16-26
- 9-16
- 4-9
- 0-4

B

Key
% Asian/Asian British: Pakistani
- 16+
- 10-16
- 6-10
- 3-6
- 1-3
- 0-1

▲ **Figure 2** Concentrations of Asian and Asian British people in London from the 2011 Census. Map A shows those with Indian ancestry, and Map B those with Pakistani ancestry.

Key word

Cultural diversity – The existence of a variety of cultural or ethnic groups within a city or society.

Cultural diversity in British cities

In London and most large British cities, **social clustering** (i.e. a preference for living close to people you wish to be with) of different ethnic communities can be seen throughout the city. This is almost always by choice, rather than by chance. In Wembley, for example, where a Hindu community has become established, one of London's two Hindu places of worship can be found – the Shree Sanatan Hindu Mandir Temple (see Figure 3). The distinct Hindu community here began to emerge when local specialist shops selling preferred foodstuffs, together with places of worship, community centres and restaurants, all offered support to the fledgling community. In the early days of migration, these community facilities offered some protection against racism from outside, and also helped to maintain a shared identity. Now, across most British cities, the economic benefits of social clustering like this are equally important, e.g. Bangladeshi restaurants in East London's Brick Lane, or the 'curry miles' in northern cities such as Bradford and Manchester.

▲ **Figure 3** *The Shree Sanatan Hindu Mandir Temple in Wembley*

Understanding immigration in the UK

The UK has a long history of immigration, both for economic reasons (e.g. EU migrants) and for those escaping persecution (e.g. the French Protestant Huguenots in the sixteenth century and Jewish refugees in the nineteenth and twentieth). South Asian migrants came to the UK after 1947 for many reasons, such as escaping war, seeking economic opportunities, or joining family members who had already settled in the UK. Many established local communities with their own clear identities.

Settlement patterns were largely determined by the easy **accessibility of key cities**. For example, many Pakistani migrants gained employment in the northern textile cities of Lancashire (e.g. Oldham and Rochdale) and West Yorkshire (e.g. Bradford); in the car manufacturing and engineering cities of the West Midlands (especially Birmingham); and in industrial estates further south (e.g. Luton). Remote rural Britain offered few job opportunities, and car ownership in those days was limited, so physical factors (e.g. distance and accessibility) were important in determining where the new migrant communities settled. Rural Britain, like East Devon (see Section 6.1), saw few of these new migrants and remains mostly White British today. Upland areas, such as the Lake District or Snowdonia, offered few employment opportunities.

▼ **Figure 4** *The Notting Hill Carnival, held annually in West London, and introduced in the 1960s to mark the positive cultural contribution made by London's West Indian community*

Many new migrant communities began as a result of deliberate **government policies**. After the end of the Second World War in 1945, labour shortages in the UK's textile factories, public transport system (e.g. London's buses), and the newly established NHS, led the government to sponsor job advertisements overseas, alongside agreed immigration targets. In response, Black Afro-Caribbean migrants began arriving from the West Indies from 1948 onwards. The first arrived on the merchant ship *Empire Windrush* to fill job vacancies in London – with most settling in Brixton, Catford, Stonebridge and Tottenham. Most of these areas are still prominent centres for London's Afro-Caribbean community, who have made a big cultural contribution to London life (see Figure 4).

Government policy and EU migration

Since 1995, the number of overseas-born people living in the UK has increased across the whole country (see Figure 5). The main reason for this increase was the UK's EU membership, as well as an expanding EU (28 countries in June 2016). EU migration started to grow after 1995, when the Schengen agreement began to allow the passport-free movement of EU citizens across most of continental Europe. The UK never signed up to Schengen and retains its border controls at all points of entry. Nevertheless, the free movement of EU citizens was a condition of EU membership. With the second largest economy in the EU, and the world's foremost business language (English), economic migration to the UK was always likely to be high.

The geography of migrants

Just as migration to the UK in the 1940s reflected the country's economic needs at the time, immigration today reflects London's domination of the UK economy. London and the South East account for half of all overseas-born people living in the UK.

Four EU countries are now in the top ten source countries for migrants living in the UK (see Figure 6), with Poland at number 2 – ahead of Pakistan and just behind India. Most migrants are working-age adults, aged 21–35. They consist of both skilled and unskilled workers, in equal numbers:

◆ Many **skilled migrant workers** gain jobs in London's knowledge economy. Skills shortages in the UK (e.g. in IT) force many UK companies to recruit overseas workers. Most migrants in this category tend to be white, highly qualified professionals from the EU, USA, South Africa and Australia.

◆ **Semi- and unskilled migrant workers** also find work easily. Many take jobs not wanted by UK workers (e.g. refuse collection), or with unsocial hours (e.g. childcare, pizza delivery). London's construction, hotel and restaurant companies would find it hard to operate without them. Many of these migrants are from the EU, but they also come from India, Pakistan, Bangladesh – and increasingly West Africa (such as Nigeria at number 7 in Figure 6).

Figure 5 The changing number of foreign-born people by region of the UK, 1995–2014

Region	1995	2014	% change
Inner London	816 000	1 374 000	68
Outer London	828 000	1 716 000	107
South East	514 000	1 109 000	116
South West	210 000	434 000	107
East of England	309 000	686 000	122
East Midlands	203 000	485 000	139
West Midlands Metropolitan County	262 000	455 000	74
Rest of West Midlands	94 000	184 000	95
South Yorkshire	49 000	105 000	114
West Yorkshire	140 000	256 000	83
Rest of Yorkshire & Humberside	56 000	119 000	112
Greater Manchester	166 000	353 000	113
Merseyside	35 000	84 000	143
Rest of North West	103 000	195 000	90
Tyne and Wear	24 000	66 000	168
Rest of North East	29 000	66 000	122
Wales	80 000	192 000	139
Scotland – Strathclyde	60 000	125 000	109
Rest of Scotland	97 000	238 000	144
Northern Ireland	53 000	129 000	143
Total	**4 129 000**	**8 371 000**	**103**

Figure 6 The top ten source countries of migrants living in the UK by country of birth in 2013

Rank	Country of birth	% share
1	India	9.2
2	Poland	9.1
3	Pakistan	6.0
4	Ireland	4.4
5	Germany	3.6
6	South Africa	2.5
7	Nigeria	2.4
8	Bangladesh	2.4
9	Romania	2.2
10	USA	2.0

In two East London boroughs – Tower Hamlets (51.5%) and Newham (52.1%) – there are more men than women.

EU migrants are less concentrated than those from Asia, Africa or the Caribbean. Although most Polish migrants, for example, live in urban areas, many also live and work in rural areas. The rural economy depends on EU migrants for many jobs, and many migrate seasonally – from daffodil to vegetable to fruit picking.

Changing fertility and mortality rates

The UK's population is ageing:

◆ Between 1974 and 2014, the UK's median age increased from 33.9 to 40.0.

◆ In 1974, 14% of people were aged over 65; by 2014, it was 18%.

Demographic change has been caused by:

◆ falling total fertility rates (see Figure 7A). Increased prosperity leads to falling birth rates as women make longer-term career choices, marry later and have fewer children.

◆ falling mortality rates and increased longevity (see Figure 7B). Improved health care and successful treatment of heart disease and cancer have improved mortality rates.

These two processes have had a major impact on population structure and the need for immigration. The UK's **dependency ratio** has increased as the size of its working population has fallen. Immigration rebalances this, because:

◆ immigrants tend to be of working age, which helps to increase the government's tax revenue to support those who are retired

◆ in the longer-term, immigrants tend to marry and have children, which increases the UK's birth and fertility rates.

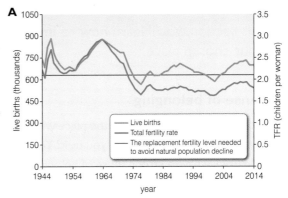
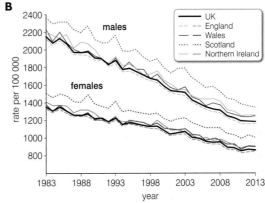

Figure 7 *Graph A: changes in the UK's live birth and total fertility rates (the average number of children born per woman). Graph B: Changes in the UK's mortality rates, 1983–2013.*

Key word

Dependency ratio – The ratio of dependents (aged under 15 or over 64) to the working-age population (aged 15-64), published as the proportion of dependents per 100 working-age population.

Over to you

1 In pairs, use Figures 1-4 to explain (a) why different ethnic groups often live in enclaves, (b) the economic, social and cultural impacts of different ethnic communities.

2 Draw a spider diagram which explains the part played by the following in establishing the UK's cultural diversity: (a) social clustering, (b) accessibility to key cities, (c) physical factors, (d) government planning policies.

3 Research one migrant community near you. Use websites, such as the ONS website, plus local resources to research (a) the numbers involved, (b) the period during which they have been migrating to the UK, (c) their economic, social and cultural contribution to your local area.

On your own

4 Distinguish between (a) 'cultural diversity' and 'ethnic enclave', (b) 'dependency ratio' and 'ageing population'.

5 Use Figure 5 to (a) describe regional variations in the numbers of overseas-born people living in the UK, (b) suggest reasons for the variations.

6 Explain how changes in the following factors have changed London's cultural characteristics: (a) fertility rates, (b) mortality rates, (c) international and internal migration.

7 Write a 750-word report on 'The geography of migrant communities in the UK'.

Exam-style questions

AS 1 Explain how ethnicity can vary within settlements in the UK. *(4 marks)*

A 2 Evaluate the factors which lead to cultural diversity in places. *(20 marks)*

In this section, you'll learn how to investigate connections that have shaped the characteristics of your chosen place.

A sense of belonging

Do you feel that you 'belong' in the place where you live or study? How attached are you to it? Yorkshire people usually have a strong sense of belonging to Yorkshire!

Two neighbouring towns in West Yorkshire – Morley (population 44 000) and Dewsbury (63 000); see Figure 1 – were once at the heart of the West Yorkshire textile-manufacturing region. However, changes experienced by both towns, help to illustrate significant economic and demographic changes within the UK as a whole. Technically, as the map shows, Morley is part of Leeds city – but Morley residents rarely identify themselves with Leeds. Figure 2 has been quoted from 'Secret Leeds', a website that records people's memories of the region.

A sense of regional identity

Like most towns in West Yorkshire, Morley and Dewsbury have similar older buildings that help to give them their character. Figure 3, a photo from 1957, shows a scene typical of both towns at the time. There was a strong sense of community and locality in West Yorkshire towns – each town even produced different textile products from its neighbours! Political geographers and sociologists argue that **centripetal forces** were at work – i.e. forces that drew people together. One of the most significant was work.

- The back-to-back terraced houses (shown in the left foreground of the photo) were characteristic of West Yorkshire's towns. They were built by the owners of the textile mills where most people worked (a mill and its chimney can be seen further down the street on the left). At its peak, Morley had over 30 mills, and their related communities lived and worked closely together.

- Family connections were also important. When someone left school (the building on the right in the photo is one of Morley's schools, with some pupils just arriving), they'd often start work in the same place as one of their parents – sons with fathers, daughters with mothers. Young people were also likely to marry somebody they grew up with and continue to live locally.

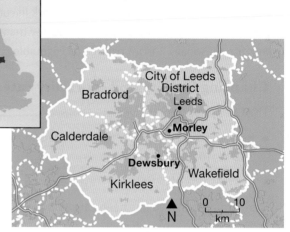

Figure 1 The locations of Morley and Dewsbury in West Yorkshire. The county is divided up into the five boroughs shown, so Morley is part of Leeds city and nearby Dewsbury is part of Kirklees.

As a Morley lad, I didn't consider myself to be from Leeds ... like many who I grew up with. I remember ... when Springfield Mill was still a textile mill and every morning old coaches full of women from South Yorkshire towns like Barnsley used to arrive to work there.

Figure 2 An online respondent recalls growing up in Morley

Figure 3 Peel Street in Morley in 1957. Scenes such as this, where school, home and workplace were all within sight – and a few metres – of each other were not unusual at the time.

All change

If 1950s Yorkshire sounds different from where you live, that's true for most places in the UK. Globalisation and immigration have brought demographic and cultural change. Globalisation has driven traditional communities apart – known as **centrifugal forces** – and immigration has changed the character of many places.

1 The impact of globalisation

By the 1980s, Morley's textile mills had all closed. Without them, younger people were forced to move away – causing **demographic change** – so that, three decades later in 2011, Morley's age profile was older than that of Leeds (see Figure 4). The old textile mills in the centre of town have now been replaced by new industrial estates on the outskirts, like the one shown in Figure 5. Morley is located near the intersection of the M1 (linking Leeds and London) and the M62 (linking Liverpool and Hull), which makes its industrial estates ideal as bases for warehousing and home-delivery companies. Most of Morley's population now works outside the town, which is an easy commute to Leeds and Bradford. Unemployment is lower than the UK average.

2 Immigration and cultural change

Like many of the UK's industrial towns, Dewsbury advertised job vacancies in India and Pakistan in the 1950s. Just as Southall (see Section 6.3) attracted mainly Indian Sikhs and Hindus, Dewsbury attracted mainly Pakistani Muslims, who now constitute 35% of the town's population. Immigration has changed the character of the town. Its mosque, the Markazi Masjid, is an important religious centre, and the town also has a sharia court for local Muslim disputes. Dewsbury has therefore undergone significant **cultural change**. Its close sense of community is helped by the mosque, by close family ties, and by one of northern England's traditional markets.

A changing economy

By the 1980s, several textile mills were owned by TNC clothing chains. However, these companies closed the mills, because clothing produced there was more expensive than that produced overseas. To replace the lost jobs, the government encouraged growth in a new '**post-industrial economy**' in tertiary (service) employment, e.g. distribution and retail like the industrial estate in Figure 5. Dewsbury still has reminders of its industrial past, as Figure 6 shows.

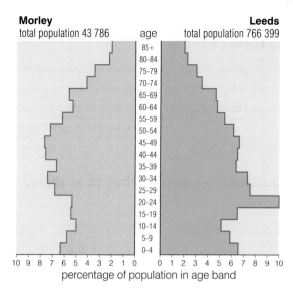

Morley
total population 43 786

age

Leeds
total population 766 399

85+
80–84
75–79
70–74
65–69
60–64
55–59
50–54
45–49
40–44
35–39
30–34
25–29
20–24
15–19
10–14
5–9
0–4

10 9 8 7 6 5 4 3 2 1 0   0 1 2 3 4 5 6 7 8 9 10
percentage of population in age band

⊘ **Figure 4** *Morley's age profile in 2011, compared to that of Leeds. The Leeds age profile is strongly influenced by the number of students at its three universities.*

⊘ **Figure 5** *Leeds 27, one of the new industrial estates in the Morley area*

▷ **Figure 6** *Dewsbury in West Yorkshire, with textile mill buildings still standing in the centre of the photo*

Investigating local demographic and cultural changes

Investigating change where you live or study is fertile territory for geographical fieldwork! There are many ways of collecting data – both quantitative and qualitative – to observe, measure and record changes that are happening locally.

Quantitative approaches to studying places

Quantitative data are sometimes called 'hard data', and are collected on scientific principles (e.g. objectivity). Most quantitative data consist of tables of figures. They can be **primary** (collected yourself, unprocessed) or **secondary** (collected and processed by another party).

1 Migration surveys

Demography is the study of changing populations: numbers, structure and characteristics. It focuses on social rather than economic processes – where people were born, their migration patterns (including at what age they moved), and their reasons for moving. Surveys such as Figure 7 are very good at recording population mobility.

2 Profiling

Profiling uses Census data to identify population characteristics. An example is shown in Figure 8 for Morley and Dewsbury. The data show how different the two towns are now, and Census data from 1951 would show what had changed. To obtain Census data, visit the Office for National Statistics (ONS) website, or use search engine phrases such as 'Census data Dewsbury'. Many county and borough councils publish Census data and GIS maps to show several indicators, e.g. multiple deprivation.

Please answer the following questions about places you've lived:

1 Where were you born? (The place and the country it's in)

_____

_____

2 Please list below each time you've moved home. (Make sure that you ask for the person's age when they moved each time.)

	1	2
Place (name of place and country)		
How old were you when you first lived there?		
What was your main reason for going to live there?		
Was it a better, similar, or worse place to live than the place before?		

3 Please tell me the number of relatives (but not parents) living in the place where you live.

No relatives	1–4	5–9	10–19	20–29	30 and over

▶ *Figure 7 A family member survey*

	Morley	Dewsbury
Ethnicity (%)		
White British	95.5	60.8
EU	1.8	1.8
Southern Asian (e.g. India, Pakistan, Bangladesh)	1.5	36.5
Black or Black British	0.1	0.01
General health (%)		
General health: not good or with a limiting long-term illness	18	19.5
Housing (%)		
Living in social-rented housing	12.9	23.7
Owner-occupied housing	69.9	59.3
Employment status: aged 16-74 (%)		
Employed full-time	47.0	40.2
Employed part-time	14.2	14.8
Permanently sick / disabled	3.8	6.2
Unemployed	3.9	2.75
Highest qualification level reached (%)		
No educational qualifications	24.9	34.2
Qualified with university degree	22.8	15.1

▶ *Figure 8 A 2011 Census profile to compare Morley and Dewsbury*

Qualitative approaches for investigating places

1 Photos

Like the scene in Figure 3, photos of places from different periods allow you to study change – and then to explore through interviews (see below) reasons for those changes, and the ways in which they affected local people. An older generation may have photos to actually show the changes taking place. For example, Figure 9 shows Machell's Mill in Dewsbury (a former textile mill which has now been converted into flats). Local research in libraries and online could trace its past history in words and pictures, and Census research could tell you which local houses were once owned by the mill owner.

2 Interviews

Interviews record people's personal experiences. They can be **structured** (where everyone is asked the same questions) or **semi-structured** (with core questions for everyone, but the freedom to go 'off script' where appropriate). Interviews can be carried out with individuals or with focus groups. Section 5.4, Figure 8 (page 207) shows a semi-structured interview used by Queen Mary University of London (QMUL) students in Canning Town, East London. You could repeat this for your own locality.

3 Social media

Different social media are increasingly being used by a wide cross-section of the population. In 2016, there were several Facebook pages about Morley and Dewsbury, which featured news, photos and memories of each place. The local newspapers for each town – the *Morley Observer* and *Dewsbury Reporter* – also feature on Twitter. These sources can be used to research change, and are important for recording people's personal experiences. Many memories are of schooldays, of changes to shops, places of work and housing – and also whether people perceive these changes to have been beneficial or not.

△ Figure 9 *Machell's Mill in Dewsbury – a former textile mill, now converted into housing*

Over to you

1 In pairs, design two spider diagrams to show (**a**) centripetal forces that made Morley and Dewsbury close communities in the 1950s, (**b**) centrifugal forces that have occurred since.

2 In groups, trial the use of (**a**) a migration survey (Figure 7), (**b**) a Census profile (Figure 8), (**c**) a photo survey, (**d**) an interview survey. Present your results and what they say about the population of your local area.

3 In pairs, search social media (such as Twitter or Facebook) to research how people relate to your local area.

On your own

4 Distinguish between the terms (**a**) centripetal and centrifugal forces, (**b**) the 'post-industrial economy' and the 'knowledge economy'.

5 Compare the photos in Figures 3 and 5. Suggest (**a**) the evidence of economic changes that have affected Morley, (**b**) how far these changes may have been beneficial.

6 Identify evidence in Figure 8 to suggest that (**a**) Dewsbury has been more affected by immigration than Morley, (**b**) Dewsbury has greater deprivation than Morley.

7 Discuss in 500 words how much your local area has been affected by (**a**) globalisation, (**b**) employment change, and (**c**) inward migration.

Exam-style questions

AS 1 Discuss how far regional and national influences have shaped the characteristics of one place you have studied. *(8 marks)*

A 2 Examine the nature of the connections which have shaped the characteristics of one place you have studied. *(20 marks)*

In this section, you'll learn how the same urban places can be perceived differently by different people, because of their personal experiences.

Manchester: a dangerous place to be ...

Cities are often portrayed as exciting places to be – full of vitality, energy and potential. But that's not always been the case. In the 1830s and 1840s, the novels of Charles Dickens gave uncomfortable accounts of life in Britain's rapidly growing industrial cities. In *Oliver Twist*, Dickens portrayed London's dark underbelly through characters like Fagin, who ran a gang of child pickpockets; Bill Sikes, a vicious murderer; and his girlfriend, Nancy, a prostitute.

London was not alone. In 1848, Elizabeth Gaskell published a landmark novel, called *Mary Barton*. It was a murder story set in industrial Manchester – but its significance was in how it portrayed the life of Manchester's industrial poor. Typically, their homes were like those shown and described in Figure 1 or even worse. The description in Figure 1, from *Mary Barton*, of the Davenport family's slum home, describes a living space that typified Britain's industrial cities in the nineteenth century. The description of the cellar where the Davenports lived is full of lurid images to convey the stench, damp and danger to health.

Friedrich Engels, a German philosopher, also described England's industrial slums in his study called *The Condition of the Working Class in England in 1844*. This was also set in Manchester, and was not a novel but a report about the conditions, poverty and filth in which working people were forced to live in England's industrial cities. He wrote:

> The streets are generally unpaved, rough, dirty, filled with vegetable and animal refuse, without sewers or gutters

The significance of these Victorian writers was that they forced their middle-class readership to have a greater awareness of the grim lives of the industrial poor in Britain's cities – most of which (like the mill in Figure 2) they would almost certainly never see themselves.

Figure 1 *Slum housing in Manchester at the end of the nineteenth century. It's worth remembering that photographic technology was developed in the second half of the nineteenth century, when the worst slums had already been cleared.*

> The smell was so foetid as almost to knock the two men down... they began to see three or four little children rolling on the damp, nay wet, brick floor through which the stagnant, filthy moisture of the street oozed up; the fireplace was empty...

Figure 2 *Murray's Mill, a survivor of industrial Manchester. The mill has been restored and now looks very different from the filth and smoke that characterised the city in the nineteenth century.*

... and also a desirable place to be!

Nineteenth-century cities were often perceived as dangerous. How images change! 15% of Manchester's population is now students – who are attracted by the city's reputation for great student life:

◆ Dedicated spaces for students, such as Manchester University's Fallowfield Campus (with its eight halls of residence), offer student-focused life (see Figure 3).

◆ Manchester has long been an attractive city for students who enjoy or study music. It has the Royal Northern College of Music, and a reputation for indie music as the base of Factory Records in the 1980s (think Joy Division, New Order). Now its venues include The Warehouse Project, Manchester Arena, and many clubs.

◆ The wider city offers sport (with two Premiership football teams), theatre, and shopping (the Arndale in the city centre, and the Trafford Centre outside the city).

◆ Manchester also offers a growing knowledge economy, with graduate employment in IT, science and engineering, and a growing financial services sector. Media employment has also increased since the BBC moved many operations to MediaCity in Salford.

The impact of these social, cultural and economic changes has transformed the age-sex profile of Manchester's population – with the largest male and female populations in the 20-30 age band, compared to that of Greater Manchester as a whole (see Figure 4).

The other side of the city

But the image of Manchester as a lively place to be a student is not all positive. The following quote comes from the online chatroom 'The Student Room':

> I have applied to Salford this year and have heard really bad things about Salford as a place, saying its really rough and you get shot n all types of things.

Everyone can think of urban places that are considered to be unsafe – or even threatening – where you might hesitate before entering alone or at night. Salford, with its run-down housing estates, has had such a reputation in the recent past.

What gives some places poor reputations? Possible reasons include poor environmental quality, e.g. litter or graffiti, or buildings that have been allowed to deteriorate prior to redevelopment. It can also – quite simply – be racism, as people perceive migrants and other ethnicities differently. Anti-Muslim sentiment in the tabloid press can encourage this.

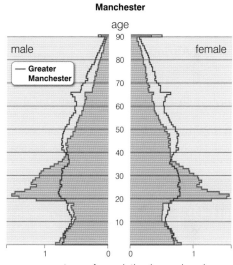

⊙ **Figure 3** *Life for Manchester University students in Fallowfield – this is the FEAST! Festival Picnic by the Lake Platt Fields Park in Fallowfield*

⊙ **Figure 4** *Manchester's age-sex pyramid, showing the impact of its large student population on the city. The red line shows the population profile for Greater Manchester as a whole (including towns such as Salford, Oldham, and Rochdale)*

Crime rates can also contribute to reputations (see Figure 5 on page 264). But are there data to support the student's anecdotal fear that she might be shot walking through Salford? In Manchester, there were 33 planned, attempted or actual homicides in the year to September 2015 (a rate of 6.3 per 100 000 population, and the sixth highest rate in England and Wales). Ten of those 33 were actual homicides. Salford was seventh with 15 (two of which were homicides) – a rate of 6.2 per 100 000 – and Rochdale was eighth with 13 (seven of which were homicides) – a rate of 6.1 per 100 000.

▼ Figure 5 *Characteristics of different parts of Greater Manchester*

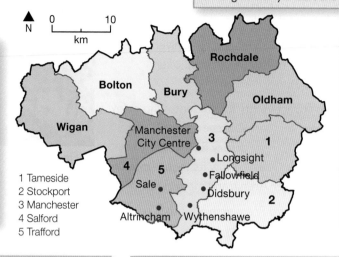

Manchester City Centre
Includes mainly commercial offices and shops, with a significant evening economy of bars and clubs. Little residential property.

Salford
Formerly older housing, with some high-crime areas, but recent regeneration has renewed some housing – and crime figures are now falling. Regeneration schemes include social and private housing, plus some up-market developments (such as Salford Quays).

Longsight
An inner-city area, consisting of mostly traditional red-brick terraced housing, with a 53% ethnically diverse population. It had a reputation as a high-crime area between gangs until 2009, when many gang members were arrested. Since then, violent crime has fallen dramatically.

1 Tameside
2 Stockport
3 Manchester
4 Salford
5 Trafford

Sale and Altrincham
A mixed, outer-suburban area, well-connected to Central Manchester via the tram network, so it's popular with commuters. It's more family oriented, with some middle-class areas of up-market housing. It has large parks and leisure spaces. Recent new-build developments include flats for young professionals.

Fallowfield
Situated 2 miles south of the city centre. It has older housing, with a mix of owner-occupied, social housing, and (increasingly) a student population with several halls of residence and a number of student houses. It's a good location for students, although residents increasingly complain about the noise and disruption of late-night student parties. Student areas are normally at high-risk for the theft of phones, stereos, TVs, etc.

Ethnicity (2011–12)	All White	All Asian	All Black British, African and Caribbean	All non-White population	House price (average 2016)	Detached house	Semi-detached house	Terraced house	Flat
Longsight	27.2%	55.3%	9.8%	72.8%	£112 272	£224 973	£150 399	£102 889	£97 504
Fallowfield	61.9%	19.6%	7.0%	38.1%	£169 982	£295 381	£189 965	£161 690	£136 328
Salford	85.6%	4.0%	2.8%	14.4%	£144 554	£297 694	£167 258	£113 690	£157 433
Sale and Altrincham	95.9%	2.5%	0.6%	4.1%	£264 182	£417 512	£275 211	£199 867	£152 920

Reported crime data (December 2015)	Total crimes	Anti-social behaviour	Violence & sexual offences	Criminal damage & arson	All forms of theft
Salford (East & West) (population 235 000)	2118	651 (30.7%)	328 (15.5%)	334 (15.8%)	398 (18.8%)
Manchester City Centre	1837	571 (31.1%)	254 (13.8%)	70 (3.8%)	577 (31.4%)
Longsight (population 15 500)	1122	330 (29.4%)	217 (19.3%)	93 (8.3%)	326 (29.1%)
Fallowfield (population 15 600)	671	152 (22.7%)	125 (18.6%)	64 (9.5%)	210 (31.3%)
Altrincham and Sale (population 60 000)	801	253 (31.6%)	111 (13.9%)	91 (11.4%)	186 (23.2%)
Total	**6549**	**1957 (29.9%)**	**924 (16.1%)**	**561 (9.8%)**	**1511 (26.3%)**

Lived experiences and 'othering'

Different groups can perceive the same urban areas differently as places to live and work. The personal views expressed in Figure 6 were an online response to someone moving to Manchester and wanting to know which areas to avoid. These views are balanced, evidenced, and are different from opinions that demean other people, using terms such as 'chavs' – a process known in psychology as '**othering**'. This process deliberately singles out and separates people thought to be different from oneself or from the mainstream. It can also encourage the development of ethnic enclaves – where those who are 'othered' prefer to live close to each other for mutual support.

With 'othering', any underlying causes are left unexplained. A good example of this is the media representation of the fictional Chatsworth council estate in Manchester, which featured in Channel 4's adult drama *Shameless* (2004–13). Chatsworth represented an example of urban council estates built in the UK in the 1960s to re-house families from inner-city slums.

During the 1980s, estates like Chatsworth suffered economic recession as a response to industrial decline. Their reputations fell as crime increased and cuts in government spending reduced housing maintenance – lowering environmental quality. The characters in *Shameless* were out of work, often at the pub, and had frequent run-ins with the police. Although the series was highly rated, journalist Owen Jones has criticised it in his book *Chavs – the demonization of the working class*, because it failed 'to address how the characters ended up in their situation, or what impact the destruction of industry has had on working-class communities in Manchester'. Ethnic communities are often similarly 'othered'.

Personally I'd add... Longsight to the list. Constant sirens, police helicopters buzzing, drunks outside the door, people shouting all night long, tyres screeching, all of my housemates had something stolen... what a place... (It's) a shame as well, 'cause the neighbours were lovely and... it was a very leafy, good-looking street and a lovely bunch of houses.

▲ **Figure 6** *Personal views about Longsight, just south of Manchester's CBD (see Figure 5), from someone who had actually lived there. This person was replying online to someone wanting to know which areas of Manchester to avoid.*

Exam-style questions

AS 1 Assess to what extent cities can be seen as 'dangerous places'. *(8 marks)*

A 2 Assess the relative importance of ways in which different urban places are perceived as places in which to live and work. *(20 marks)*

Over to you

1 In pairs, use newspapers to explore how media images of cities today compare with those of Victorian cities as dangerous places to live.

2 a Use the website police.uk to obtain crime data for three urban places that you know.

  b Devise a six-slide presentation to show variations in crime between the three places.

  c Now use the website mouseprice.com to compare how far crime rates affect house prices in the three places.

3 a Design and carry out interviews with people of two contrasting age **or** ethnic groups about their perceptions of inner-city and outer-suburban areas as places to live and work in one urban area that you know. This could include your student colleagues about their perceptions of their university choices.

  b To what extent have you found examples of 'othering' (e.g. of different age groups or ethnicities) in your interviews?

On your own

4 Distinguish between the terms 'perception' and 'othering'.

5 a Use Figure 5 to compare the five identified areas of Greater Manchester in terms of (**a**) data for different crimes, (**b**) house prices, (**c**) ethnicity.

  b To what extent do the data identify Salford as somewhere that is 'really rough and you get shot'?

6 Suggest reasons for the following:

- Inner-city areas are seen as desirable by younger age groups and some ethnic communities.
- Outer-city areas, such as Sale and Altrincham, are preferred by older age groups.

In this section, you'll learn how rural places can be perceived differently by different people because of their experiences.

Cornwall – the best place to be!

Just as cities can be perceived differently – providing opportunity and excitement for some, but seen as threatening and risky to others – rural places can also be viewed in different ways. The *Metro's* headline for 27 March 2015 (see Figure 1) had no doubts – rural Cornwall was the best! Among the 15 reasons given to justify this statement, it listed the following:

◆ Superfast broadband. Not perhaps what most people would think of first, but Cornwall was the first place in the UK to trial superfast broadband. By 2015, 92% of Cornish people and businesses were connected, which is well above even London's average! It's a good way of attracting new businesses to the county.

◆ After work drinks on the beach. The *Metro* claimed that you could leave work at 5pm and be on the beach at 5.30! Cornwall is a long narrow county, so you're never far from the sea.

◆ Its coast and scenery. Figure 2 says it all!

◆ Local food and drink. To the existing list of Cornish pasties, over 60 cheeses, ice cream, clotted cream, and fresh seafood, can now be added award-winning wines. Camel Valley wines in North Cornwall have won gold medals for sparkling wines in competition with French champagnes!

Cornwall's rural geography

With 700 km of coastline, an area of 3560 km², and a population of 540 000, Cornwall's population density of 155 per km² makes it predominantly rural. It's a long county – two hours by train from Penzance in the west to Plymouth on the Devon/Cornwall border. The largest town, St Austell, has a population of just 35 000, and its county town, Truro, is one of the smallest county towns in the UK.

Cornwall has historically been isolated from the rest of the UK. There are no motorways; the nearest is the M5 at Exeter in Devon. Rail is also slow west of Exeter, and Cornwall's only airport, at Newquay, has few flights outside the summer tourist season. This presents a problem for business people, who spend a lot of time and money travelling, which adds to their costs. That's one of the reasons why having superfast broadband is seen as so important for Cornwall.

15 reasons why Cornwall is the best place to live in the UK

Forget official statistics – a quick straw poll amongst Cornwall-based friends has affirmed that us Cornish inhabitants are a pretty happy crew.

With miles of coastline, a sub-tropical climate, and the home of Poldark, there are plenty of reasons why Cornwall is considered to be the best place to live.

Economically we may be classed as one of the poorest areas in the UK – but the 15 examples we've listed show that we're the richest in many other ways.

Figure 1 *The Metro's view of Cornwall in March 2015 – the best place to be!*

Figure 2 *The classic view of Cornwall in summer – the rural idyll*

Some like it that way!

Cornwall's remoteness may put some people off, but there are many others who like its isolation – it's led to a sense of Cornish 'identity' that many want to share, and which forms a strong part of its attraction for tourists.

◆ The Cornish Nationalist Party, Mebyon Kernow, has councillors on Cornwall Council.

◆ The Cornish language – Celtic in its roots (like Gaelic and Welsh) – only has 4000 speakers, but they are powerful enough to campaign for bi-lingual road signs.

◆ Some of Britain's earliest settlements can be found in remote West Cornwall.

◆ The Cornish coastline thrives on stories of past smuggling, protected by its small sheltered harbours, as well as dramatic shipwrecks and fishermen's tales.

◆ Cornwall's coast – and its clear Atlantic light – has also attracted generations of artists, particularly to enclaves such as St Ives in the far west of the county.

Like other remote counties with a strong heritage – both real and fictional – Cornwall is never short of tourists. Just as many people visit Dorset to see Thomas Hardy's literary Wessex, so too many tourists visit Cornwall to see the setting for Daphne Du Maurier's novels, such as *Jamaica* Inn or *Rebecca*, or Winston Graham's *Poldark* novels, dramatized by the BBC in recent years. '*Poldark* country' lies in the west of the county, where Cornwall's tin- and copper-mining heritage (shown in Figure 3) provides the dramatic scenery of abandoned mine workings clinging to the rugged Atlantic coastline. These are landscapes to which people become easily attached.

▲ **Figure 3** *Relics of tin and copper mines in West Cornwall – the setting of the* Poldark *novels, dramatized on BBC television*

Living in Cornwall

Actually living in Cornwall presents different challenges to simply visiting for a week or two on holiday (see Figure 4). Cornwall has one of the UK's fastest-growing populations – in spite of an ageing population and the departure of many young people for university and/or work elsewhere (see Section 6.1). The recent population increase has largely been caused by inward migration. Traditionally, Cornwall has been a popular choice for retirees. However, more recently, many families have decided to move to Cornwall, looking for a less-stressful rural lifestyle. Many incoming migrants are actually 'returners', i.e. they originally came from the county.

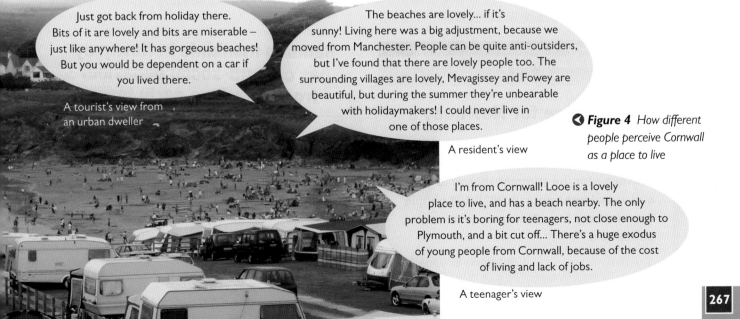

Just got back from holiday there. Bits of it are lovely and bits are miserable – just like anywhere! It has gorgeous beaches! But you would be dependent on a car if you lived there.

A tourist's view from an urban dweller

The beaches are lovely... if it's sunny! Living here was a big adjustment, because we moved from Manchester. People can be quite anti-outsiders, but I've found that there are lovely people too. The surrounding villages are lovely, Mevagissey and Fowey are beautiful, but during the summer they're unbearable with holidaymakers! I could never live in one of those places.

A resident's view

◀ **Figure 4** *How different people perceive Cornwall as a place to live*

I'm from Cornwall! Looe is a lovely place to live, and has a beach nearby. The only problem is it's boring for teenagers, not close enough to Plymouth, and a bit cut off... There's a huge exodus of young people from Cornwall, because of the cost of living and lack of jobs.

A teenager's view

The downside of Cornwall

However, there is a downside to the 'rural idyll' of Cornwall. Several factors stand out:

◆ **Its remoteness**. Cornwall is distant in terms of travel times from other parts of the UK. In winter 2014, coastal damage at Dawlish in Devon (see page 108) cut the only rail link into Cornwall. Many Cornish hotels suffered for several months afterwards, because the national media gave the impression that the South West was 'closed for business'.

◆ **Lack of commuting opportunities**. Only Saltash and parts of east Cornwall have easy access to Plymouth and the higher salaries and urban lifestyle available in cities.

◆ **The climate** (see Figure 5). In spite of its higher sunshine hours and warmer winters, Cornwall has some big climatic disadvantages, including frequent storms and high winter rainfall.

◆ **Limited social opportunities**. Teenagers notice this (see Figure 4 on page 267), but so do older people who have retired to the county – away from old friends and family.

◆ **A limited range of services**. Many village shops are under threat, and Exeter and Plymouth (both in Devon) are the only large cities available for certain types of shopping. However, shopping online has begun to ease the problem. Access to health care (see opposite) is also a problem.

◆ **High transport costs**. Cornwall has high levels of car ownership, but commuting or shopping are both costly in terms of fuel consumption.

◆ **Tourism**. Some parts of Cornwall, such as Newquay, have a poor reputation. Many local people feel that its popularity for drunken stag and hen parties has damaged the town's reputation (see Figure 6).

Climate data	Cornwall	London	Manchester
Sunshine hours per year	1640	1620	1420
Rainfall per year (mm)	1000	625	810
Days of rainfall (>1mm) per month	12–13	10	12–14
Days of air frost per year	14	34	41
Average winter temperatures (°C)			
Max	9.0	8.0	7.5
Min	4.0	2.0	1.5
Average summer temperatures (°C)			
Max	19.5	24.0	20.5
Min	14.0	14.0	13.0

◀ **Figure 5** *Cornwall's climate, compared to that of London and Manchester*

▼ **Figure 6** *News headlines about Newquay*

Home Business World

Extra police for Newquay's summer season

Extra police officers are patrolling in Newquay for the summer season to cut anti-social behaviour. In 2011, 104 people were banned from Newquay town centre, police said.

(BBC News 30 June 2012)

Banning mankinis in Newquay has cut anti-social behaviour from notorious party town

Officers in the Cornish town said a 'robust' attitude to inappropriate behaviour in public has helped to shed its 'Wild West' image as a haven for stag and hen parties.

(*Daily Mirror*, 28 May 2015)

Cornish health services

Access to health facilities can be poor in rural areas. West Cornwall (see Figure 7) is the county's remotest area. Its population is elderly. In recent health surveys, it was ranked as one of the UK's most deprived areas, as measured by:

♦ its shorter life expectancy

♦ the greater likelihood of serious illness and disability

♦ the proportion of adults under 60 who suffer anxiety disorders and periods in hospital, and who claim health benefits.

Only 38% of west Cornish villages have a doctor's surgery, and even then their opening hours vary – St Just has a surgery every weekday, but the villages of Sennen, St Buryan and Polgigga have surgeries just one morning a week.

♦ Access to transport for more serious medical appointments is difficult. Buses to Penzance and St Ives operate in 70% of villages, but there may be only 3-4 a day. People rely on cars or taxis.

♦ The Royal Cornwall Hospital in Truro provides wide-ranging treatments, including emergency care. But living in Penzance means a 26-mile journey to Truro, which affects survival rates for severe heart attacks and strokes. In the tourist season, traffic congestion can also be a problem.

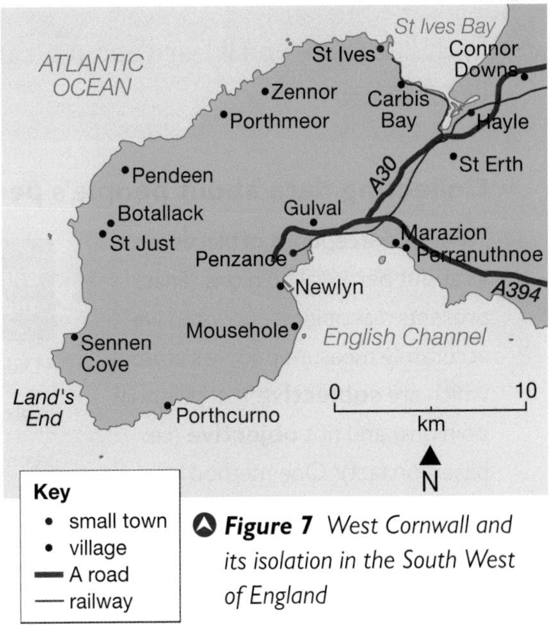

▲ **Figure 7** West Cornwall and its isolation in the South West of England

Key
- small town
- village
- A road
- railway

Roseland Parc retirement village

Roseland Parc is a new type of rural enclave. It's a retirement village built *within* the village of Tregony in South Cornwall, yet completely separate from it. It consists of one- and two-bedroom flats, which are 'sheltered' with a warden. There is a minimum age of 55. On site there are facilities, such as a restaurant, and there are nurses available, plus care home and full nursing facilities for the frail and for end-of-life care. It's not cheap; flats cost about £200 000, plus service charges, and care provision costs £4000 per person per month. These enclaves are likely to be more common in future as the UK's rural population ages.

Exam-style questions

AS 1 Examine to what extent rural places can be described as idyllic. (8 marks)

A 2 Discuss the evidence base to explain why rural locations are sometimes perceived as undesirable. (20 marks)

Over to you

1 In pairs, use the information in this section to identify the characteristics of Cornwall that people might see as 'idyllic'.

2 a Research online advertising materials for Cornwall (e.g. tourist promotions material, art exhibitions, holiday cottages) to show how this 'idyll' is significant in promoting Cornwall.

   b Research online the phrase 'What's Cornwall like to live in'. Use the material and Figure 4 to compare visual and text images you find with those in the tourist information. Explain any similarities and differences.

3 a In pairs, draw a table headed 'Quantitative evidence' and 'Qualitative evidence'. Complete it to show the evidence base for the negative side of Cornwall.

   b Decide on balance whether the positives of Cornwall outweigh the negatives for (i) teenagers, (ii) families, (iii) retirees.

On your own

4 Identify and explain those features of Cornwall regarded as 'undesirable'.

5 Explain in what ways rural areas often suffer poor health services.

6 To what extent is the damage to Newquay's reputation described in Figure 6 something that could affect Cornwall as a whole?

7 For a rural place that you know well, compare it with Cornwall in terms of its (a) physical landscape, (b) cultural identity, (c) upsides and downsides.

In this section, you'll learn about a range of ways by which to evaluate how people view their living spaces.

Collecting data about people's perceptions of places

People's perceptions of places are all about personal opinions. That presents a problem – how can we accurately measure people's attitudes, which are **subjective** (i.e. personal opinions) and not **objective** (i.e. based on fact)? One method involves studying house prices! This is a blunt instrument, because houses vary from small flats to country mansions, but basically a house (like anything) is worth what someone is willing to pay for it (house price = objective fact). However, the price paid reflects the desire of someone to live in a particular house in a particular area (desire = subjective opinion) – so, house prices in an area are a fairly accurate reflection of people's perceptions of that area.

Boroughs and districts	Population (2011 Census)	Average house price	Reported ASB incidents	Total reported crimes	Crime rate per 1000 people
Didsbury	26 788	£294 084	118	492	18.4
Fallowfield	15 211	£169 982	152	671	44.1
Longsight	15 429	£112 272	330	1122	72.7
Manchester East	82 943	£80 658	359	1101	13.3
Manchester North	98 709	£99 861	396	1276	12.9
Oldham	98 609	£135 557	429	2439	24.7
Salford	224 900	£144 554	651	2118	9.4
Trafford North	94 310	£165 687	188	773	8.2
Trafford South (includes Sale & Altrincham)	117 490	£264 182	253	801	6.8
Wythenshawe	74 200	£137 276	219	685	9.2

Figure 1 *Population, house prices, and some reported crime data for selected Greater Manchester boroughs and districts in December 2015. See Figure 2 for their locations. ASB stands for 'anti-social behaviour'.*

To understand this further, it helps to analyse factors that influence house prices, such as crime statistics. Data from the national crime database (www.police.co.uk) show how far crime statistics correlate with house prices. The database uses police neighbourhoods within particular geographical areas, and zooms in to crimes recorded in specific streets or houses on certain dates.

Figure 1 includes some crime statistics for selected areas of Greater Manchester. When examining these statistics, it's important to take account of population size. Salford, for example, has one of the largest reported crime *totals* in Figure 1, but it also has a much larger population than the other areas referred to! Therefore, it's important to calculate a '**crime rate** per 1000 people' (the last column in the table). To do this, divide total crimes by the population and multiply by 1000. This changes the result for Salford significantly – from a place that (on the surface) has many crimes, it actually emerges as an area of Greater Manchester with one of the lowest crime rates in the table.

Among the categories of crime listed in the table is anti-social behaviour (ASB), which accounts for 30% of all UK reported crime. It's the largest category in most places – even in rural areas. Given that ASB is a nuisance for those on the receiving end, a working hypothesis might be that 'House prices are related to the number of ASB incidents'. The relationship between the two can then be explored using two techniques – a scatter graph (see Figure 3) and Spearman Rank (see page 307).

1 Tameside
2 Stockport
3 Manchester
4 Salford
5 Trafford

Figure 2 *Selected Greater Manchester boroughs and districts*

BACKGROUND

Researching your own data

If you want to replicate similar data to those in Figure 1 for your own area, the process is mostly straightforward.

◆ **Crime data**. Use www.police.uk to obtain crime data:
  • Simply click on 'Find your neighbourhood' to enter a place name or postcode. This will take you to a map of your policing neighbourhood.
  • You can then obtain police data from any scale by clicking on 'Explore the crime map'.
  • The drop-down menu will give you crime data by category.
  • Clicking on the broad map allows you to zoom in, so that particular instances of crime in streets – or even individual houses – are obtainable.

◆ **House prices**. You can research average house prices via www.zoopla.co.uk. Use 'House prices', then 'UK area stats', which will allow you to enter the name or postcode of any area.

◆ **Census data**. This can give you everything from population data to deprivation indicators. Use the ONS website (www.ons.gov.uk), though local council data are often easier to use, because your local council will have processed the data already! A simple phrase on Google, such as 'population of Oldham', will often get quicker results.

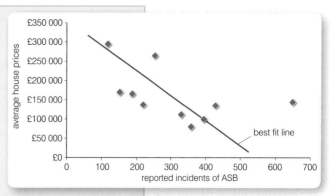

▲ *Figure 3 A scatter graph showing the relationship between average house prices and reported incidents of anti-social behaviour (ASB) in December 2015 in selected parts of Greater Manchester*

Investigating how people feel about their local area

Several geographers and polling companies have investigated what affects people's attitudes about where they live. IPSOS-MORI is a large polling company; it publishes opinion polls at election time, but usually it carries out commissioned surveys about all aspects of human behaviour – the results of which it then sells to organisations. The sample sizes used are enormous, so the results are significant.

Many of IPSOS-MORI's findings are useful. Its 2009 report *People, Perceptions and Place* contains data about crime and how it affects people's feelings about their area. However, it's important to understand how the data were collected. The Satisfaction data shown in Figure 4 are percentages of how well people thought that seven ASB indicators (e.g. graffiti) were being handled (i.e. how satisfied they were with the actions being taken). Each indicator had a three-point score – so the total was out of 21. The higher the score, the higher the satisfaction was with its handling. Scoring 14 out of 21 would therefore be 66.7%. This graph shows the results for all 323 local authorities studied, so the data are significant! The result is not surprising; it's a negative relationship, with the greatest satisfaction in those areas with the lowest incidence of ASB.

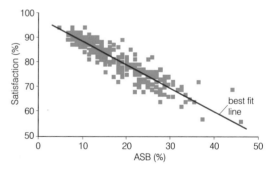

▲ *Figure 4 The relationship between how satisfied people feel about the area in which they live and the amount of reported anti-social behaviour (ASB) there*

▷ *Figure 5 Anti-social behaviour – graffiti in Hackney Wick in East London*

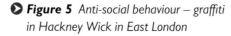

Collecting primary data about attitudes and perceptions

It's worth carrying out your own versions of IPSOS-MORI studies – a process known as **replicating**. Surveys which replicate others are useful; the methods have already been established, and your own findings can then be compared with national patterns. Figure 6 shows one such survey, called a **'Satisfaction with life survey'**. Its purpose is to find out how people feel about the area in which they live. By collecting data from a large sample, it's then possible to identify the **drivers** (key factors that drive whether or not people like the area in which they live).

Studying the questions shows that:

◆ 1 to 6 deal with people, housing and a sense of neighbourliness

◆ 7 to 10 deal with safety

◆ 11 to 14 deal with aspects of environmental quality

◆ 15 to 20 are about the local council, police and transport.

By collating the results from a large sample, it's then possible to add up the scores, express them as percentages, and then analyse which factors/drivers produced the highest scores. The result will be a list of factors that work either towards strong communities (positive drivers – i.e. things which score highly) or against them (negative drivers – i.e. things which score poorly). Figure 7 shows an example of this.

Ask recipients to score their local area for the following qualities. Each statement should be scored between 1 and 5 (where 5 is best).

In the area where you live, how much do you feel ...

1 a sense of belonging to the area? ☐

2 that people here treat each other with respect? ☐

3 that it's a quiet neighbourhood without noisy neighbours? ☐

4 that people from different backgrounds get on well together? ☐

5 that your area has good-quality housing (well kept, maintained)? ☐

6 that housing here creates a mix of different ages (families, couples, the elderly)? ☐

7 safe going out during the day? ☐

8 safe to go out after dark? ☐

9 that the area is safe for children? ☐

10 that elderly people can get about easily on the pavements? ☐

11 that the area is kept litter-free? ☐

12 satisfied with local parks and open spaces? ☐

13 that your area has plenty of leisure amenities (e.g. swimming pool, pub)? ☐

14 that the area has little vandalism? ☐

15 that local public services work well to keep the area litter-free? ☐

16 satisfied with the way the local council runs things? ☐

17 satisfied with the local police force? ☐

18 that there are drug problems in the area? ☐

19 that there are safe road-crossing points? ☐

20 that you have good local transport? ☐

Finally: Add up a total score out of 100 = ☐

▲ **Figure 6** *A satisfaction with life survey to find out how people feel about the area in which they live*

▼ **Figure 7** *Positive and negative drivers/factors that affect whether or not people are satisfied with their local area (according to polling group IPSOS-MORI)*

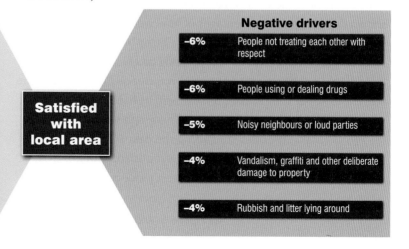

Positive drivers	
Belong to immediate neighbourhood	15%
Satisfaction with the way the local council runs things	14%
Safe to go out during the day	11%
Satisfaction with parks and open spaces	8%
Safe to go out after dark	7%
Keeping public land clear of litter and refuse	5%
Local public services working to make the area cleaner and greener	5%
Satisfaction with local police force	5%
People from different backgrounds get on well together	5%

Satisfied with local area

Negative drivers	
–6%	People not treating each other with respect
–6%	People using or dealing drugs
–5%	Noisy neighbours or loud parties
–4%	Vandalism, graffiti and other deliberate damage to property
–4%	Rubbish and litter lying around

Environmental Quality Surveys (EQS)

Sometimes it's possible to identify factors about the environment that might contribute to people's enjoyment of, or lack of satisfaction with, an area. In this case, an **Environmental Quality Survey** (see Figure 8) can be used to measure people's feelings about the environment. It's a bi-polar survey (a survey which uses terms at opposite extremes within which to place people's opinions). It consists of indicators that record people's feelings. These can then be used to score – either individually (such as what people think of building design or maintenance), or as a whole by adding up a total score. These results can then be plotted geographically to identify how satisfaction with the environment varies.

Name of place assessed: _____ Brief description of environment (e.g. housing, industry, offices): _____

Qualities being assessed		Agree +2	Generally agree +1	Av. 0	Generally disagree −1	Disagree −2	
Building design and quality	**1** Well designed / pleasing to the eye						Poorly designed / ugly
	2 In good condition – e.g. paintwork, woodwork						In poor condition
	3 Houses well maintained or improved						Poorly maintained / no improvement
	4 Outside – land, gardens are kept tidy / in good condition						Outside – no gardens, or land / open space in poor condition
	5 No vandalism; graffiti has been cleaned up						Extensive vandalism or visible graffiti in large amounts

▲ **Figure 8** *Sample statements from an Environmental Quality Survey (EQS)*

Creating images of places

Media (music, photography, film, art and literature) can expand the view given by statistics about representations of places. These include:

- Music, e.g. The Specials' 'Ghost Town', a negative view of the condition of Britain's inner cities in 1981.
- Photography, e.g. Figure 5 of Hackney Wick – an image that could give a very negative image of Hackney.
- Film, e.g. *East is East*, a portrayal of a multicultural community in Salford, which presents some of the challenges facing such communities.
- Art, e.g. paintings by L. S. Lowry of industrial Manchester, which can present either a romantic view of Manchester, or alternatively a very depressed view of small 'matchstick' people dwarfed against industrial buildings and cities.
- Literature, e.g. Ian McKewan's novel, *Saturday*, about one day in London in 2005.

Over to you

1 Using Figure 1 as a guide, research similar data for your area (using the web resources explained in the Background box).

 2 **a** Describe the relationship in the scatter graph (Figure 3).
   b Using the data in Figure 1 and the guidance in Section 7.6, carry out a Spearman's Rank correlation on the two variables in Figure 3.
   c Test your correlation for significance and explain its strength.
   d Repeat questions **b** and **c** for any other pair of variables in Figure 1.

3 **a** In pairs, research and prepare a brief presentation on how your area, city or county is portrayed in music, photography, film, art and literature.
   b Discuss and explain how and why your researched sources portray different images of the area.

4 Test out the survey in Figure 6 in your local area. Discuss its strengths and weaknesses.

On your own

5 Distinguish between the following pairs of terms: **(a)** positive and negative drivers, **(b)** subjective and objective data.

6 Using Figure 1, explain **(a)** the value, and **(b)** potential problems in using crime data to assess people's perception of a place.

Exam-style questions

 AS 1 Explain how different media provide contrasting images of particular places. *(6 marks)*

 A 2 Evaluate the use of statistical evidence in determining the image that people may have of a particular place. *(20 marks)*

In this section, you'll learn how the UK's culture and society are becoming increasingly diverse.

Head versus heart?

Ben works in Leeds. He graduated from Leeds University and already it feels like home. But like many graduates in northern cities, he's finding it difficult to get a well-paid job. His close friends are in Leeds, and he's had plenty of offers to work in call centres at about £15 000 salary, but feels that's not why he went to university. Many of his university colleagues have moved to London, and now earn over £35 000 a year. What should be do – leave, or stay? It's a head versus heart decision.

If Ben moves, he will join others who are contributing to London's fastest-ever population growth (see Section 6.2). Each year, over 2.5 million people move around the UK, which has one of the highest internal migration rates in Europe. The popular hotspots are in London and South East England. However, internal migration is just one explanation of why different parts of the UK are growing in population at different rates. The process of population change is a complex series of inflows and outflows of people, as shown by the system diagram in Figure 1. It has five **variables** (or components) – flows in (i.e. births and migrant inflow), flows out (i.e. deaths and migrant outflow), plus temporary inflows and outflows (e.g. Armed Forces personnel leaving or returning from service overseas).

The data in Figure 1 are for 2013, when the UK's net overall population increased by just over 400 000. The two largest contributors to this growth were:

◆ births minus deaths, which contributed a net 212 100 in 2013 (56% of the growth)

◆ immigration minus emigration, which contributed a net 183 400 in 2013 (44% of the growth).

◀ **Figure 1** The UK's population dynamics in 2013, displayed as a system diagram with five variables.

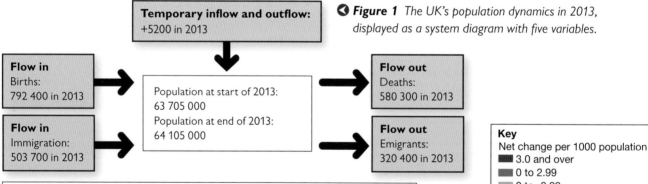

Population change = (births minus deaths) +/- (immigration minus emigration) +/- (temporary inflow minus temporary outflow)

London's population dynamics

London's population dynamics are interesting! Figure 3 shows that its overall population growth in 2013 was greater than the UK's other standard economic regions. In 2013, London had the UK's largest:

◆ natural increase (82 900, 40% of the UK's increase)

◆ net international migration (79 500, 43% of the UK's net total).

However, one further population dynamic is at work. In spite of a high inward migration to London, caused by people such as Ben, an even greater outflow from London is taking place at the same time! As Figures 2 and 3 show, in 2013 London had the greatest outward internal migration – a deficit of 55 000 (-6.5 per 1000 population).

▶ **Figure 2** Net change per 1000 population as a result of internal migration in each of the UK's 12 standard economic regions

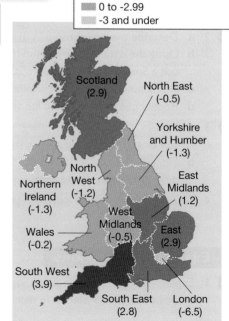

Standard economic region	Overall population increase	Births minus deaths (natural increase)	Net international migration	Net internal migration	Temporary inflow and outflow
North East	8.2	2.6	6.6	-1.4	0.4
North West	18.9	17.4	9.8	-8.5	0.3
Yorkshire and The Humber	21.0	14.6	12.7	-6.8	0.5
East Midlands	31.0	11.7	12.7	5.6	1.0
West Midlands	32.1	19.9	14.0	-3.0	1.3
East of England	46.8	19.0	12.6	17.5	-2.3
London	108.2	82.9	79.5	-55.0	0.7
South East	67.9	26.5	16.1	24.9	0.4
South West	38.0	5.1	10.9	21.0	0.9
Scotland	14.1	0.9	2.1	7.9	3.2
Wales	8.3	1.8	7.3	-0.7	-0.1
Northern Ireland	6.1	9.6	-0.9	-1.5	-1.1
Total	400.6	212.1	183.4	0.0	5.2

◀ **Figure 3** *The population components of each of the UK's 12 standard economic regions in 2013 (all figures are in thousands, and please note that the data may not add up perfectly due to rounding)*

The move away from cities

Why are so many people leaving London? There are two basic reasons for the outflow of people:

◆ Whilst young graduates – like Ben – are moving into London, a greater outflow is occurring amongst those of middle and retirement age. Their most frequent destinations are the South East (caused mainly by families wanting more space for children) and the South West (caused mainly by those retiring). This is actually a national trend – since 2008, there has been an increase in net internal migration away from cities to more rural settlements. As a result, London has become a younger city, while the other regions have aged, especially the South West (see Section 6.1).

◆ London's rapidly rising property prices since 2000 have also given many older age groups significant financial assets – allowing them to sell a flat or small house within London in return for a much larger house outside. These are usually longer-term residents, who have seen their properties substantially increase in value since they bought them.

What is left behind is a capital city with an increasing proportion of young professionals, as well as immigrant and ethnic communities (see Figure 4). In 2011, the year of the last Census, Newham in East London became the first London borough to register a minority White British population.

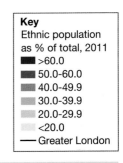

Key
Ethnic population as % of total, 2011
- ■ >60.0
- ■ 50.0-60.0
- ■ 40.0-49.9
- ■ 30.0-39.9
- ■ 20.0-29.9
- □ <20.0
- — Greater London

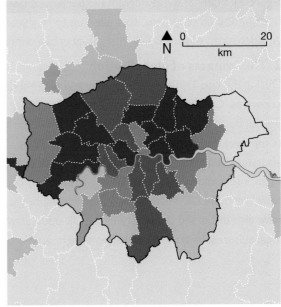

▶ **Figure 4** *Ethnic population (excluding White British and Irish) as a percentage of the total population in each London borough in 2011*

The outflow of people from London has a long history. In the 1960s, geographers in both the UK and the USA used to talk about the 'white flight' of middle-class white families from cities such as London or New York for more spacious suburbs. The term 'white flight' suggested an underlying racism. This process had several stages:

◆ Those living in the outer suburbs often left the city when, for example, they could afford a property in a suburbanised village.

◆ When they left, a re-shuffling of residents would occur – with better-off migrant communities dispersing from inner-city areas towards the outer suburbs that the white middle classes had left.

Now migration is more complex, but the same kind of process occurs – as, for example, Asian Indian families living in Southall (see Section 6.3) move further out to suburban towns, such as Iver in Berkshire.

Understanding UK immigration

International migration has changed UK culture and society. Since the end of the Second World War in 1945, inward flows have come from two sources – Europe and beyond Europe. Those born beyond Europe mainly consist of people from former British colonies, e.g. India, Pakistan and the West Indies.

Post-war immigration has had two major phases:

1 Post-colonial migration

The British government promoted large-scale immigration from former British colonies from the 1940s to the 1960s. They gave these migrants the right to settle in the UK, and also sponsored them to fill particular job vacancies (such as in the newly founded NHS; see Section 6.3). Once established in the UK, a migrant could apply for family members from their home country to join them. These migrants formed the majority of non-UK-born residents in 2014 (see Figure 5). Many of these migrants experienced discrimination in both housing and employment, so – even up to fifty years later – their grandchildren often still live in areas of urban deprivation.

2 The impacts of globalisation

During the late 1990s, a period emerged when the UK became known as 'Cool Britannia'. This term has long since fallen into disuse, but it branded the attraction of the UK for its global links to companies (for work), and its global culture (for a varied and liberal lifestyle). As a result, the UK has become a destination of choice for many international migrants. People from every country in the world now live in the UK. Many are London-based, but substantial numbers have also dispersed elsewhere in the UK upon arrival – influencing local cultures, languages and religions right across the country.

UK economic regions	UK-born	Non-UK born
North East	2477	140
North West	6340	569
Yorkshire and The Humber	4862	479
East Midlands	4061	430
West Midlands	4836	612
East of England	5284	635
London	5099	2853
South East	7580	1011
South West	4862	416
Wales	2854	163
Scotland	4842	353
Northern Ireland	1690	120

... of which EU	... of which non-EU
47	93
200	369
169	310
159	271
169	443
267	368
788	2065
365	646
185	231
71	92
171	182
83	37

 Figure 5 *Residents of the UK's 12 economic regions by category of birth in 2014 (all figures in thousands)*

As a result of globalisation, the UK's population has risen rapidly for three reasons:

◆ Many TNCs encourage immigration from the USA, Australia, South Africa and New Zealand to work in the knowledge economy (particularly in London). Their demand for specific skills can only be satisfied by overseas recruitment.

◆ After 1992, the 'Maastricht Agreement' allowed free movement of workers between EU member states. The expansion of the EU in 2004 led to increased economic migration to the UK. Polish migrants now represent the UK's largest-ever inward migration – over 600 000 Poles have arrived in the UK since 2004.

◆ Younger international migrants have helped to combat an ageing UK workforce. As well as in the knowledge economy, international migrants also provide skills and workers for the caring professions, hotels, retail and catering, agriculture, and construction projects (e.g. the 2012 Olympic Park and London Crossrail).

Migrants in rural areas

Rural parts of the UK, not just urban areas, are affected by migration. Lincolnshire, in eastern England, is a farming county; its fertile soils make it the UK's most productive area for growing vegetables, such as cabbages, peas and broccoli (see Figure 6). Vegetable picking and packing there is now done largely by Eastern Europeans for low wages.

The arrival of the Eastern Europeans has had a huge impact:

◆ Boston, one of Lincolnshire's largest towns, saw its overseas-born population increase by 470% between 2001 and 2011. Some locals (who like diversity) have welcomed the migrants, but others (who believe that migrants undercut local labour rates) have not welcomed them.

◆ In August 2015, the *Financial Times* reported that Boston's Latvian community was suffering abuse from contractors supplying vegetable pickers to farms (known as gang-masters). Many Latvian pickers were being paid below the legal £6.21 farm minimum wage, and they also had excessive costs for accommodation and transport deducted from their wages. The *FT* reported that 'far from draining the public purse of tax credits, or taking jobs from native workers, some migrants are being ruthlessly exploited by gang-masters, and enriching criminals who prey on their poor language skills and ignorance of local labour laws.'

◆ *Figure 6* Vegetable pickers in Lincolnshire overseen by a gang-master

Exam-style questions

AS 1 Explain one challenge and one opportunity offered by rural locations for migrants. *(4 marks)*

A 2 Assess the significance of international migration on diversity in the UK. *(20 marks)*

Over to you

1 Using the formula below Figure 1, devise an equation for London's population in 2013.

2 **a** Using Figures 2 and 3, describe the pattern of net internal migration in the UK.

 b In pairs, research one region outside London with a **net gain** and one with a **net loss** from internal migration. Research the likely economic and social causes for each.

3 **a** In pairs, brainstorm reasons why rural counties may not accept immigration as readily as London.

 b Research the arguments for and against using migrant labour on farms.

On your own

4 Define the terms 'population dynamics' and 'population variables'.

5 Explain how population dynamics have left London younger, and more ethnically and culturally diverse, than any other region of the UK.

6 Using Figure 5, outline the relative significance of EU and non-EU migration in different parts of the UK.

In this section, you'll learn how levels of segregation reflect cultural, economic and social variation and change over time.

A snip at the price

If, in early 2016, you had a spare £32.5 million, the house shown in Figure 1 was for sale. The asking price reflected both its size and its location – overlooking Regents Park. With 1100 square metres of living space (think ten average-sized three-bedroom houses), it included a library, dining room, breakfast room, kitchen, drawing room, study, five bedroom suites, games room, roof garden, guest suite, gym, utility room, and garages. If you needed a mortgage, the agents quoted one costing over £150 000 a month for 25 years!

In recent years, Central London has attracted large numbers of 'super-rich'. In 2016, residents living at one of its most prestigious developments, One Hyde Park in Knightsbridge (see Figure 2), included the Minister of Foreign Affairs of Qatar, a Ukrainian business **oligarch**, two Russian property developers, and the President of Kazakhstan. It is deliberate **segregation** – a decision to live in self-contained, security-guarded enclaves.

Why choose London?

In 2014, the *Financial Times* reported that wealthy Russians were buying up London property to escape economic punishments – known as **sanctions** – which had been imposed against Russia, and also to protect their financial assets by purchasing overseas property in a safe haven. In 2013, estate agents Knight Frank placed Russians top of the list of overseas buyers of London homes – estimating their total spending that year at £500 million. At that point, an estimated 100 Russian billionaires were living in London. Most of them had made their fortunes by buying up the assets of the former Soviet Union after its collapse in 1991. They were mostly concentrated in segregated enclaves in Central London, such as Kensington, Mayfair and Westminster (see Figure 3).

⬙ **Figure 1** *The house in Cornwall Terrace, Central London, that was put for up sale in early 2016 for £32.5 million*

Key word

Oligarch – A member of a small group of wealthy, powerful people who control a country or organisation.

Segregation – The separation of people of different backgrounds, wealth, cultures or nationalities.

⬙ **Figure 2** *One Hyde Park in Knightsbridge, where even the smallest apartment cost several million pounds in 2016*

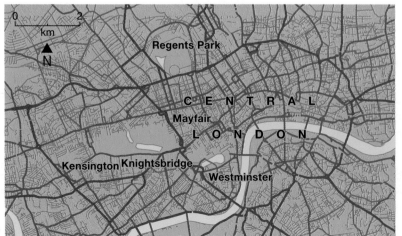

⬑ **Figure 3** *The areas of Central London where many of the 'super-rich' live*

For the wealthy, London's attractions are mostly economic:

◆ Property in London is a good investment. A smaller property, near that in Figure 1, was bought for £595 000 in 1996 and sold for £3.7 million in 2014 – an average price increase of 10.4% per year.

◆ The British pound holds its value more consistently than the Russian rouble, due to Russia's volatile economy and its dependency on oil and gas.

◆ London is also the main capital-raising centre for Russian companies, and offers financial respectability. In 2013, 28 Russian firms, worth £260 billion, were listed on London's Stock Exchange.

◆ The UK makes it easy for the wealthy to migrate. It grants three-year visas to those investing over £1 million in government bonds. Provided they keep the bonds, they can buy residency for £10 million after two years. Between 2008 and 2013, 433 such visas were granted to Russians, and 419 to Chinese.

Other reasons for the wealthy to move to the UK include the following:

◆ One third of Russians buying property in the UK, do so to educate their children in private schools. In 2013, over 8% of non-British pupils in independent schools were Russian – earning those schools £60 million a year in school fees.

◆ Crime rates in the UK are also low; in 2014, London had fewer than 100 murders, and the UK under 600, compared to Russia's 16 000.

Ethnicity and culture in London

The segregation of different **ethnic** groups in London is a common feature of many large cities. The availability of cheaper rental property in big cities makes it easier for migrants to settle there, and local work can then make these communities permanent (such as the Hindu and Sikh communities in Southall; see Section 6.3). Many West Indian immigrants in the 1950s were recruited to drive London's buses, so many of them settled in areas near to London Transport bus garages, e.g. in Brixton. Many descendants of those early immigrants still live in the same areas. Once established, the growth of specialist retail outlets (see Figure 4), places of worship and leisure – to reinforce their **culture** – help to make different ethnic communities more permanent. London's Greek Cypriot community (see the next page) is another example of this. All are examples of **ethnic segregation**.

Key word
Ethnic – A social group identified by a distinctive culture, religion, language, or similar.
Culture – The ideas, beliefs, customs and social behaviour of a group or society.
Ethnic segregation – The voluntary or enforced separation of people of different cultures or nationalities.

▶ *Figure 4 A greengrocer's shop in Brixton, South London – small independent shops like this are often a good indicator of local ethnicity or culture*

279

London's Greek Cypriot community

The majority of London's original Greek Cypriot community arrived in the 1960s and 1970s (many escaping the Turkish invasion of Cyprus in 1974). Including Greeks from the mainland, there are now about 180 000 Greek speakers in London, and Greek is the twelfth most commonly spoken language in the capital. The Greek Orthodox Church, of which there are several in London, plays a significant role in Greek language teaching (supported by the Greek and Greek Cypriot governments). This ensures that as well as their **assimilation** into the UK, the **cultural distinctiveness** of those with Greek heritage is maintained.

The Greek Cypriot community has traditionally lived in North-West London (see Figure 5). It has established:

◆ several places of worship, with Greek Orthodox Churches in North London and Paddington, as well as a number of Greek shops and other businesses (see Figure 5) and a dedicated part of the public cemetery in New Southgate

◆ a Greek newspaper, the *Parikiaki*, which serves the Greek Cypriot community, and also a London Greek radio station.

Key word

Assimilation – The gradual integration of an immigrant group into the lifestyle and culture of the host country, sometimes at the expense of their own distinctiveness.

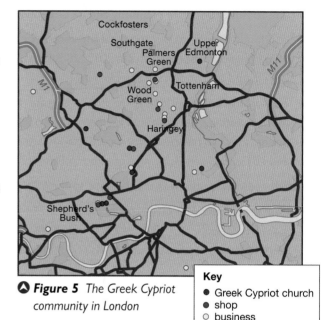

⊘ **Figure 5** *The Greek Cypriot community in London*

Key
● Greek Cypriot church
● shop
○ business

BACKGROUND

Measuring cultural diversity

Several cities and countries have very diverse populations. Inner London is Europe's most diverse place, followed by Luxembourg, Brussels and Outer London. But not all places are so diverse. Figure 6 shows global diversity on a map of **cultural fractionization**. This approach was devised by Klaus Desmet and others at UCLA in 2015. Cultural fractionization measures how diverse countries are, by measuring people's attitudes towards, for example, religion, democracy and the law. Using people's

responses, the authors devised an index between 1 (total diversity) and 0 (no diversity), as shown in Figure 6 – the higher the figure, the more diverse the culture. The global average was 0:53. Darker-shaded countries in Figure 6 are more culturally diverse than those shaded lighter. As examples, France and India have high cultural diversity, whereas Egypt has low diversity (i.e. a high degree of cultural conformity).

⊘ **Figure 6** *Cultural fractionization (or diversity), using an index devised by Klaus Desmet and others in 2015*

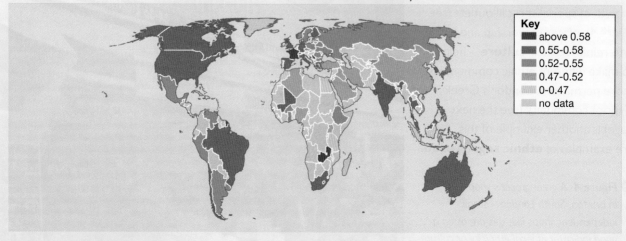

Key
■ above 0.58
■ 0.55-0.58
■ 0.52-0.55
■ 0.47-0.52
▦ 0-0.47
☐ no data

How people experience change

As inward migration changes the character of places and communities, people's views about those places change. Over decades, former residents see their former locality alter – and they may eventually cease to identify with it completely once they leave and a new cultural identity emerges. For example, Brick Lane in East London (Figure 7) has seen waves of migrants since the seventeenth century. The earliest migrants were escaping persecution, e.g. the French Protestant Huguenots in the seventeenth century and Jewish refugees from Russia and Germany in the late nineteenth and twentieth centuries. The synagogue shown in Figure 7a (established in a period when violent anti-Semitism in London was a reality) is hidden behind what looks like a plain terraced house.

Over time, London's main Jewish community moved to Golders Green, and Brick Lane has now become home to a large Bangladeshi community. Restaurants, local shops (see Figure 7b) and mosques reflect Bangladeshi culture. However, Brick Lane also borders the City of London, and the established Bangladeshi community is now under threat as **gentrification** drives out all of those unable to afford the higher property prices paid by City workers (Figure 7c).

Key word

Gentrification – A change in social status, where former working-class inner-city areas are increasingly occupied and renewed by the middle classes.

Over to you

1 a In pairs, outline factors that attract the super-rich to London.
  b To what extent do such migrants live in distinctive places?

2 In pairs, list arguments in favour of migration policies that welcome the super-rich.

3 a Using Figure 6, describe the global pattern of cultural diversity.
  b Research one country with an index above 0.58 and one with 0.52 or less to explain the level of cultural diversity in each.

4 Using Figures 7a–7c, outline the changes that can take place as one community replaces another over time.

On your own

5 Distinguish between **(a)** ethnicity and culture, **(b)** cultural fractionization and diversity, **(c)** assimilation and cultural distinctiveness.

6 Explain why some countries are more culturally diverse than others.

7 a Research one other culture which is present in London. Find its location(s) within London, its specialist shops, places of worship and lifestyle features.
  b Write 750 words to discuss the economic and cultural contributions of ethnic communities to London.

8 Outline possible conflicts caused by gentrifying parts of cities, such as Brick Lane.

Exam-style questions

AS 1 Explain why international migrants tend to live in distinctive places. *(4 marks)*

A 2 Assess the ways in which levels of segregation reflect cultural, economic and social variation and change over time. *(20 marks)*

△ **Figure 7a** Sandys Row Synagogue in East London. Originally a French Huguenot church in the seventeenth century, it became a synagogue in the late nineteenth century as a new Jewish refugee population moved into the area.

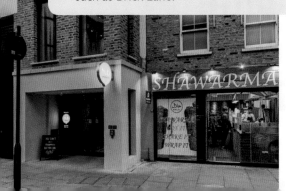

▷ **Figure 7b** Bangladeshi shops and restaurants in Brick Lane

◁ **Figure 7c** In 2016, this new hotel opened on Brick Lane, where previously a BanglaCity supermarket was located. Critics say that this shows how Brick Lane is changing as property is upgraded and gentrified.

In this section, you'll learn how changes to diverse places can lead to tension and conflict.

Out with the old, in with the new

Every day, huge container ships arrive at Britain's newest container port – London Gateway in Essex (see Figure 1). This new port is 30 km east of Central London, and it can cope with the world's largest container ships. In the 1970s, the development of container ships signed the death warrant for the original Port of London (to the east of Tower Bridge). The Thames simply wasn't deep enough that close to Central London to accommodate them. From being Europe's largest port in 1900, London's dock facilities had to shift further and further downstream, and the last of the original East End docks closed in 1981.

The closure of London's original docks was devastating for the dockworkers:

◆ Between 1978 and 1983, 12 000 jobs were lost. In the 1981 Census, over 60% of adult men were unemployed in some parts of East London.

◆ The Thames riverside east of Tower Bridge consisted of abandoned docks and derelict wharves – not a good image for a major city.

◆ Many nearby industries in East London's Lea Valley also closed, because they needed the port facilities to import raw materials and export finished products.

The population of the East End fell by 100 000 between 1971 and 1981, as people left to find work.

London's changing East End

The closure of the docks left a legacy that was hard to tackle:

◆ Economically, East London was left in a spiral of decline.

◆ Socially, race relations and alleged heavy handedness by police had already raised tensions in some parts of London, and increased unemployment tested the situation even further.

◆ Environmentally, 21 km² of riverside property was left derelict or run-down, like the building shown in Figure 2.

Faced with these problems, the Conservative government of the time reacted by attempting to rebrand inner cities. Planners began envisioning a different image for inner cities – a process known as **re-imaging** (in other words, how the image of a place could be altered, such as how it's portrayed in the media).

⬙ *Figure 1* A container ship unloading at the new London Gateway container port in Essex

⬙ *Figure 2* Derelict buildings, like this former flour mill, lined the Thames downstream from Central London in the 1980s

Regenerating London Docklands

The job of regenerating 21 km² of potential building land, close to Central London, went to the LDDC (London Docklands Development Corporation) – a central government agency. Formed in 1981, it was highly controversial:

◆ The LDDC's focus was to encourage economic growth, so it brought together key **players**, e.g. private property developers keen to purchase public land (the former Port of London was owned by the government), architects and private investors. This process was known as **market-led regeneration** – leaving the private sector (i.e. the free market) to decide on the future of Docklands, rather than local councils or communities. The LDDC's plans focused on economic growth, profitable new housing, and prestige infrastructure like the new London City Airport to service Canary Wharf (see below).

◆ Local borough councils had no say in the regeneration of the area, and there were tensions between the developments they wanted (affordable housing, work for local people) versus those approved by the LDDC. The Docklands regeneration was a **top-down** project, which ignored community groups who wanted social housing as part of a different vision for East London – and certainly did not want an airport (see Figure 3)!

Figure 3 *A poster from the 1980s protesting against the construction of a new airport in London Docklands. That airport is now called London City Airport.*

Economic growth

The LDDC's flagship project was Canary Wharf, now London's second CBD (see Figure 4). A huge transformation took place, with high-rise offices attracting TNCs (especially banks such as Barclays and HSBC) and stimulating quaternary employment to replace the docks and former secondary industries. The aim was to create high-earning jobs. High earners, it was argued, would generate other jobs as wealth 'trickled down' to poorer communities. Companies based in Canary Wharf now operate as part of the knowledge economy (see Section 6.1).

◆ In 2015, Canary Wharf's annual output was over £6 billion – greater than that of the entire principality of Monaco.

◆ 100 000 commuters travel there every day, and the average salary in 2015 (including those on low pay, such as shop workers, who bring down the average!) was £100 000 – plus generous annual bonuses.

But even in the twenty-first century, the decision to create Canary Wharf is controversial. Although employment has grown, most of those who work in Canary Wharf commute in from other areas. The East End is no longer one of the UK's most deprived areas, but poverty is still present. In 2012, 27% of Newham's working population earned less than £7 per hour – the highest percentage of any London borough.

Figure 4 *Canary Wharf after redevelopment*

The gentrification process

The changes to London Docklands have created both challenges and opportunities for different people. The regeneration process has transformed the population of the area (known as Eastenders):

◆ Regeneration has increased the area's housing supply – but of luxury housing in riverside locations, rather than social housing. This has led to **gentrification** – what has been called 'the march of the middle classes into East London'. New riverside properties in converted warehouses suddenly became desirable and very expensive.

◆ As a result, a divide emerged between the new and old Eastenders, and the spaces occupied by each. For example, the new young professionals drink and eat in renovated 'foodie' pubs and restaurants, while the traditional East End pub often struggles to survive.

▼ Figure 5 *People's lived experiences of the changing East End*

Changes to social housing

Before regeneration, most East London housing was rented from local councils at low-cost, which reflected the low average incomes in the area. Then two major changes occurred:

◆ The Conservative government introduced the **Right to Buy** scheme, which gave those living in council housing the right to buy it at a reduced price. Between 1980 and 2015, over 127 000 council properties were sold nationally, which reduced the available supply of social housing. Then, as London's house prices soared, low-income groups were increasingly excluded from living in East London – and the list of applicants for the reduced pool of social housing lengthened (for example, 20 000 people were on the council house waiting list in Tower Hamlets in 2016).

◆ Many of those traditional East End residents who bought their council houses in the 1980s have now sold up and left the area – many retiring to the Essex coast, e.g. Southend-on-Sea. Their old properties have now been taken over by a younger generation. In 2011, the average age of the Docklands population of Newham and Tower Hamlets was 31 (the UK average is 40). As a result, traditional East End communities have been broken up, and there has been a major change in how the East End 'feels' (as the opinions in Figure 5 show).

> Before, people used to stay and you got to know them … the trouble is now I hardly know anybody any more down this street.

> In cases where the properties are let and the turnover of residents is quite high, you lack that close community – because people tend not to stay in the same place for very long.

> My great-grandmother came from Bethnal Green, and moved from working in London to Tilbury Docks. I think it's a good thing their East End has gone – it was unremitting poverty.

> Born and bred in the East End (Poplar and the Isle of Dogs). Moved away but like a magnet it keeps drawing me back. Unfortunately I just can't afford to move back here.

> Essex is now the home of the East End, as all the original East End people have moved out – Romford Market is where you will find true East End people. Not Newham or Tower Hamlets!

Migrant experiences in East London

East London has always been ethnically diverse, but increased immigration since 2000 has made Newham and Tower Hamlets London's most ethnically diverse boroughs (see Sections 6.1 and 6.2). Most people in London perceive that migration brings benefits, both economic and cultural.

However, immigration is not problem-free. While most immigrants settle in quickly, some experience social exclusion – mostly caused by family isolation, but also due to hostility from those claiming that migrants threaten traditional 'British' culture.

▼ *Figure 6* *The East London Mosque and London Muslim Centre, in Tower Hamlets, has won awards for community integration*

◆ In 2015, the BBC reported that hate crimes against Muslims in London had risen by 70% since 2014. 'Tell MAMA', an organisation monitoring Islamophobic attacks, claims that 60% of victims were women and that those wearing a face veil were more likely to experience aggression. Many incidents involve far-right-wing groups, some of whom run anti-mosque campaigns. Yet, in spite of these threats and tensions, some East London mosques have won awards for community integration and their work against extremism (see Figure 6).

◆ Eastern European migrants experience less aggression, but even so they are not always welcomed openly. One research project in East London found that 9% of Eastern Europeans said that they never spoke to local people (e.g. in shops), whereas 26% of long-term Eastenders reported that they never talked to Eastern Europeans.

Over to you

1 Using Google Maps, produce a map of East London to show the places mentioned in this section. Show the location of London's original docks and the changes that have taken place since the 1980s.

2 In pairs, draw a spider diagram to show how the following events have affected local people and their experience of living in East London: **(a)** the Port of London closure, **(b)** the regeneration of Canary Wharf, **(c)** the Right to Buy, **(d)** the out-migration of traditional Eastenders, **(e)** gentrification, **(f)** large-scale immigration.

 3 In pairs, discuss what the quotes in Figure 5 reveal about **(a)** different people's values, **(b)** how people's experiences of East London depend on who you are.

On your own

4 Distinguish between **(a)** regeneration and re-imaging, **(b)** regeneration and gentrification.

5 Explain whether you believe that the regeneration of London Docklands produced more winners or losers.

6 Explain why different ethnic groups may experience different levels of integration within local communities.

Exam-style questions

AS 1 Explain how changes to land use locally can create challenges and opportunities for people. *(4 marks)*

 A 2 Assess the ways in which changes to diverse places can lead to tension and conflict. *(20 marks)*

In this section, you'll learn how the management of cultural and demographic issues can be measured.

Cities as unequal spaces

Although their bright lights lure many people from a wide range of backgrounds, large cities are actually very unequal places. Rich and poor live in close proximity, but lead different lives. Income inequalities exist in all major urban areas; although there are exceptions in Figure 1, such as Tokyo, the general relationship is nonetheless that the bigger the city, the greater the inequality.

▲ **Figure 1** *The relationship between city size and inequality, as measured by the Gini index*

Inequality is measured using an indicator known as the **Gini index**. It measures how far income distribution within a country deviates from perfect equality. An index rating of 0% = perfect equality, and a rating of 100% = perfect inequality. The larger the percentage, the greater the concentration of wealth amongst a few. Figure 1 shows the relationship between the Gini index and city size.

Inequality in London

London is far from being the world's most unequal city, with a Gini value of about 43% (as Figure 1 shows). But inequality is certainly present there – and it's growing. In 2011, the London School of Economics (LSE) published a report into London's poverty. It showed that:

◆ of all 12 standard UK economic regions, London had the highest proportion of households in the top 10% of incomes – but also the highest proportion in the bottom 10%.

◆ inequality in London grew during the economic boom between 1997 and 2008, when the incomes of its wealthiest residents grew but London's overall poverty level remained unchanged.

As London has become more diverse, so its inequality has increased. This is due to three factors: economic activity among different ethnic groups, low pay, and housing costs.

1 Economic activity and ethnicity

Most people earn their income through employment, so economic inactivity is a cause of poverty – whether through unemployment, disability, or being a student. Although most international migrants to the UK are economic, the employment level varies between different ethnic groups. For example, in 2011, women in Bangladeshi (54%) and Pakistani (52%) families had the highest levels of economic inactivity among all adults, because of their traditional family roles.

▼ **Figure 2** *Two-bedroom flats in a block such as this can cost well over £350 000 in East London – high prices like these are unaffordable to those on low or even average pay*

2 Poverty and low pay

Low pay is also important as a cause of poverty in certain types of employment:

◆ 37% of employed men worked in low-skilled, low-pay occupations in 2011. However, this percentage was higher among certain ethnic groups; 50% of men of Pakistani, Black African and Bangladeshi ethnicity worked in such jobs, e.g. in restaurants or hotels.

◆ 59% of employed women were in low-skilled jobs in 2011, particularly Bangladeshi (67%) and Caribbean (66%) women, who were commonly employed in the NHS and social care.

Low-income households are therefore more likely to contain people from certain ethnic backgrounds. About 40% of people from ethnic minorities live in low-income households, which is twice the poverty rate of White British people (see Figure 3). While only 10% of White British people are in low-income employment, the percentage rises to 65% (Bangladeshis), 50% (Pakistanis) and 30% (Black Africans).

3 Poverty and housing costs

Although average incomes in London are higher than in the rest of the UK, poverty there has increased – largely due to the high cost of housing. Between 1995 and 2015, general inflation in the UK (e.g. of food and services) rose by 73%, but house prices in many parts of London rose by up to 1000%! Households in London now spend up to 60% of their income on housing, compared with a UK average of 25%. After housing costs are deducted, inner London now has the UK's highest poverty rate.

Many people can no longer afford to live in large areas of London (see Figure 4). This situation has worsened since 2010, due to government cuts to housing benefit, which particularly affects low-income households – and, therefore, more ethnic minorities.

Tackling inequality

Since the 1980s, the UK government has increasingly used regeneration projects to deliver improvements in wealth. The theory is that they create jobs – both in the initial rebuilding work, and then through secondary jobs created as more people are attracted to the regenerated area. In theory, therefore, regenerated areas should see increased incomes and falling inequality. This should be the case particularly in the three East London boroughs of Barking and Dagenham, Newham, and Tower Hamlets, which have shown historically high levels of deprivation, and which have been through both the Docklands regeneration and also the regeneration necessary to host the 2012 Olympic and Paralympic Games. As a result, there should be obvious improvements to incomes, deprivation and health in these boroughs.

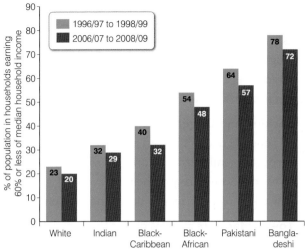

🔺 **Figure 3** *The proportion of people earning 60% or less of median household income, after housing costs have been deducted – by ethnic group*

Key word

Low-income households – Those earning 60% or less of median household income, after housing costs have been deducted.

🔻 **Figure 4** *Those areas of London which, by 2016, had become unaffordable to people receiving Local Housing Allowance (a form of housing benefit)*

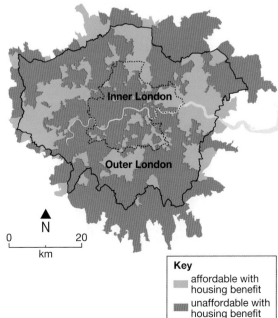

Levels of income

Figure 5 shows the number of jobs and median weekly incomes in the following three East London boroughs in 2014. These three boroughs have all undergone regeneration at different times in the recent past, with mixed results:

London borough	Number of jobs	Median weekly income (2013)	Annual % change
Barking and Dagenham	41 000	£517.40	6.0
Tower Hamlets	217 000	£804.90	5.0
Newham	63 000	£475.70	-5.8
UK	**24 385 000**	**£403.90**	**0.0**

⬆ **Figure 5** *Jobs and median incomes in three East London boroughs*

- **Barking and Dagenham**. Until recently, this was one of London's biggest manufacturing boroughs – until Ford (motor vehicles) and Sanofi (a large pharmaceutical company) closed many of their production facilities there. Regeneration is now taking place, mainly to deliver housing on industrial brownfield sites.

- **Tower Hamlets** remains one of London's poorest boroughs, in spite of the fact that it contains Canary Wharf (see Section 6.10). Most of those working in Canary Wharf live outside the area, and the redevelopment has generated few local jobs.

- **Newham** hosted the 2012 Olympic Games, and contains most of the new Queen Elizabeth Olympic Park. It also contains the new Westfield Shopping Centre, which generated 10 000 jobs. However, most of the construction work was completed by 2011, and retail jobs tend to be part-time, seasonal and low paid.

Levels of deprivation and health indicators

Although East London remains London's most deprived area, Figures 7A and 7B show reducing deprivation levels between 2010 and 2015. Health indicators have yet to follow this pattern, as Figure 6 shows. It therefore seems as if regeneration has brought some success. However, deprivation data should be treated with caution:

- In 2010, the UK – and London, in particular – was recovering from the financial banking crisis. Economic improvements had begun, and were maintained, so deprivation would have been expected to fall between 2010 and 2015 anyway.

- Unemployment fell nationally in the same period, and is one of the indicators by which deprivation is measured.

- Many health indicators were not linked to regeneration but to targets in the NHS, e.g. in diagnosing and treating cancer. These would count as reductions in deprivation, but would not be due to regeneration.

Indicator	Barking and Dagenham		London	
	2010	2015	2010	2015
Male life expectancy (years)	76.4	76.3	78.2	78.2
Female life expectancy (years)	80.6	81.2	82.7	83.4
Infant mortality rate per 1000 live births	4.9	5.3	4.6	4.4

⬆ **Figure 6** *Changing life expectancy and infant mortality, 2010–2015*

⬇ **Figure 7** *Deprivation levels across London in 2010 (map A) and 2015 (map B)*

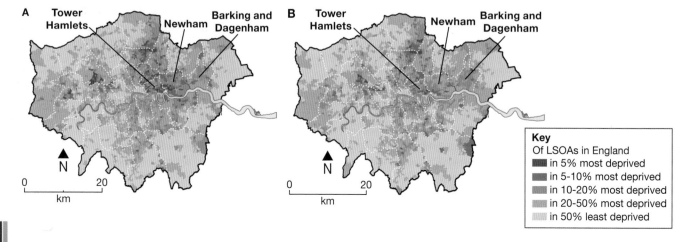

A Tower Hamlets, Newham, Barking and Dagenham

B Tower Hamlets, Newham, Barking and Dagenham

0 — 20 km

0 — 20 km

Key
Of LSOAs in England
▨ in 5% most deprived
▨ in 5-10% most deprived
▨ in 10-20% most deprived
▨ in 20-50% most deprived
▨ in 50% least deprived

The assimilation of different cultures

Although London is extremely diverse, the degree to which different ethnic groups become assimilated varies considerably. One of the ways in which cultural assimilation can be measured is by measuring voter turnout during elections. Historically, General Elections have higher turnouts than Local Elections. Nationally, for example, there was a 66.1% turnout in the 2015 General Election, but only about 30% for Local Elections.

Ethnicity and deprivation affect voting behaviour – to quote the news website, East London Lines, 'the poor don't vote'. However, ethnicity can sometimes actually *increase* the likelihood of voting! Ed Fieldhouse, Professor at Manchester University, found that voting in the 2001 General Election was greater within East London's Bangladeshi community than amongst Londoners as a whole – even though the area was poorer. He believes that people vote where they have faced prejudice, or exploitation at work, and that traditions of community organisation have transferred there from Bangladesh.

Figure 8 *The East London Mosque, from which several community initiatives have taken place to help improve community relations*

Community engagement

Crime has fallen across London almost every year since 2000, which makes it amongst the world's safest global cities. However, people are still targeted with hate crimes because of who they are, or who they are perceived to be. One particular target has been the Muslim community in London, which has tried to respond and improve community relations. Figure 8 shows the East London Mosque and London Muslim Centre in Tower Hamlets. This borough contains the UK's largest Muslim community, and it holds community open days (an initiative encouraged by the Muslim Council of Britain), so that people of other faiths can see a mosque and understand its part in Muslim life.

Over to you

1 Describe the general relationship between city size and inequality, as shown in Figure 1.

2 a In pairs, draw a spider diagram to show the three factors which cause inequality in London.

  b Using coloured highlighters, show how far ethnicity, low pay, and housing costs contribute to poverty in London.

3 a Compare the pattern of deprivation in 2010 and 2015 in Figure 6.

  b Identify possible reasons for any changes that you observe in particular parts of London.

4 a 'Regeneration always brings benefits'. Using Figures 5 and 7, identify data to (i) support, and (ii) contest this claim

  b Suggest other factors that might explain changes in deprivation.

On your own

5 Distinguish between 'inequality' and the 'Gini index'.

6 a Explain why cities are unequal places.

  b Suggest some of the implications for London if the patterns shown in Figure 4 continue.

7 Explain why voting is a good indicator of cultural assimilation.

Exam-style questions

 1 Explain two ways in which reductions in deprivation can be measured. *(4 marks)*

 2 Discuss the extent to which attempts to manage cultural and demographic inequalities have proved successful. *(20 marks)*

In this section, you'll learn how different urban players have different criteria for assessing the success of change in diverse urban communities.

Diversity and changing communities

During the 2012 Olympic and Paralympic Games, London was praised internationally for its tolerance and successful integration of migrants representing almost every culture. British athletes such as Mo Farah (Figure 1), who moved to the UK from Somalia as a child, and Jessica Ennis-Hill, from a multicultural family in Sheffield, were seen as proof of the UK's ability to bring together people from different ethnicities – as well as evidence that the UK's **multicultural** society was working well.

Yet beyond London, where there are often fewer migrants, immigration is contentious:

◆ Almost every opinion poll since 2000 has found that immigration features in the top five most important issues for the British public.

◆ In one poll, over 50% of respondents considered it one of the top three causes of tension in the UK.

◆ The tabloid press frequently report in hostile ways about immigration and immigrants.

But for those actually living in multiethnic areas, the reality is often different. A report by British Future, a think tank about migration, suggested that – while people are anxious about immigration at a national level (see Figure 2A) – they do not observe or experience racism that much in their own locality (see Figure 2B), and are generally positive in their own communities about living with people of mixed ethnic backgrounds. How can such differing views exist simultaneously?

▶ **Figure 2** *How people in different countries viewed immigration in 2013 (Graph A), and how often people from ethnic minority backgrounds in the UK experienced racism in 2015 (Graph B)*

National strategies towards multicultural issues

British attitudes towards multicultural communities have developed significantly since 1968. In that year, Conservative politician Enoch Powell made a speech about immigration, in which he predicted 'rivers of blood' on Britain's streets from rising tensions between people of different ethnicities. Of course there have been some problems – the inner-city riots of the early 1980s in London, Leeds and Birmingham, for example – but these riots and disturbances were only partly caused by racism, with some started by extreme right-wing groups. Often, the riots were linked to protests by ethnic minorities about their perceived harsh treatment at the hands of the police. However, it seemed at the time that **centrifugal forces** were at work (see Section 6.4) – i.e. those that forced people apart.

> **Key word**
>
> **Multicultural** – The existence of people of different ethnic backgrounds in society.

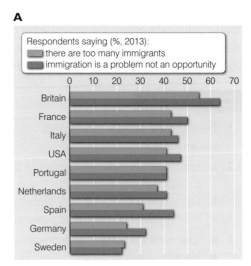

▲ **Figure 1** *Mo Farah, one of the UK's most successful athletes, winning gold at London's 2012 Olympic Games*

A

Respondents saying (%, 2013):
▪ there are too many immigrants
▪ immigration is a problem not an opportunity

Britain
France
Italy
USA
Portugal
Netherlands
Spain
Germany
Sweden

B

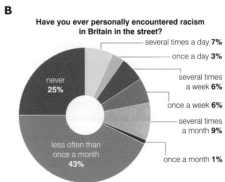

Have you ever personally encountered racism in Britain in the street?

- several times a day **7%**
- once a day **3%**
- several times a week **6%**
- once a week **6%**
- several times a month **9%**
- once a month **1%**
- less often than once a month **43%**
- never **25%**

Different UK governments, and other interested parties, have pursued changing policies to try to integrate people of different ethnicities and reduce any tensions between them:

◆ Early immigrants in the 1950s lived detached from the majority white community – **segregated** into separate enclaves. There was little understanding between the different communities, and events such as London's annual Notting Hill Carnival (see page 255) were started in an attempt to bring people together.

◆ Later policies sought **assimilation**, where immigrant communities were expected to integrate into the culture of the host country – giving up their customs and losing their cultural distinctiveness. If people came to live in the UK, it was argued, they should fully adopt 'Britishness'. In 2015, 78% of people still believed this.

◆ Now, governments try to lead public opinion by adopting **pluralist** policies which encourage ethnic cultures to flourish in a diverse society, where everyone recognises the strengths of each group and the contribution it makes.

Local strategies in Slough

Slough is a large town in Berkshire, with a population of 140 000. It has never been seen as an attractive place – like a lot of towns located between 20 and 40 miles from London, it consists of many uniform suburban estates. The town lies along the 'M4 corridor' (see Figure 3), which extends west of London between Heathrow and Newbury.

Demographic and ethnic groups in Slough

Slough shows how ethnic tensions can be managed successfully. In 2011, there were 18 significant ethnic groupings in Slough, with no single group dominating:

◆ 34.5% were White British (the UK average is 80.5%)

◆ Over 75% claimed to be 'British', even though only 60% were actually born in the UK – so many migrants had adopted British nationality.

◆ There were no distinct ethnic enclaves, unlike many other British towns and cities (such as Oldham in Greater Manchester).

Slough has long-established migrant communities. In the 1920s and 1930s, Welsh and Scots migrants sought work there. A Polish community settled after 1945, followed by Indian immigrants in the 1950s. The 1980s and 1990s were difficult times. Tensions spread to the UK from overseas, for example between Hindus and Sikhs following the Amritsar Temple Massacre in 1984 in India. People feared that ethnic tensions might spread, especially after racist crimes such as the murder of Stephen Lawrence in South London.

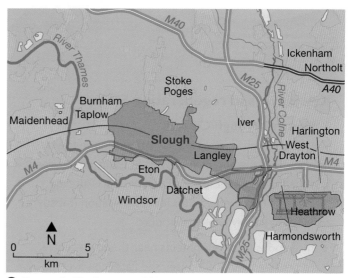

Figure 3 *The location of Slough in the Thames Valley*

Figure 4 *Wellesley Road in Slough, where the older housing has been sub-divided into rented flats (typical of the sort of housing that newly-arrived migrants settle in)*

'It's a very ugly place, but there's plenty going on, plenty of work.'

Adesh Singh

'Good people, from everywhere in the world. You can have a good life here.'

Jarek Miller, 29, from Poland and now married to a local woman

'Indigenous White British are getting more tolerant ... Racial tensions are almost non-existent.'

Charlie McGreachie, head teacher of a local primary school.

'From the outside, Slough might be seen as a joke – *The Office* and all that. But actually it's a well-kept secret success story, a multicultural success.'

Rob Deeks of Aik Saath

▲ **Figure 5** *How four people feel about Slough*

Key players in regenerating Slough

1 Slough Borough Council

Slough Borough Council is a key **player** in the town's future – with the aim of improving housing, employment and the environment for local people. It has already identified 39 brownfield sites for regeneration (mostly for housing). Early projects include:

◆ three housing developments near the town centre, to provide over 200 houses and flats (however, no affordable housing has been included)

◆ 29 new council-rented houses and a community centre on different sites

◆ the Curve, a new cultural learning centre, with a library, adult education facility, cafe, and performance centre

◆ a new sports stadium for Slough Town Football Club at Arbour Park, with football pitches, a multi-use games area, clubhouse and accommodation block, athletics track, dance academy, and 89 new homes.

In addition, the Council is working with Network Rail on a new rail link into Heathrow Airport.

2 Community groups

In the early 1990s, an organisation called **Aik Saath** began to tackle tensions between gangs of young Muslims, Hindus and Sikhs. Aik Saath (meaning 'Together As One') is a charity that focuses on young people in

schools and places of worship. Its aim is to increase **centripetal forces** in Slough (see Section 6.4) – to bring people together. Aik Saath works with teachers and young people on themes such as extremism, anti-racism, and a programme called 'Saying No To Knives'. It interacts with primary and secondary schools, and also with Thames Valley Police, using discussions and workshops aimed at mutual understanding. Aik Saath trains young people, teachers and youth workers to deal with sensitive issues. The charity has also engaged with recent Eastern European migrants, who have settled in well. Figure 5 shows some of the positive views about Slough as a result.

3 Stakeholders in the local economy

Slough is prosperous. The town's proximity to Heathrow has led many TNCs to establish offices there, e.g. BlackBerry, O2, Nintendo, Mars and Dulux. However, Slough does have its problems. The greatest need is for affordable rented housing, e.g. the local council has 7000 properties and a waiting list of 7000. Pressure is growing as cuts to housing benefit force many low-wage earners to move out of London to commuter towns such as Slough. The Slough Regeneration Partnership was formed in 2012 between Slough Borough Council and a local construction company to begin regenerating the town (see Figure 6).

Aspire Southall is another partnership between companies in Slough and Slough Borough Council. It aims to develop employability skills among the local population. Programmes involve young people, teachers, parents and training advisors in building skills – using professional trainers from the Council and a local College to encourage young people from Slough's multicultural communities. It has its own training centre, and the programme is being managed by local companies (such as Mars).

4 Environmental stakeholders

Slough Borough Council is also responsible for community health, through improving the local environment (e.g. air quality), transport, and safety (e.g. pavements, street lighting). However, air quality is hard to manage so close to the M4 and Heathrow (see Figure 3 on page 291).

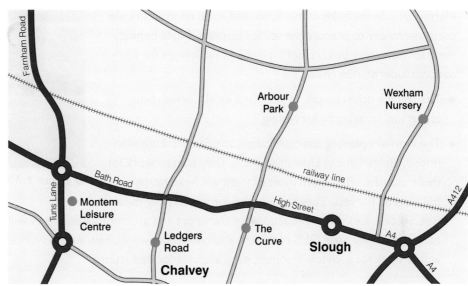

▲ **Figure 6** *The plan for Slough approved by the Slough Regeneration Partnership*

Over to you

1 In pairs, suggest possible reasons for the attitudes and experiences shown in Figures 2A and 2B.

2 a Research the causes, duration and impacts of one UK inner-city riot in the early 1980s.

   b In pairs, draw an annotated diagram to show the centrifugal forces in the 1980s that helped to cause the inner-city riots.

   c Now draw a second diagram to show how Aik Saath has tried to improve centripetal forces in Slough.

3 a Copy and complete Figure 7 to show which players might **(a)** agree, **(b)** disagree with each other about regeneration proposals for Slough, and why.

   b To what extent does your conflict matrix agree with the comment in Figure 5 that Slough is 'a well-kept secret success story'

On your own

4 Distinguish between segregation, assimilation and pluralism.

5 Suggest reasons why the people in Figure 5 might feel as they do about Slough.

6 Write a 500-word report showing to what extent your local area **(a)** could learn from Slough, **(b)** might want a different future from that of Slough.

	Player A				
Player A		Player B			
Player B			Player C		
Player C				Player D	
Player D					Player E
Player E					

Key

++ Strong agreement

+ Some agreement

— Strong disagreement

– Some disagreement

▲ **Figure 7** *A conflict matrix for analysing attitudes between different players*

Exam-style questions

AS 1 Explain the success of recent changes to one diverse urban living space. *(4 marks)*

A 2 Assess the improvements to an urban living space for different demographic and ethnic groups. *(20 marks)*

In this section, you'll learn how different rural players assess the success of change in diverse rural communities.

It's planting time again

March 2016. In the fields of mid-Cornwall, local contractors are using machinery to prepare the soil for planting. Close behind them follow two teams on foot – planting cauliflowers for a national supermarket chain.

◆ The tractor drivers work all year for a local Cornish firm: ploughing, spraying or harvesting.

◆ The workers planting the cauliflowers are Eastern European (mostly Bulgarian and Lithuanian men). They plan to work for three months, after which most of them will head home until the autumn harvest. They are working for a local gang-master (see Section 6.8). After deductions for transport and accommodation, they will earn just £2 per hour (25% of the UK adult minimum wage). Few British workers would work for such low incomes, even allowing for deductions.

⬤ **Figure 1** *Migrant workers harvesting broccoli in Cornwall*

Diversity in Cornwall

South West England is one of the UK's least ethnically diverse regions – 95% of the population is White British (see Section 6.1). The Census only records permanent residents, and most agricultural migrant workers (see Figure 1) are temporary. Nevertheless, these workers have increased the range of ethnic backgrounds in Cornwall. Many live on caravan sites that would otherwise close in winter and spring.

Cornwall – image versus reality

Cornwall is the UK's top tourist destination for its scenery (see Figure 2), and is also a favoured destination for retirement. For both groups – tourists and retirees – place image is vital. In recent years, Cornwall has tried to **re-image** itself to attract a broader demographic (e.g. young adults, who might visit Cornwall off-season).

But the reality for other people contrasts sharply with this image. Jobs in tourism are mostly seasonal, low-paid, and part-time. Cornwall's biggest economic problem is that it now lacks a year-round economy for adult employment.

◆ The county's 'old economy' consisted of primary sector employment in farming, fishing, tin mining and china clay quarrying. These jobs were year-round and permanent.

◆ Until the 2000s, most farmland in Cornwall was used for livestock farming. However, for years, supermarkets drove down the price they paid for milk, so that, by 2016, milk cost 34p a litre to produce – but dairy farmers were paid just 28p. Therefore, many Cornish dairy farmers have now switched to arable or vegetable farming.

> **Key word**
>
> **Re-imaging** – How the image of a place is changed and portrayed in the media. This term is used when rebranding new images of places, and by tourist agencies to promote images (like the photo in Figure 2).

▼ **Figure 2** *Cornwall's scenery*

- However, these new farming practices have had environmental consequences. Because milk prices are so low, dairy pasture is increasingly being ploughed up and planted with crops. With Cornwall's hilly slopes and wet winters, there has been a big increase in soil erosion (see Figure 3).

The post-production countryside

Rural areas now produce less, so geographers talk about a **post-production countryside**. As well as in agriculture, employment has fallen sharply in the china clay industry – from 10 000 people in the 1960s to just 800 in 2015. A decline in employment prospects has led to a lack of opportunity, so Cornwall's young, well-qualified population leaves – causing a 'brain drain'. In 2011, Cornwall had England's lowest full-time average annual earnings of £25 155 (77% of the UK average). Its problem is therefore how to build a higher-income economy that will provide year-round employment. At the same time, it has to bear the cost of derelict land from centuries of mining and quarrying (see Figure 4).

National and local strategies for rural areas

Recent regeneration in Cornwall has involved grants from local and national government, plus the EU. Between 1999 and 2007, an EU funding programme – called Objective One – aimed to boost rural development. Its principle was simple; it matched individual capital funds to reduce the risk of businesses failing in their first year. Its most successful investment has been the Eden Project (see page 296).

Since 2010, Cornwall has also qualified for government **Regional Aid**, which is given to companies investing there. Within this 'Regional Aid' programme are **Enterprise Zones** (specific locations which attract particular forms of local aid). In 2015, there were 44 of these (all shown in Figure 5). Most of them are urban, but some rural areas have also been designated as Enterprise Zones – of which Cornwall is one. Enterprise Zones incentives focus investment on a small area, which is then 'branded' to attract particular companies. The incentives in Cornwall include:

- council business tax discounts (up to £160 000 for five years)
- a planning-free environment in which no planning permission is needed for building
- county-wide superfast broadband (see page 296).

Figure 3 Soil erosion caused by ploughing hilly land on a farm in mid-Cornwall in 2016

Figure 4 Wasteland produced by Cornwall's china clay industry

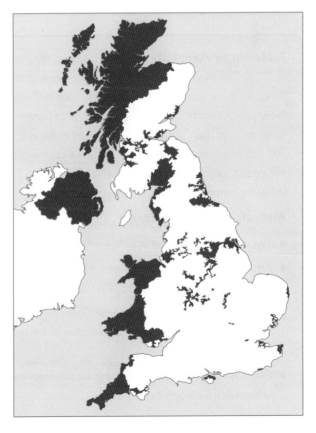

Figure 5 Areas (in red) qualifying for Regional Aid in the UK

Case studies of investment in Cornwall

1 The Eden Project

The Eden Project opened in 2001 (see Figure 6). It consists of two large conservatories (called 'biomes'), which exhibit major plant types, plus an education centre about sustainable living and a youth hostel. It has transformed the landscape – from a former china clay quarry to a completely re-imagined environment. The cost was £140 million – three-quarters of which came from the EU Objective One funding programme and the National Lottery.

▶ **Figure 6** *The Eden Project*

Benefits

In its first ten years, The Eden Project:

* generated £1.1 billion for Cornwall's economy (seven times its cost)
* attracted 13 million visitors
* directly employed 650 people (many of whom were previously unemployed) and also helped to sustain 3000 related jobs (e.g. supplying food)
* increased employment in other tourist-related businesses in Cornwall
* extended tourism as a year-round sector, by providing tourists with a destination for a rainy day (not unknown in Cornwall!)

2 Superfast broadband

By 2016, over 95% of Cornwall had access to fibre broadband (the first county to achieve this), and it also had the greatest take-up as a percentage of population. Cornwall now has the world's largest rural fibre network. It cost £132 million: £53.5 million from the EU Regional Development Fund and £78.5 million from BT. This superfast broadband network encourages businesses, particularly knowledge-economy companies and those working from home. One evaluation showed that 2000 jobs have been created with an annual economic impact of around £200 million.

3 Newquay Aerohub

In 2014, Cornwall Council obtained Enterprise Zone status for Newquay Aerohub Business Park (adjacent to Newquay Airport). The aim was to attract investment to an aviation 'hub' (or focus) that would generate 700 skilled permanent jobs by 2015. It's a partnership between Cornwall Council and private investors. The following companies were established in the first year:

Aircraft-related industries

* Two helicopter and aircraft training centres
* Bristow Helicopters (operates coastal Search and Rescue operations)
* British International Helicopters (operates offshore operations for the Ministry of Defence and Royal Navy)
* Cornwall Air Ambulance Trust (two emergency ambulance helicopters)
* Skybus (operates passenger flights between Cornwall and the Isles of Scilly)

Others

* Ainscough Wind Energy (wind turbine maintenance)
* Bloodhound SSC (a research and development centre for the Bloodhound Super Sonic Car, which will attempt a land speed record of up to 1000 mph)

The above list, achieved in a year, is impressive. But not all are 'new' jobs. For example, Bristow Helicopters and British International Helicopters took over jobs formerly done within Cornwall by the Ministry of Defence. These were displaced from the public sector into the private, and are therefore not new as such.

Key players in managing change in Cornwall

1 The EU
- Convergence funding has been granted to Cornwall since 1999.

2 UK central government agencies
- The South West Regional Development Agency (SWRDA) made investment grants before it was abolished in 2010.
- Most grants have been cut and are now given out by central government.

3 Local government
- The public sector (NHS and Cornwall Council) is Cornwall's largest employer.
- Since 2010, Cornwall Council has had no start-up funding. It offers rebates on business taxes as part of its Enterprise Zone at Newquay.

- A Local Enterprise Partnership supports business growth, but little funding is available – only 'help with exports, training or start-up advice'.

4 Stakeholders in the local economy
- The biggest industry is tourism. It wants better infrastructure (roads, rail and air travel)
- Other industries include food and farming. They want greater economic expansion.

5 Environmental stakeholders
- Cornwall's biggest asset is its scenery and environment.
- Many environmentalists are concerned about wasteland created by clay extraction and also the way in which soil is eroded from new arable land.

6 Stakeholders in people
- **Education**, e.g. Combined Universities in Cornwall, Cornwall FE Colleges.

Over to you

1 In pairs, draw up a six-slide presentation to show images of Cornwall that tourists, migrant workers, retired residents, young people, and working adults may have.

2 In pairs, draw up a table listing the advantages and disadvantages to the Cornish economy of **(a)** migrant workers, **(b)** supermarket policies towards the prices they pay to farmers.

3 **a** In groups of three, select each regeneration project opposite – one each. Then copy the table on the right, and assess how successful you think your project has been.

  b Present an agreed rank order with reasons – which is the most successful project?

4 Based on your evaluations of the projects in question 3, rank the importance of each of the players in regenerating Cornwall. Which actions have been most significant in bringing benefits to Cornwall?

Does it, or will it:	
Economically	
• improve prospects for farmers?	
• generate year-round employment?	
• help to develop a 'knowledge economy'?	
Demographically and socially	
• prevent further 'brain drain'?	
• provide opportunity for young people?	
• alter Cornwall's ageing population?	
Environmentally	
• make good use of Cornwall's environment?	

On your own

5 **a** Create a conflict matrix like Figure 7 in Section 6.12. Then complete the players for the examples in this section from Cornwall.

  b Identify which groups may be in **(a)** agreement, **(b)** disagreement, and why this might be the case.

6 'People's image of Cornwall contrasts with its reality.' Explain in 400 words how far you think the three regeneration projects opposite will change the reality of Cornwall for many people.

Exam-style questions

 1 Using examples, examine the impact of national and local strategies for change on a rural area. *(8 marks)*

 2 Assess the success of managing change in rural communities for different stakeholders. *(20 marks)*

Having studied this topic, you can now consider the three synoptic themes embedded in this chapter. 'Players' and 'Attitudes and Actions' were introduced on page 245. This page focuses on 'Futures and Uncertainties', as well as revisiting the four Enquiry Questions around which this topic has been framed (see page 245).

3 Futures and Uncertainties

People approach questions about the future in different ways. They include those who favour:

- **business as usual**, i.e. letting things stand. This might involve leaving diversity as a process from which the government might stay away, except to protect communities and individuals from racism.

- **more sustainable strategies** towards diversity, particularly the engagement of local communities in managing diversity. Sections 6.12 and 6.13 explore the legacy left by different approaches.

- **radical action**, which might involve directly engaging with diverse communities.

Working in groups, select two examples of regeneration from this chapter and then discuss the following questions in relation to each one:

1 How far does population diversity add to the UK's population economically and socially?

2 Should governments foster or supress diversity?

3 Should population diversity be left to 'see how it goes', or does it need managing with the full control of the law to confront racism, etc.?

Revisiting the Enquiry Questions

These are the key questions that drive the whole topic:

1 How do population structures vary?

2 How do different people view diverse living spaces?

3 Why are there demographic and cultural tensions in diverse places?

4 How successfully are cultural and demographic issues managed?

Having studied this topic, you can now consider answers to these questions.

Discuss the following questions in a group:

4 Consider Sections 6.1-6.4. How and why are populations so varied between one place and another?

5 Consider Sections 6.5-6.7. Why are cities and rural areas perceived so differently by different groups of people?

6 Consider Sections 6.8-6.10. How significant are demographic and cultural tensions between different places in the UK?

7 Consider Sections 6.11-6.13. How well do different places cope with diversity and its challenges?

Books, music, and films on this topic

Books to read

1. *Small Island* by Andrea Levy (2004)
   This novel is based on the migration of West Indian people to England after 1948, from British colonies in the Caribbean. It focuses on the challenges faced by the migrants in England and how they adapted.

2. *Pigeon English* by Stephen Kelman (2011)
   This novel is narrated by an 11-year-old Ghanaian immigrant living in London with his mother and sister.

3. *Segregation and Mistrust: Diversity, Isolation, and Social Cohesion* by Eric M. Uslaner (2012)
   This book assesses the view that trust in communities is boosted when societies are more diverse, based on networks in the USA, Canada and the UK.

Music to listen to

1. 'Black or White' by Michael Jackson (1991)
   A song released about racial harmony around the world, which attempts to address the issue of racism.

2. 'Ghetto Gospel' by 2Pac ft. Elton John (2004)
   This song addresses racism in the world and how it's viewed.

3. 'One Love' by Bob Marley (1977)
   This song celebrates the equality and diversity of all people.

Films to see

1. *East is East* (1999)
   This comedy film is based around a modern British family with mixed Pakistani and English heritage.

2. *My Big Fat Greek Wedding* (2002) & *My Big Fat Greek Wedding 2* (2016)
   These two romantic-comedy films address the role of the family in modern society, as well as the pressure placed on people in the family because of cultural norms in Greek families.

3. *My Brother the Devil* (2013)
   This British crime drama tells the story of two immigrants from Egypt growing up in London and how their paths to adulthood differ.

This skill is of particular use for the following topic:
- Coastal landscapes, Section 3.2

Drawing a sketch

Field sketching is an important skill for geographers, because it highlights the very features in the physical and human environment that are significant to them. Instead of a photo, labelled features on a well-drawn field sketch highlight the 'stuff' of Geography – landforms, human features, or processes. It omits irrelevant detail and focuses on the subject.

Don't worry, you don't need to be a good artist to draw an effective field sketch – but you do need to represent the landscape. The following tips might help you:

1. Use a pencil, not a pen. If you want to draw a pen-and-ink sketch, use pencil first so that you can easily correct your work!

2. Sketch the horizon first to give some idea of relative scale. Hold a pencil at arms length to work out the proportions of the features you want to draw.

3. Divide the landscape below the horizon into two or three parts. It's easier than trying to take in the whole thing. Then complete each part.

4. If you are drawing physical features, use shading and lines to show slope (e.g. field boundaries to show slope angle).

Labelling your sketch

Identify the main features that you are trying to show. For example, the Figure 1 photo of Beachy Head on the East Sussex coast shows considerable mass movement. Note how the Figure 2 field sketch of this same scene has picked out the movement on the cliff face and labelled it.

Notice that the field sketch contains the following elements:

- A location (a place name or grid reference will do), so that the reader can see where the sketch is of, and from which direction it has been viewed.

- Good labelling, to draw attention to points of interest for geographers.

- An estimated height, to give some idea of the landform's scale.

- The lighthouse, to emphasise the height of the cliffs.

⬤ **Figure 2** *Beachy Head in East Sussex*

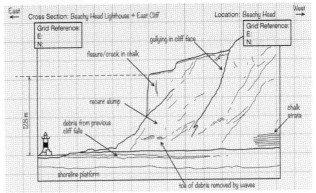

▶ **Figure 2** *A field sketch of Beachy Head to show its key features*

Carrying out an EIA

The law requires that an Environmental Impact Assessment (EIA) is carried out before any development proposals are approved. An EIA is a quantitative means of estimating the environmental changes arising from a proposal – to assess the development's potential problems and benefits.

The criteria on each EIA vary, depending on the proposal. But they tend to be fairly standard – using indicators that are familiar to geographers:

1. Environmental, i.e. those impacts that affect the scenic value of a place, or noise.

2. Economic, i.e. those impacts that affect the local economy.

3. Social, i.e. those impacts that affect the local quality of life.

The EIA is completed as follows:

1. The existing environment is assessed.

2. The proposed development is drawn up.

3. Any potential impacts of the proposal are considered in relation to their effect on the:

   • natural environment, e.g. habitats, potential pollution and visual impact

   • human environment, e.g. health and quality of life, design and local employment.

4. These potential impacts are then scored in a table, using a scale from -3 (negative impact) to +3 (positive impact) – see Figure 1.

⬇ **Figure 2** *A sample Environmental Impact Assessment (EIA)*

	Positive impact			No impact	Negative impact		
	strong +3	general +2	slight +1	0	slight -1	general -2	strong -3
1 Environmental factors – impacts of the development on:							
1.1 overall scenic quality							
1.2 local architecture (will it blend in?)							
1.3 general height (will it stand out?)							
1.4 wildlife habitats							
1.5 noise and traffic levels							
2 Economic factors – impacts of the development on:							
2.1 economic land uses							
2.2 local employment							
3 Social factors – impacts of the development on:							
3.1 land used for social purposes, e.g. housing							
3.2 the local quality of life							
3.3 the amenity value of the area							
Total score = _____ on a scale between +30 and -30							

Certain additional points need to be considered:

◆ The above method of assessment assumes that each factor is equal in terms of its importance.

◆ However, in reality, some impacts are more important than others. For example, recreation and employment may be more important to residents than visual appearance.

◆ Therefore, if needed, particular factors can be weighted in the EIA table. For example, if you feel that scenic value is doubly important (e.g. in a National Park like the Lake District), you could double the scores for Criterion 1.1 above, i.e. -6, -4, etc. through to +6. This would then affect the total score available at the end.

This skill is of particular use for the following topic:
- Glaciated landscapes, Section 2.9

Till fabric analysis is the study of the orientation and dip of particles within glacial till deposits. Stones lie in a particular way (known as **orientation**), linked to former ice movement (see Figure 1). They usually lie either parallel to ice movement or (very occasionally) at right angles. Measurements are normally made of the orientation of their long axis in degrees from magnetic north, using a compass, as well as their size and roundness.

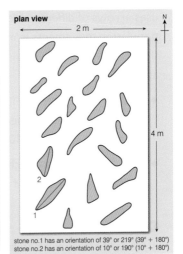

⬇ **Figure 1** *Glacial till aligned along a direction of flow along which they were deposited*

stone no.1 has an orientation of 39° or 219° (39° + 180°)
stone no.2 has an orientation of 10° or 190° (10° + 180°)

Measuring sediment size

- Most simply, measure the longest axis – the 'a axis' in Figure 2. However, to calculate the volume and size of each stone, measure its width (the 'b axis' in Figure 2) and its height (or depth – the 'c axis' in Figure 2).

- For each measurement, use a tape measure or (preferably) sliding callipers.

⬆ **Figure 2** *Measuring sediment size along three axes, a, b and c*

Measuring sediment roundness

Roundness refers to the sharpness of edges and corners. Visual estimates, using a roundness chart known as the Power's Scale (see Figure 3), are normal. Use the Cailleux Index, which you should research online, for more detailed measurements; this is normally used in universities by researchers.

class 1	class 2	class 3	class 4	class 5	class 6
very angular	angular	sub-angular	sub-rounded	rounded	well rounded

⬆ **Figure 3** *A scale of roundness known as Power's Scale*

Measuring orientation

Measure the orientation of the 'a axes' of pebbles. Gently remove stones from the till and position the 'a axis' with a non-magnetic object, (e.g. a knitting needle or plant support) into the imprint left by the stone. Then measure its bearing using a compass.

Presenting results using a rose diagram

A rose diagram is used to portray either numbers or proportions of pebbles in different azimuthal classes (i.e. angles of bearing from magnetic north). It is easy to plot and uses numbers of sediment (the radial lines in Figure 4) plotted on points of a compass reading from North clockwise.

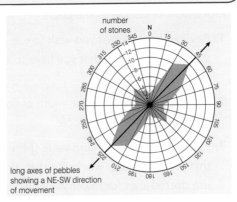

long axes of pebbles showing a NE-SW direction of movement

⬆ **Figure 4** *A rose diagram showing the axes of pebbles plotted on a compass rose – in this case, showing that most pebbles were deposited along a NE-SW direction of movement*

This skill is of particular use for the following topic:
♦ Globalisation, Section 4.7

Using flow lines and arrows

Flow-line maps show movements, e.g. of people or goods. For example, Figure 1 uses arrows of different widths to show population movements from London to the other economic regions in England and Wales.

♦ The width of each flow line represents a different number of people, using a set scale (e.g. 1 mm = 1000 people).

♦ The arrowheads on each flow line indicate the direction of movement.

▶ **Figure 2** *Flows of people from London to the eight other English economic regions, plus Wales, in mid-2012. The width of each arrow is in proportion to the number of people moving out of London to that region, so the widest arrow is to the South East region and the narrowest to the North East.*

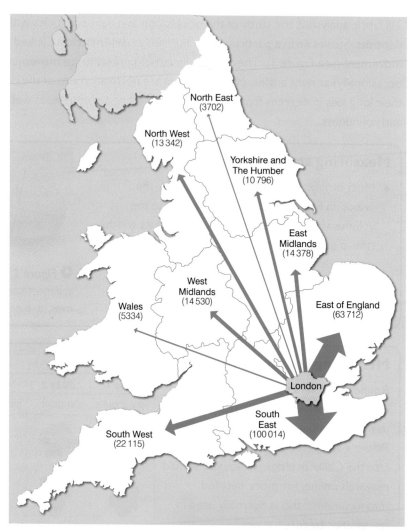

Follow these steps to construct a flow-line map showing the movement of people to the UK:

1. Review the data that you wish to use (e.g. Figure 2) and identify the highest number that you have to show – in this case, India (734 000).

2. Decide on an appropriate width of flow line to show this, e.g. 1 mm = 100 000 people.

3. Now check the minimum value (Nigeria; 181 000) and check that your scale also works for that. (In some cases, flow maps use dotted lines for lower values where the width of the flow is so narrow that it could not be drawn accurately.)

4. Plot the flows on a world map, with arrows leading to the UK (the host country) from the ten major source countries in Figure 2.

Rank	Country of origin	Population
1	India	734 000
2	Poland	679 000
3	Pakistan	502 000
4	Republic of Ireland	376 000
5	Germany	297 000
6	South Africa	221 000
7	Bangladesh	217 000
8	USA	199 000
9	China	191 000
10	Nigeria	181 000

⬣ **Figure 2** *The top ten source countries for international migrants living in the UK in 2013*

This skill is of particular use for the following topics:

◆ Glaciated landscapes, Section 2.6
◆ Coastal landscapes, Section 3.3

What is 'central tendency'?

The term 'measures of central tendency' refers to a group of statistical tests that analyse a particular data set and then compare it with others. Because this skill explores data distribution, it has widespread applications (e.g. analysing how rainfall varies between years).

Central points of data sets

There are three different statistical measures to show the central point of a data set: mean, mode, and median.

1 The **mean** is calculated by adding up the total and dividing by the number of items. However, it can be influenced by any extreme values (known as **outliers** or anomalies). So, for example, mean rainfall (see Figure 4 on page 304) might be influenced by a few unusually wet years. The mean also gives no indication of how data within a set are spread around the average; two completely different sets of data can have the same mean. However, 'mean' can be useful for geographers – such as when comparing beach data collected at different locations in order to identify trends (see Figure 1).

2 The **mode** – or modal value – is the most commonly occurring value within a data set.

3 The **median** is the mid-point value of a rank order – half of the data set lie above it and half below. The data must be ranked first. Using a dispersion diagram (see Figure 2 on page 304) makes this easier.

  • Where there are an odd number of data items, the median is always a whole number (e.g. the eighth value from 15 in Figure 2).

  • Where there are an even number of data items, the median lies across the two items at the mid-point (e.g. between the 10th and 11th in a data set of 20), where it is the average of those two values.

Like the mean, two data sets with the same median can have completely different distributions.

⬆ *Figure 2* *A Level students using quadrats to collect beach data*

Data variance

How data vary – called **data variance** – can be crucial in an investigation (e.g. why beaches vary considerably along a stretch of coast). Identifying variations leads to the question 'why are they varied?' – and so promotes further investigation.

1 Range

The range describes the span of data across a set, within which all data lie. It's a simple calculation (subtracting the lowest from the highest value). The range describes the spread of the data, but gives no idea about how the data are distributed (e.g. evenly spaced or clustered). It can also be affected by any outliers (i.e. data lying well beyond the range of the majority).

2 Dispersion

Dispersion means what it says – how data are distributed, or dispersed, within the range. A dispersion diagram (see Figure 2) reveals how far data are dispersed or clustered.

3 Inter-quartile range

The inter-quartile range (see Figure 2) is a statistical value to show where the middle 50% of the data lie within any given set. It is based around the median, and takes all data into account (discounting the effect of any anomalies).

The inter-quartile range is calculated as follows:

- Find the median. (In Figure 2, this would be the eighth of 15 items.)

- Find the 25% and 75% quartiles. These are points representing the limits of the middle half of the data set. To find these, count the number of data items either side of the median (seven on each side of the median in Figure 2). Calculate the median of each half (i.e. the fourth value of seven in Figure 2). The median of the upper half is called the **upper quartile**, and that of the lower half is called the **lower quartile**.

- The difference between the upper and lower quartiles is called the **inter-quartile** range (see Figure 2).

The inter-quartile range can be used in data presentation, using **box and whisker plots** (see Figure 3). This shows the median, upper and lower quartiles (forming the 'box'), the tail either side of the inter-quartile range (the 'whisker'), and the range. This is useful when comparing two data sets with similar means, medians or ranges.

4 Variance and standard deviation

Figure 4 shows rainfall data. Notice that no two years are the same. How can geographers make sense of such data? Page 305 explains.

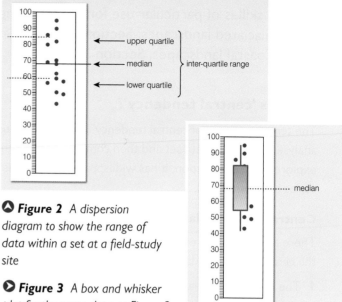

Figure 2 A dispersion diagram to show the range of data within a set at a field-study site

Figure 3 A box and whisker plot for the same data as Figure 2

Figure 4 Rainfall values over a 25-year period

Year	Rainfall (mm)	Variance from mean (x-x̄)	Variance (x-x̄) squared
9	1266	127.6	16281.76
17	1246	107.6	11577.76
3	1242	103.6	10732.96
13	1236	97.6	9525.76
4	1231	92.6	8574.76
12	1212	73.6	5416.96
25	1190	51.6	2662.56
1	1183	44.6	1989.16
10	1177	38.6	1489.96
14	1175	36.6	1339.56
19	1154	15.6	243.36
22	1146	7.6	57.76
16	Median 1133	-5.4	29.16
21	1120	-18.4	338.56
7	1117	-21.4	457.96
15	1104	-34.4	1183.36
24	1093	-45.4	2061.16
18	1092	-46.4	2152.96
11	1073	-65.4	4277.16
2	1069	-69.4	4816.36
23	1068	-70.4	4956.16
20	1055	-83.4	6955.56
6	1054	-84.4	7123.36
8	1021	-117.4	13782.76
5	1003	-135.4	18333.16
Total Σ	28460		Σ 136360
Mean x̄	1138.4	**Square root**	369.3

Figure 5 shows a normal distribution curve for any group of people or objects. As well as the mean value, the graph shows the **variance** and **standard deviation** from the mean (i.e. the amount within which most items are expected to fall). See the explanation in the text boxes to understand how these are calculated.

Some years have above-average rainfall, and some below. **Variance** (also known as **deviation**) shows how far each year varies from the average. The majority of years fall within a certain range from the average. **Standard deviation** is a value which shows **by how much** most years vary from the mean. It's like quartiles (see above), except that it quantifies a figure.

Standard deviation is widely used in data analysis to show how dispersed data are from the mean. It aids prediction, such as the statistical likelihood of drought or flooding. To calculate standard deviation, use either Excel (find the SD function in 'Formulae' on the toolbar), or the longhand method using the procedure and formula outlined in the box on the right.

The result on the right means that the amount by which rainfall is likely to deviate either side of the mean is 73.85 mm. The years to which this applies are shown in bold in Figure 4, i.e. in 16 of 25 years (64%), and in Figure 5 as 'μ+1'/'μ−1'.

♦ It means that in 64% of years, rainfall is likely to be within 73.85 mm of the mean – similar to the normal distribution shown in Figure 5.

♦ In most data sets, twice the standard deviation (in this case, 147.7) will normally encompass 95% of data (between 1280.7 and 985.3). In the case of the rainfall in Figure 4, only one year out of 25 (i.e. 4%) lies outside this range, so the rainfall data have a close-to-normal distribution.

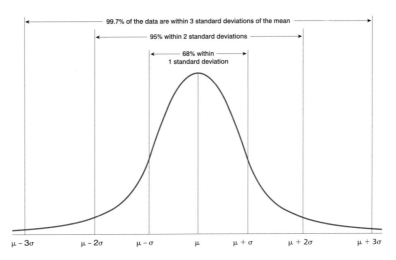

♦ **Figure 5** *A normal distribution curve. In the diagram, μ is the mean and σ is the symbol for standard deviation (see text for explanation).*

Variance is calculated as follows:

♦ Find the mean.

♦ Calculate, in a separate column, how much each value differs from the mean. In Figure 4, this is shown in the column headed 'Variance from mean (x-x)'.

How to calculate standard deviation (using the data from Figure 4)

♦ Calculate the mean ($\dot{x}$), variance and variance-squared values as shown in Figure 4. (Because some are positive values and others negative, squaring them makes all values positive.)

♦ Total the squared values, as shown in the column 'Variance (x-x) squared' in Figure 4.

♦ Use the formula below.

It shows:

▸ the standard deviation value sigma (σ) =

▸ the square root of $\sum$ (the sum of) variance squared values

▸ n (the number of examples)

It is written: $\sigma = \dfrac{\sqrt{(\sum(x-\dot{x})^2}}{n}$

♦ In this case, the number of examples is 25; the mean value x is 1138.4, the value $\sqrt{(\sum(x-x)^2}$ is 136360. The equation therefore reads:

$$\dfrac{136360}{25} = 5454.4, \text{ of which the square root} = 73.85$$

♦ Therefore σ = 73.85.

This skill is of particular use for the following topics:

◆ Glaciated landscapes, Section 2.8
◆ Diverse places, Section 6.7
◆ Regenerating places, Section 5.7

Relationships between sets of data

The purpose of investigating relationships between sets of data is to identify any potential statistical correlation between them. For example, the data in Figure 1 compare GNI per capita in 13 African and Asian countries against life expectancy in years from birth – to see whether, statistically, a higher income leads to a higher life expectancy.

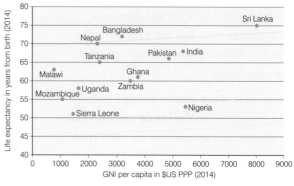

◬ **Figure 1** *A scatter graph showing the relationship between GNI per capita and life expectancy in years from birth in 13 African and Asian countries*

Scatter graphs

Scatter graphs show two sets of data, known as **variables**. One will be the 'dependent variable' and the other the 'independent variable'. The dependent variable is the one you are seeking to explore; the independent variable is one you know is fixed. For instance, when plotting population over time, 'population' is the dependent variable – the one you are seeking to explore over time. Time is the independent variable. Consider why the variables have been shown in Figure 1 in the way they have.

Follow these steps to draw a scatter graph:

◆ Plot a frame for the graph – and a scale.

◆ Plot the data as a number of points. Do not join them.

◆ The result will be a graph of scatter points, where a trend or pattern may emerge.

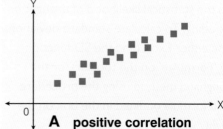

A positive correlation

Best fit line

Having drawn scatter points, you can now investigate any possible relationship between them by drawing a best fit line.

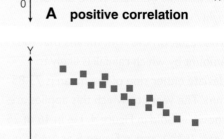

B negative correlation

Follow these steps to draw a best fit line:

◆ Calculate the mean of each data set that you are testing.

◆ Plot this as a 'mean point' (a point with a circle drawn around it).

◆ Place a ruler on the graph, passing through the mean point. Rotate it until it is aligned with the trend of the points you have drawn. Depending on the number of points plotted, your ruler should ideally pass through about 2, and leave half of the remainder on either side. This is a 'best fit' line.

◆ When you are confident of a 'best fit', draw it as a line.

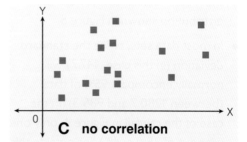

C no correlation

Then use Figure 2 to interpret your result. Is it a positive correlation (Graph A, where one variable increases in line with the other), negative (Graph B), or no correlation (Graph C, with a random pattern)?

◬ **Figure 2** *Scatter graphs showing three possible correlations*

Spearman's rank correlation

Unlike scatter graphs, which are interpreted by eye, Spearman's rank correlation is a statistical method used to test the strength of the relationship between two variables. It uses ranked data to test the relationship, and gives a fixed figure (between -1 and +1) to show the strength of the relationship:

◆ +1 indicates a perfect positive correlation

◆ -1 a perfect negative correlation

◆ 0 indicates that there is no correlation (see Figure 3).

In order to test the strength of the correlation, complete Steps 1-5 in Figure 4.

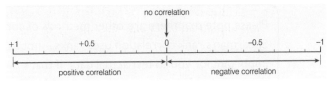

▲ **Figure 3** *Assessing the strength of any correlation between two variables, using the Spearman's rank correlation method*

▼ **Figure 4** *The five steps used when completing a Spearman's rank correlation.* **Note:** *When two numbers or more tie, rank them equally (e.g. 2=) – but remember to skip a number when dealing with the next rank down (i.e. 1, 2, 2, 4, etc.)*

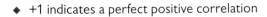

| 1. Rank median household income from '1' highest to '20' lowest | 2. Rank level 4 qualifications from '1' highest to '20' lowest | 3. Subtract the first rank from the second | 4. Square each difference 'd' |

Ward in Newham	Median household income (£), 2012-13	Rank order	% Level 4 qualifications or above (2011)	Rank order	Difference in ranks (d)	d²
1. Beckton	34 100		27.9			
2. Boleyn	31 630		28.8			
3. Canning Town North	28 910		23.9			
4. Canning Town South	32 870		33.3			
5. Custom House	31 840		24.2			
6. East Ham Central	32 380		32.6			
7. East Ham North	32 340		31.1			
8. East Ham South	31 430		24.1			
9. Forest Gate North	34 270		33.0			
10. Forest Gate South	34 550		32.7			
11. Green Street East	31 570		30.0			
12. Green Street West	31 080		29.8			
13. Little Ilford	29 730		24.6			
14. Manor Park	31 580		30.2			
15. Plaistow North	30 740		26.8			
16. Plaistow South	31 750		26.2			
17. Royal Docks	38 580		42.7			
18. Stratford and New Town	35 840		40.6		5. Add up all the d²	
19. Wall End	32 330		29.6			
20. West Ham	32 310		29.8			
					TOTAL (or Σ) d²	

Next, use the following formula to calculate the Spearman's rank correlation coefficient value R(s) – substituting data from Figure 4:

$$r_s = 1 - \frac{6 \times \sum d^2}{n^3 - n}$$

Where

◆ R(s) is the correlation figure

◆ n is the number of paired values (in this case, the number of wards – 20)

◆ $\sum d^2$ is the sum of the squared differences, shown in Step 5 above.

Finally, interpret your result using Figure 3.

Please note that there are other methods of correlating two sets of data. Spearman's rank correlation uses ranks – not actual data. However, it may be important for you to test actual data. If that's the case, use Pearson correlation in the statistical function menu in a spreadsheet such as Excel.

Testing for significance

For greater precision, any correlation coefficient should be tested for significance. Significance is a mathematical term, which allows the user to get a measure of confidence in the correlation they have calculated. For example, a correlation of + or - 0.5 does not suggest a definite link between sets of data, and the result could be due to random chance. To test whether this is likely, use the **significance table** in Figure 5.

▼ **Figure 5** *A significance table*

To test for significance follow the steps below:

♦ Note how many paired data items you used in your correlation. This is labelled as 'degrees of freedom' in Figure 5. Degrees of freedom are a statistical convention, arrived at by subtracting 1 from the number of paired items being correlated. (For example, if you were testing the data in Figure 4, there are 20 wards in Newham, so 20-1 = 19 degrees of freedom.) Find this number on the horizontal scale in Figure 5.

♦ Now find your correlation coefficient (ignoring the + or -) on the vertical scale in Figure 5.

♦ Finally, find where the two points meet. If the point occurs above the 5% line, you have found a low probability (under 5%) that your correlation was caused by chance. If it occurs above the 1% line, the likelihood of chance is even less, and even better if it occurs above the 0.1% line!

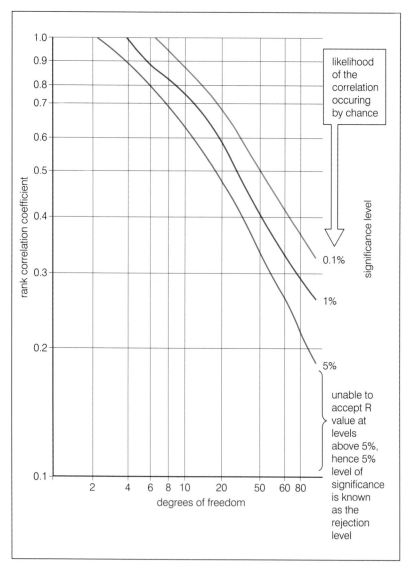

Significance tables allow a correlation to be expressed as a percentage level of **confidence limits** (or **intervals**), usually 95% or 99%. This means that 95 or 99 times out of 100, the relationship is unlikely to have been caused by chance – and is therefore significant. The greater the 'n' value, the lower the likelihood will be that any correlation was caused by chance.

This skill is of particular use for the following topic:
 ◆ Globalisation, Section 4.9

The Lorenz curve

Measuring inequality is one of the fundamentals of geography. The Lorenz curve was developed to show and measure any inequality in graphical form. It assumes that – in an equal world – 10% of the population would have 10% of a country's wealth, 20% would have 20% of the wealth, and so on. As Figure 1 shows, this would result in a straight line – i.e. perfect equality. Plotting a Lorenz curve shows how much inequality actually exists in a particular situation. Look at Figure 1. The more the Lorenz curve line bends away from the straight line of perfect equality, the greater the inequality represented will be.

Figure 1 uses income distribution as a measure of inequality. By dividing the percentage of different groups of people by their share of national income, the Lorenz curve shows how relatively wealthy or poor certain sections of the population were. When Figure 1 was prepared in the 1970s, South Africa's 'Black African' population made up over 70% of the total, yet they received just 19% of national income. Meanwhile, the 'White' population represented just 14% of the total, yet they received 73% of national income.

Ratios of advantage

By dividing a particular group's share of national income by its percentage of the total population, a **ratio of advantage** can be calculated for each group, which is then used to plot a Lorenz curve.

The graph in Figure 1 was calculated as follows:

 ◆ The share of national income of 'Black Africans' in South Africa (the poorest group) was 19% in the 1970s. Divided by their share of the population (75%), this gave a ratio of advantage of 0.25.

 ◆ This was then added to for other groups in the population – with the 'White' population added last. Their 73% share of national income was divided by their 14% share of the population – giving a ratio of advantage of 5.2.

Any figure higher than 1 shows advantage, whereas any figure below 1 shows disadvantage. The higher the figure is above 1, the greater the advantage will be.

Drawing a Lorenz curve

A Lorenz curve is plotted on a graph using data showing ratio of advantage. Follow the steps described on page 310 (using a graph outline copied from Figure 2) and plot the data for the UK shown in Figure 3. Notice this time that, unlike the South African racial groups used in Figure 1 (which were unequal in size), the data for the UK are divided into equal-sized groups (known as deciles, or tenths).

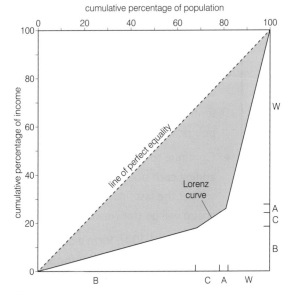

🔺 **Figure 1** A Lorenz curve for the distribution of population and national income in South Africa in the 1970s. In the diagram, 'B' means 'Black African', 'C' means 'Coloured' (a term used to describe inter-racial or unregistered people), 'A' means 'Asian', and 'W' means 'White'. The terminology used to describe each population group was a legal classification used in South Africa until the ending of apartheid in 1994.

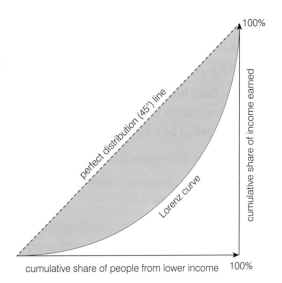

🔺 **Figure 2** A Lorenz curve graph outline. To use this to plot your own data, scale each of the axes 0-100% in graduations of 10%.

- Draw a graph outline and plot a perfect distribution line (as shown in Figure 2).

- Ratios of advantage have already been calculated and ranked in Figure 3, by taking each population decile and its share of national wealth.

- Begin by plotting the group with the lowest ratio – the 'Poorest tenth' in Figure 3. This will be point 1 on your graph.

- Now take the next lowest group. Add its share of wealth and population to those for the 'Poorest tenth' that you have already plotted, and plot these points. In this way, you are plotting cumulative figures.

- Continue to plot the remaining groups in order, adding each one onto the previous cumulative figure – so that the last figure you plot reaches 100%. The last group will be the wealthiest, or most advantaged.

- Join points 1-10 to form a curve, known as the Lorenz curve. The area between the straight line and the curve shows how unequally income is distributed. The more curved the line, the greater the inequality.

Decile of population	% wealth (2012-14)	Cumulative %
Poorest tenth %	0.05	0.05
2nd poorest	0.49	0.54
3rd poorest	1.28	1.82
4th poorest	2.63	4.45
5th poorest	4.25	8.70
6th poorest	6.23	14.93
7th poorest	8.81	23.74
8th poorest	12.52	36.26
9th poorest	18.93	55.19
Wealthiest tenth	44.81	100.00

Figure 3 Data showing wealth distribution in the UK, 2012-14. Wealth is defined here as the sum total of people's property and pension wealth, added to the total of their other assets (e.g. savings) – minus any debts.

The Gini index and Gini coefficient

The **Gini index** measures the inequality of wealth distribution. It measures the area between the Lorenz curve and a line of absolute equality (see Figure 2), expressed as a percentage of the maximum area under the line. On graph paper, this would be done by counting small graph squares. It is shown as a value between 0 and 100:

- A lower index indicates more equal distribution, with 0 as perfect equality.

- A higher index indicates more unequal distribution, with 100 corresponding to perfect inequality (where one person has all the income).

The World Bank uses this method to show inequalities globally (the index ranges from 24.9 in Japan to 70.7 in Namibia).

The **Gini index** is sometimes shown as the **Gini coefficient**, which is the same thing except that it's shown as a value between 0 and 1 (i.e. the Gini index divided by 100). An index of 24.9 for Japan therefore becomes 0.249.

The main advantages of either the Gini index or the Gini coefficient are that they:

- measure inequality, rather than presenting a uniform picture of an entire population (such GDP or GNI)

- can be used to compare inequalities of different kinds within a country, e.g. urban areas compared to rural, or comparing different sections of the population

- are easy to use to detect trends over time. In an ideal world, the Gini index would increase as GDP increases, showing that the population shares in increased wealth. If this is not happening, then wealth is becoming more concentrated in the hands of a few.

7.8 Chi-squared

This skill is of particular use for the following topic:
- ◆ Coastal landscapes, Section 3.5

Measuring distributions

Distributions are an essential part of geographical studies, e.g. the distribution of particular plants across a sand dune. Measuring and analysing distributions is the purpose of a statistical test known as **chi-squared**.

Study Figure 1, which shows the number of marram grass plants at various points in an area of sand dunes. In the chi-square test, these are known as the 'Observed' – or actual – numbers that exist. These will be represented by [O].

Distance from high-water mark	10m	20m	30m	40m	50m	60m	70m	80m	Total
Number of marram plants observed within a plant quadrat [O]	1	11	15	14	9	5	1	0	56

At first glance, it seems as if distance from the high-water mark is important in determining the number of marram grass plants present. But **how** important? The **chi-square** test shows how clustered distributions are – it compares how clustered a distribution is with how it might be if everything were evenly distributed.

⬥ **Figure 1** *Changes in the number of marram grass plants observed within a plant quadrat as the distance from the high-water mark increases*

Performing the test

1 Establish a hypothesis. For this exercise, we will assume that the hypothesis is as follows: 'There is no difference in the distribution of marram grass plants with increasing distance from the high-water mark'.

2 Decide what the hypothesis means. For example, a total of 56 plants were recorded in Figure 1. If the hypothesis in Step 1 was true, marram grass plants would be distributed equally – i.e. 7 plants for each of the eight sampling points. In the chi-square test, these are called 'expected' – or theoretical – numbers, which would exist if marram grass were evenly distributed to match the hypothesis. These are shown by [E].

3 Now compare Observed values, [O] in Figure 1, with expected values, [E] in Figure 2 on page 312, and calculate the value known as chi-squared. This is done by:

- calculating the difference between each observed [O] and expected [E] frequency

- squaring it in order to remove minus numbers

- dividing by the expected frequency [E], as shown in Figure 2.

Distance from high water mark	10m	20m	30m	40m	50m	60m	70m	80m	Total
Number of marram plants observed within a plant quadrat [O]	1	11	15	14	9	5	1	0	56
Expected number [E]	7	7	7	7	7	7	7	7	
O-E	-6	4	8	7	2	-2	-6	-7	
$[O-E]^2$	36	16	64	49	4	4	36	49	
$\dfrac{[O-E]^2}{E}$	5.14	2.29	9.14	7	0.57	0.57	5.14	7	36.86

4 The chi-square figure itself – x^2 – is calculated by adding up the final row (i.e. 36.86). This figure means nothing by itself – it has to be referred to using significance tables to see whether or not it is statistically significant. The standard table for significance testing for chi-squared is shown in Figure 3.

5 The value (36.86) is read off against the appropriate degrees of freedom in Figure 3. Degrees of freedom are a statistical convention (see page 308). They are arrived at by subtracting 1 from the number of categories tested; in this case, 8 plant quadrats. Therefore, 8-1 = 7 degrees of freedom.

6 In this case, the value 36.86 (plotted against 7 degrees of freedom) suggests that there is a 0.1% probability of this value occurring by chance, i.e. once in 1000 occasions! Therefore, the hypothesis is rejected. It means that chance alone is unlikely to have caused the observed distribution of marram grass, and there is a 99.9% probability that the observed differences are statistically highly significant. In this case, the distribution of marram grass is closely linked to distance from the high-water mark.

⬥ **Figure 2** *A completed table showing the calculations needed for chi-squared. The final value is shown in the bottom-right hand cell – i.e. 36.86.*

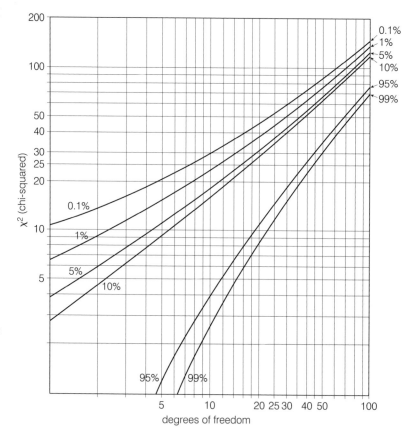

⬥ **Figure 3** *Degrees of freedom for chi-squared*

7.9 Using an Index of Diversity

This skill is of particular use for the following topic:
- ◆ Coastal landscapes, Section 3.5

What is an Index of Diversity?

Distributions and diversity are essential concepts for geographers. For example, in the study of sand dunes, plant communities can be measured over time – in order to study plant succession. An **Index of Diversity** measures diversity by counting both the number of plant species within a community, and also the number of plants present from each species. The final calculated Index produces a single figure – the higher the figure, the greater the plant diversity. On sand-dune profiles, diversity normally increases with distance inland (see Figure 1). However, climatic climax communities outnumber, and eventually out-shade, other species, so reduced diversity can be expected as dunes age.

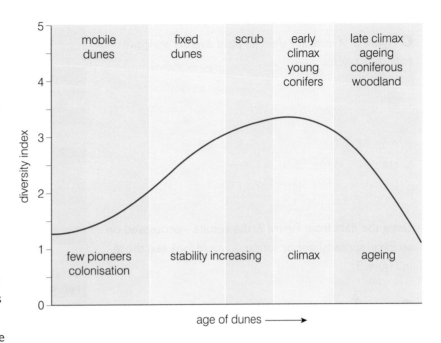

↗ **Figure 2** A theoretical Index of Diversity across a sand-dune profile

---

There are several ways of calculating an Index of Diversity. One of the most straightforward is the Simpson Yale Index. It has the following two advantages:

- ◆ No familiarity is needed with individual plant names – only the ability to identify different species (see Figure 2 on page 314, where the species are simply recorded as A, B, C, etc.).

- ◆ The Index is simple to calculate – although fiddly, unless you use a spreadsheet.

The formula for calculating the Index (D) is:

$$D = \frac{N(N-1)}{\sum n(n-1)}$$

Where:

N = the total number of individual plants recorded. In Figure 2, this value is 112.

n = the number of individuals within each species

∑ = the sum of

Collecting and recording information

Figure 2 shows a simple recording sheet for two sand-dune profiles – **1** and **2**.

▶ **Figure 2** A recording sheet for plant diversity along two sand-dune profiles

Sand-dune profile 1		Sand-dune profile 2	
Plant species	Number within each species	Plant species	Number within each species
A	4	A	2
B	12	B	6
C	19	C	67
D	8	D	5
E	12	E	43
F	4	F	2
G	3	G	9
H	2		
I	23		
J	11		
K	9		
L	5		
Total plants counted	**112**		**134**
Total number of species	**12**		**7**

Using the data from Figure 2, the results – processed on an Excel spreadsheet or similar – would look like those in Figure 3.

▼ **Figure 3** The processed field results of plant communities along sand-dune profiles **1** and **2**

Sand-dune profile 1				Sand-dune profile 2			
Plant species	Number within each species (n)	n-1	n(n-1)	Plant species	Number within each species (n)	n-1	n(n-1)
A	4	3	12	A	2	1	2
B	12	11	132	B	6	5	30
C	19	18	342	C	67	66	4422
D	8	7	56	D	5	4	20
E	12	11	132	E	43	42	1806
F	4	3	12	F	2	1	2
G	3	2	6	G	9	8	72
H	2	1	2				
I	23	22	506				
J	11	10	110				
K	9	8	72				
L	5	4	20				
Total counted	112			**Total counted**	134		
Total species	12			**Total species**	7		
Value of N(N-1)	12432	$\sum$n(n-1)	1402	**Value of N(N-1)**	17822	$\sum$n(n-1)	6354

Using Figure 3, the Index of Diversity (D) is 8.87 for sand-dune profile **1**, and 2.81 for sand-dune profile **2**. Therefore, it is clear that diversity is greater in sand-dune profile **1**.

Student t-test

This skill is of particular use for the following topic:
- ◆ Coastal landscapes, Section 3.5

What is the student t-test?

One problem with collecting fieldwork data is that it's not always possible to collect sufficient data to reach a significant sample. In mathematical terms, 'significant' means reliable – and the smaller the sample, the less reliable it's likely to be. For example, in sand-dune studies, you may not record as many as 30 species in a survey (particularly with a small quadrat). The problem is that – in terms of statistical reliability – a mathematical mean (see Section 7.5) is only regarded as reliable if the sample is greater than 30.

However, in cases where the number of samples is lower than 30, it's still possible to test for reliability using the student t-test. Figure 1 shows how significance can even be assessed using sample sizes as low as six!

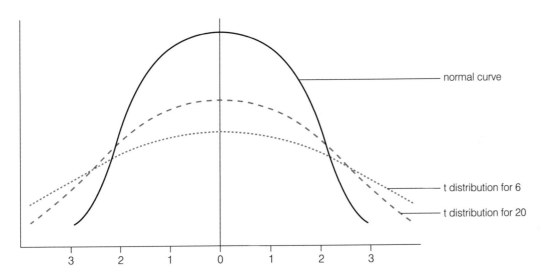

Figure 1 *The t distribution for different sample sizes compared to the normal distribution curve. To see a fully annotated normal distribution curve, see Figure 5 in Section 7.5.*

Calculating the t-test

Figure 2 on page 316 shows the total number of plant species collected using quadrats at ten points along two sand-dune profiles, X and Y. Profile Y has only eight points – perhaps because the dunes were narrower than they were for Profile X. Using the student t-test, it's possible to gauge the reliability of data for each profile.

Quadrat location number	Total species along dune profile X (x in the formula)	Total species along dune profile Y (y in the formula)
1	6	14
2	8	12
3	9	6
4	4	11
5	7	15
6	11	14
7	7	17
8	6	8
9	8	
10	7	

Calculating the coefficient t for the student t-test uses a formula that looks more complex than it really is.

$$t = \frac{\dot{x} - \dot{y}}{\sqrt{\dfrac{(\sum x2/n_x) - \dot{x}^2}{(n_x - 1)} + \dfrac{(\sum y2/n_y) - \dot{y}^2}{(n_y - 1)}}}$$

Where:

$\dot{x} - \dot{y}$ are the means of each sample

n_x and n_y are the number of individuals in each sample X and Y respectively

$\sum$ means 'the sum of', or total.

⬤ **Figure 2** *Plant species collected at different quadrat recording points along two sand-dune profiles – **X** and **Y***

The data from Figure 2 work out as shown in Figure 3 – simply add two more columns and three extra rows to Figure 2 and carry out the calculations.

Quadrat location number	Total species along dune profile X (x in the formula)	Total species along dune profile Y (y in the formula)	x^2	y^2
1	6	14	36	196
2	8	12	64	144
3	9	6	81	36
4	4	11	16	121
5	7	15	49	225
6	11	14	121	196
7	7	17	49	289
8	6	8	36	64
9	8		64	
10	7		49	
Totals	$\sum x = 73$	$\sum y = 97$	$\sum x^2 = 565$	$\sum y^2 = 1271$
	$\dot{x} = 7.3$	$\dot{y} = 12.13$		
	$n_x = 10$	$n_y = 8$		

◀ **Figure 3** *Calculating the student t-test*

Using the formula from the tint box above gives a value of t = 3.38. To test this value for significance, do the following:

1. Calculate the degrees of freedom. In this case, there are two data sets, X and Y, so 'n-1' must be calculated twice and added to give a total. With samples of 10 and 8, n-1 becomes 9 and 7 respectively (making 16 when added).

2. Find an online critical-values table by searching using the phrase 'Critical values student t-test'. By using the table, you will see that the value 3.38 is greater than the critical value of t at p = 0.05. This means that the result is significant – there is a significant difference between the plants along the two sand-dune profiles.

In this chapter, you'll learn about the fieldwork skills needed for the AS course, and also be guided about using them in the AS exam.

What fieldwork is needed at AS level?

Geographical study is incomplete without fieldwork! Through fieldwork, geographers gain a sense of the realities of the world outside the classroom, and learn the skills needed to investigate questions and issues at first hand (see Figure 1).

AS level students must complete a minimum of two days of fieldwork (compared with four days for the full two-year A Level course). This fieldwork must focus on processes in both physical and human geography, so it's likely that you will spend one day developing your skills in each. These skills will then be assessed in your AS exams.

Many of those taking the AS exams will go on to complete an Individual Investigation for A Level, which is a piece of written coursework. The skills you develop at AS – almost certainly as part of a group, in locations like Lulworth (see Figure 2) – will therefore be essential in helping you to develop this independent piece of work. Your school or college must submit a written fieldwork statement to Pearson about the fieldwork that you have undertaken.

▲ **Figure 1** *Fieldwork – investigating issues first hand!*

Which topics are assessed?

Although fieldwork for A Level can be based on any part of the specification, the AS exams assess fieldwork on the following topics:

- **Either** Glaciated landscapes **or** Coastal landscapes (Chapters 2 and 3 in this textbook)

PLUS

- **Either** Regenerating places **or** Diverse places (Chapters 5 and 6 in this textbook).

Remember – you are only restricted to these topics for the AS exams. If you intend to do the full two-year course, you may study any topic you wish that is related to the specification. The purpose of the AS year is to build your skills base.

How is fieldwork assessed?

Fieldwork is assessed at AS in both Papers 1 and 2 – each within optional sections B and C. You have to answer one fieldwork question in each exam. The questions will assess your:

♦ knowledge and understanding of investigating geographical questions

♦ skills in interpreting, analysing and evaluating data collected through fieldwork

♦ ability to construct arguments and draw conclusions about your fieldwork.

▲ **Figure 2** *Lulworth Cove in Dorset – a favourite location for fieldwork. What questions could be investigated here?*

Practical geographical fieldwork

To get the most out of your geographical enquiry, follow the three steps below:

1 Preliminary work

a Identify a question

What possible, practicable fieldwork opportunities are presented by the place?

b Contextualise fieldwork

- Research relevant secondary background information (from the Internet, journals, books) about the place and the theme (e.g. population change, coastal processes). Research theories and models.

- Develop hypotheses (see the box on the right) and/or suitable key questions.

2 Primary data collection in the field

a Design: where, how many?

- Plan how many data-collection sites are practicable for group or individual observations.

- Choose sampling procedures (systematic, random or stratified) and sample sizes.

- Consider health and safety factors and undertake risk assessments.

b Equipment and recording

- Design data-collection methods that will help you investigate your hypotheses. Obtain appropriate equipment to ensure accuracy and reliability. Develop recording sheets to record your results.

- Read any sections of this book which you think could help you, e.g. using questionnaires from Section 5.4 in Chapter 5 Regenerating places.

3 Presentation, analysis, conclusions and evaluation

a Data processing and presentation

- Consider how you will use ICT to manage, collate and process information (e.g. shared spreadsheets), and also present your findings.

- Plan what statistical analysis you will use from Chapter 7, e.g. measures of central tendency, correlation techniques.

Hypothesis testing

A study normally begins with a broad question, such as 'How far has regeneration in X been a success?'. However, for more-focused elements of a study, especially those gathering quantitative data, it is normal to develop hypotheses. A hypothesis is a prediction or idea that states a relationship between two variables, using scientific method. It is then tested using data. On that basis, it may then be accepted, rejected, or be inconclusive.

A null hypothesis

A null hypothesis is an opposite, i.e. it assumes that no relationship exists between variables. Its purpose is scientific – it intends to provide an alternative, so that the investigator does not prefer one prediction over another, just because they expect a relationship to exist.

As an example:

Hypothesis (written as H_1): 'There are wide differences in pebble size between different points along a beach'.

Null hypothesis (written as H_0): There is no difference in pebble size between different points along a beach.'

In this sense, we accept that there is no relationship until evidence proves otherwise.

b Analysis and conclusions

- Review any patterns shown by your results and then draw them together into a conclusion – using evidence and reasoned argument.

- Consider any relevant statistical analysis from Chapter 7, e.g. significance tables, student t-test.

c Reflect critically on the results

Evaluate your enquiry, including the accuracy and appropriateness of the data and the methods used, as well as the validity and reliability of your conclusions.

Some ideas to get you started

The following four outlines have been written to help you meet the AS requirement of 2 days fieldwork, as well as giving you some idea of the level expected in the exam questions that assess the fieldwork.

Field study topic 2: Glaciated landscapes and change

These can be relict or active glaciated landscapes, in either upland or lowland areas. The following themes could be appropriate:

◆ Changing glacial and/or fluvio-glacial sediments.

◆ Glacial and/or fluvio-glacial landform morphology.

◆ The impact of human activity on fragile glaciated landscapes.

Planning an investigation into glaciated landscapes

1 Pre-visit

Use the following sources to help you gather information about the area you will visit:

◆ **The British Geological Survey website** (bgs. ac.uk) – for more about the geology of the area. Remember – solid geology describes solid rock, while drift geology is the surface (e.g. glacial deposits).

◆ **Climate data**, e.g. Met Office data for information about upland or lowland areas.

◆ Background reading on the glaciated past of the area, and how geographers know about the track of past glaciers.

◆ **GIS data**, from packages such as ArcGIS, plus Google Earth and satellite images to identify the main features of glacier types.

2 The field trip itself

Depending on your focus, a wide range of fieldwork data could be collected, such as:

◆ the mapping of erosional features, e.g. slope mapping of corries, or scree slopes. Investigations of secondary data could include analysis of corrie orientation.

◆ till fabric analysis (see Section 7.3).

◆ recording slope form and shape of depositional features, e.g. drumlins (see Figure 3), fluvio-glacial deposits, ice-contact features (e.g. eskers), or pro-glacial deposits (e.g. varves).

◆ investigations into drumlin morphometry and an orientation survey to measure the correlation of height, length and elongation ratio.

◆ photographs and sketches.

Figure 3 *A series of drumlins in the north-eastern Lake District in Cumbria*

3 Follow up

Consider how to analyse the data collected. For example:

◆ sediment analysis; analyse data for central tendency (see Section 7.5), Spearman's rank (see Section 7.6). Test for significance (see Section 7.6) or use the Student t-test (see Section 7.10).

◆ the numerical analysis of mean rates of glacial recession in different global regions, using secondary evidence.

◆ a statistical comparison of two data sets from contrasting drumlin 'swarms'.

Warning!

Never take risks of any kind in any upland environment, even in the height of summer. **NEVER** work in high-level country alone. Always go prepared for sudden changes in the weather, and wear strong footwear for steep and/or high ground. Be aware that landscapes in upland areas can be fascinating, but also very dangerous.

Field study topic 3: Coastal landscapes and change

This could take place along a single stretch of coast, or at more than one coastal location on the following themes:

◆ Changing coastal sediments.

◆ Changing coastal profiles.

◆ An assessment of coastal management approaches.

Planning an investigation into coastal landscapes

1 Pre-visit

Use the following sources to help you research the coast that you will visit:

◆ **The British Geological Survey website** (bgs. ac.uk) to find out more about the geology of that particular stretch of coast. Remember – solid geology describes solid rock, while drift geology is the surface (e.g. glacial deposits).

◆ **Climate data**, e.g. wind rose data – search for 'wind rose …' including the name of the area you are researching.

◆ **Historical maps** and **land use surveys** to identify land uses. You could include the First OS map series (1846 onwards) and First Land Utilisation Survey (1930s) online at: www.visionofbritain.org.uk

◆ **GIS data**, from packages such as ArcGIS, or Google Earth and satellite images to compare recent land use with historical data.

◆ Research **ICZM** and **SMP** for the coast.

2 The field trip itself

Depending on your focus, a wide range of fieldwork data could be collected, such as:

◆ the mapping of physical features, e.g. cliff heights, cliff material, landslides (plus coastal defences). More 'pure' physical techniques could be carried out on wave type and period, longshore drift, beach profiles and sediment.

◆ biodiversity mapping and surveys – either physically to assess plant succession or dune morphology (like that in Figure 4), or to assess the ecological 'value' of the area.

◆ EIAs of sea defences to show environmental impacts. Collect cost-benefit data as part of your research using secondary sources.

◆ personal interviews about the experience of living in an area where erosion is a risk.

◆ photographs and sketches.

◆ the effectiveness of strategies being used in ICZM and SMPs.

3 Follow up

Consider how to analyse the data collected. For example:

◆ sediment analysis; analyse data for central tendency (see Section 7.5), Spearman's rank (see Section 7.6). Test for significance (see Section 7.6), or use the Student t-test (see Section 7.10).

◆ plant analysis along dune transects; analyse data using chi-square (see Section 7.8) and a diversity index (see Section 7.9).

⬤ **Figure 4** *Sand dune plant succession at Daymer Bay, Cornwall*

Warning! ⚠️

Never take risks of any kind in **any** coastal environment, even in the height of summer. **NEVER** work near the base of any cliff – especially where slumping is frequent and dangerous, or where you might get cut off by the tide. Be aware that coastal landscapes can be fascinating, but also very dangerous.

Field study topic 5: Regenerating places

This fieldwork could take place in a single urban or rural area on the following themes:

◆ Evidence of regeneration.

◆ Public opinions about local regeneration.

◆ An analysis of the key players, winners and losers from regeneration.

Planning a field trip to a regenerated area

1 Pre-visit

Use the following sources to research the city or rural area that you will visit:

◆ **Historical maps** and **land use surveys** (e.g. GOAD) to identify previous settlement size or economic activities. The first OS map series (1846 onwards) and Land Utilisation Survey (1930s) are both available online at: www.visionofbritain.org. uk. GOAD maps of past urban land-use patterns in your fieldwork area can establish land-use change associated with rebranding.

◆ Historical material could be combined with **GIS data** from packages such as ArcGIS, Google Earth or satellite images to compare recent land use with historical data.

◆ Research the role of councils in rebranding by googling city planning departments – use phrases such as 'Manchester regeneration' or 'Dorset regeneration'.

◆ **Newspaper articles** can be used to research local issues and conflicts. Use local papers, e.g. *Yorkshire Post* for Leeds and Sheffield, and local BBC News websites for your chosen city – e.g. news.bbc.co.uk/ *name of county or city*'

2 The field trip itself

Depending on your focus, a wide range of data could be collected, such as:

◆ land-use and services mapping, and 'shopping basket surveys' in urban retail centres to identify the basis of the local economy, and any influence by students and/ or migrants.

◆ Environmental Quality Surveys (see Section 6.7) to identify costs and benefits of (a) regeneration, (b) gentrification, and (c) studentification in different areas, and of regeneration programmes.

◆ questionnaires to assess local opinions and views about re-imaging, e.g. the Eden Project, etc.

◆ photographs and sketches. These can either be at ground level or oblique aerial (e.g. of aspects of regeneration, such as the Olympic Park in Figure 5), or vertical aerial from Google Maps.

◆ EIAs of buildings in the regeneration area, or of planned regeneration schemes.

⊙ **Figure 5** *London's Queen Elizabeth Olympic Park – site of a major regeneration project from what was once a run-down industrial area*

3 Follow up

In class, further sources could be accessed at the 'writing up stage':

◆ The Census website, www.neighbourhood.statistics. gov.uk can be used to assess population trends, social conditions and economic trends.

◆ Environmental Impact Assessment (see Section 7.2). This could be drawn up to bring together all of the pros/cons of developing regeneration schemes.

◆ Comparing regenerated and un-regenerated areas through EQS scores and annotated photos.

Warning!

Never take risks when using questionnaires or conducting interviews. **ALWAYS** design your questionnaires carefully. If some questions concern sensitive issues (e.g. migration), discuss them with your teacher first. Prepare interviews by arrangement beforehand, so that you know who the interviewee will be. Phone interviews can be just as useful as face-to-face interviews. Be aware of traffic and your personal safety at all times.

Field study topic 6: Diverse places

This could take place in a single urban or rural area. The following themes would be appropriate for study:

- Attitudes towards geo-demographic change.
- The extent of deprivation in an area.

Planning a field trip to an area to study diversity

1 Pre-visit

Use the following sources to research the city or rural area that you will visit:

- **Historical maps** to identify previous settlement size. The first OS map series (1846 onwards) are available online at: www.visionofbritain.org.uk.
- Data from past censuses can identify aspects of population diversity and change. Combined with details of church, synagogue or mosque attendances and membership, population composition can be traced more easily.
- Historical material could be combined with **GIS data** from packages such as ArcGIS about population change.
- **Newspaper articles** can be used to research current population issues. Use local papers, e.g. *Yorkshire Post* for Leeds and Sheffield, and local BBC News websites for your chosen city – e.g. news.bbc. co.uk/*name of county or city*'

2 The field trip itself

Depending on your focus, a wide range of fieldwork data could be collected, such as:

- land-use and services mapping, which record population diversity (such as the photo of East London in Figure 6). This can be used to assess any influence by migrants on the local economy.
- Environmental Quality Surveys (see Section 6.7) to identify environmental quality as part of an overall deprivation survey.
- crime and design vulnerability surveys to assess potential hotspots. Simple recording sheets can be used from the national police database.

- questionnaires to assess local opinions and views about changes in population and migrants, together with interview techniques in Section 6.4.
- photographs and sketches. These can either be at ground level (e.g. of aspects of regeneration), or oblique aerial, or vertical aerial from Google Maps.

3 Follow up

Back in the classroom, a range of further sources could help the 'writing up stage':

- The Census website, www.neighbourhood.statistics. gov.uk can be used for population and social trends.
- Mapping crime data and appearances of crime, e.g. using the police and crime website.
- Comparing the impact of diverse populations on the social and environmental appearances of parts of cities.

🔺 **Figure 6** *East London has some of the UK's most diverse populations, which can be recorded on land use surveys in streets like this*

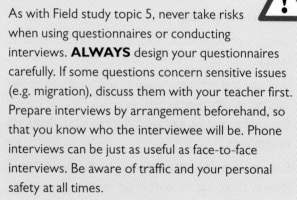

Warning!

As with Field study topic 5, never take risks when using questionnaires or conducting interviews. **ALWAYS** design your questionnaires carefully. If some questions concern sensitive issues (e.g. migration), discuss them with your teacher first. Prepare interviews by arrangement beforehand, so that you know who the interviewee will be. Phone interviews can be just as useful as face-to-face interviews. Be aware of traffic and your personal safety at all times.

The AS course is a lively and interesting one, full of contemporary topics, such as regeneration, and old favourites like tectonics. But achieving exam success is also important: How will you be examined? What kinds of questions might appear in the exam? How will you earn marks? That is where this chapter can help.

AS versus A Level?

Remember that the AS Geography qualification does not count towards A Level. The Topic content is the same as one half of the A Level course, but it's assessed at a lower level. The AS course exists for the following three reasons:

1. You may not want to study a full A Level, but you still want to take Geography further than GCSE because you enjoy it.

2. You might prefer an interim exam half way through the course, because it will give you some exam practice.

3. Some universities believe that AS grades already achieved are a good indicator of future success at A Level.

Like the full two-year A Level course, AS Geography is balanced between physical, human and environmental geography. It provides you with knowledge, understanding and skills to prepare you for higher education or employment.

What does the AS specification include?

The AS specification consists of six topics, of which you must study four. They are grouped into two themes, known as Areas of Study.

Area of Study 1: Dynamic Landscapes

Topic 1: Tectonic Processes and Hazards (Chapter 1 in this textbook). This topic is compulsory.

Topic 2: Landscape Systems, Processes and Change – you must study one of

◆ **either** Option 2A: Glaciated Landscapes and Change (Chapter 2 in this textbook)

◆ **or** Option 2B: Coastal Landscapes and Change (Chapter 3 in this textbook)

Area of Study 2: Dynamic Places

Topic 3: Globalisation (Chapter 4 in this textbook). This topic is also compulsory.

Topic 4: Shaping Places – you must study one of

◆ **either** Option 4A: Regenerating Places (Chapter 5 in this textbook)

◆ **or** Option 4B: Diverse Places (Chapter 6 in this textbook)

Other elements of the course

Three other elements feature in the exam: geographical skills, fieldwork, and synoptic themes (see page 324).

Geographical skills (Chapter 7 in this textbook). Many geographical skills have been integrated into each chapter, but often those in Chapter 7 are used in more than one topic.

Fieldwork (Chapter 8 in this textbook). There is no coursework at AS, but questions will be included in the exam about the fieldwork you complete as part of your AS course.

Synoptic themes and concepts

The three synoptic themes

The three synoptic themes are entitled: 'Players', 'Attitudes and Actions', and 'Futures and Uncertainties'. They are designed to help you make links between different topics.

1. Players

These are the people, groups and organisations with a stake (or a viewpoint) in a geographical issue or decision. A stake may be financial (e.g. a property owner or investor) or environmental (e.g. a pressure group). Some players have more influence than others, e.g. international players such as IGOs, NGOs, national and local government, large and small businesses, TNCs, and pressure groups.

2. Attitudes and Actions

These are the attitudes or views that people or organisations may have about a geographical issue, and the actions that they are prepared to take in response. Such actions might include policies, strategies and management methods. The views held may vary, depending on the person's or organisation's identity and links to a place, their political and religious views, or the importance they give to promoting profit, social justice, equality or the natural environment.

3. Futures and Uncertainties

Different approaches can be taken towards making decisions about geographical issues that will affect people in the future. These approaches include 'business as usual', as well as prioritizing more sustainable strategies and radical alternatives. For example, debates about coasts and climate change involve mitigation and adaptation. The impacts of choices made now can be uncertain for reasons that include scientific, demographic, economic and political uncertainties.

14 key geographical concepts

The AS and A Level Geography courses have been designed with **content** in mind (e.g. tectonics), but they are also underpinned by geographical **concepts**. These concepts are threaded through areas of geographical study. For example, understanding causality – that is, exploring causes – is as important to human topics (e.g. urban regeneration) as it is to physical topics (e.g. the causes of earthquakes).

The 14 key geographical concepts are:

1. causality
2. systems
3. equilibrium
4. feedback
5. inequality
6. representation
7. identity
8. globalisation
9. interdependence
10. mitigation and adaptation
11. sustainability
12. risk
13. resilience
14. thresholds.

Figure 1 shows how the four AS topics can be related to the 14 key concepts.

🔻 **Figure 1** *How topics can be linked to the 14 key concepts that underpin the AS and A Level Geography courses. The table is only partly completed – you should be able to think of far more links!*

Concept	Tectonics	Glaciated / Coastal Landscapes	Globalisation	Rebranding Places / Diverse Places
1. Causality	✓	✓	✓	✓
2. Systems		✓		
3. Equilibrium		✓		
4. Feedback		✓		
5. Inequality			✓	✓
6. Representation				
7. Identity				
8. Globalisation				
9. Interdependence				
10. Mitigation and adaptation				
11. Sustainability				
12. Risk				
13. Resilience				
14. Thresholds				

How will you be assessed?

There are two exams for AS Geography. Unlike A Level, there is no coursework – but you must complete two days of AS fieldwork. Figure 2 explains what the AS exams will involve.

▼ **Figure 2** *A summary of the AS exam requirements*

Topics assessed	Assessment information
Paper 1	
There is a total of 90 marks for this exam, worth 50% of the AS qualification. **Compulsory Section A** (Assesses Topic 1: Tectonic Processes and Hazards in Question 1) Answer all parts of this question. **28 marks** **Optional:** **Either** Section B (Assesses Topic 2A Glaciated Landscapes and Change) Answer • Question 2 (Glaciated Landscapes and Change) – **28 marks** • AND Question 3 (Fieldwork) – **16 marks** • AND Question 4 (Core Synoptic) – a single **16-mark** question. **Or Section C** (Assesses Topic 2B Coastal Landscapes and Change) Answer • Question 5 (Coastal Landscapes and Change) – **28 marks** • AND Question 6 (Fieldwork) – **16 marks** • AND Question 7 (Core Synoptic) – a single **16-mark** question.	**Time:** 1 hour 45 minutes **Sections:** Consists of three sections of which you do two. You will have a resource booklet. The examination may include multiple-choice questions, short open, open response, calculations and resource-linked questions. You will need a calculator. The exam includes 9-, 12-, and 16-mark extended-writing questions. • Question 1 (Tectonic Processes and Hazards), Question 2 (Glaciated Landscapes and Change) and Question 5 (Coastal Landscapes and Change) will include a **12-mark** question. • In optional Sections B or C, **one** question will assess fieldwork (Question 3 in Glaciated Landscapes and Change or Question 6 in Coastal Landscapes and Change). These will include a **9-mark** question. • Sections B and C each end with a single **16-mark** synoptic question that assesses your understanding of one of three synoptic themes (see opposite) and of one of the 14 key geographical concepts (see opposite).
Paper 2	
There is a total of 90 marks for this exam, worth 50% of the AS qualification. **Compulsory Section A** (Assesses Topic 3: Globalisation in Question 1) Answer all parts of this question. **28 marks** **Optional:** **Either** Section B (Assesses Topic 4A Regenerating Places) Answer • Question 2 (Regenerating Places) – **28 marks** • AND Question 3 (Fieldwork) – **16 marks** • AND Question 4 (Core Synoptic) – a single **16-mark** question. **Or Section C** (Assesses Topic 4B Diverse Places) Answer • Question 5 (Diverse Places) – **28 marks** • AND Question 6 (Fieldwork) – **16 marks** • AND Question 7 (Core Synoptic) – a single **16-mark** question.	**Time:** 1 hour 45 minutes **Sections:** Consists of three sections of which you do two. You will have a resource booklet. The examination may include multiple-choice questions, short open, open response, calculations and resource-linked questions. You will need a calculator. The exam includes 9-, 12-, and 16-mark extended-writing questions. • Question 1 (Globalisation), Question 2 (Regenerating Places) and Question 5 (Diverse Places) will include a **12-mark** question. • In optional Sections B or C, **one** question will assess fieldwork (Question 3 in Regenerating Places or Question 6 in Diverse Places). These will include a **9-mark** question. • Sections B and C each end with a single **16-mark** synoptic question that assesses your understanding of one of three synoptic themes (see opposite) and of one of the 14 key geographical concepts (see opposite).

How are the exam papers marked?

Examiners have to know what it is that they are assessing you on, so they use **Assessment Objectives**, or AOs for short. There are three AOs at AS level:

◆ **AO1 Knowledge and understanding**. Worth 40% of the AS exam marks. These questions test what you know.

◆ **AO2 Application of understanding** (e.g. in interpreting, analysing and evaluating geographical information and issues). Worth 35.6% of the AS exam marks. These questions test what you understand about new situations or data that you are presented with.

◆ **AO3 Geographical skills** (e.g. in handling and processing information when investigating geographical questions and issues, such as interpretation, analysis and evaluation of data and evidence), as well as constructing arguments and drawing conclusions. Worth 24.4% of the AS exam marks. These questions test your skills.

Understanding the mark schemes

Examiners use two types of mark scheme, which depend on the number of marks allocated for shorter questions (up to 4 marks) and those for extended written answers (6 marks or more).

Shorter answers (worth up to 4 marks)

Questions carrying up to 4 marks are **point marked**. For every correct point that you make, you earn a mark. Sometimes these are single marks for a 1-mark question. Others require the development of a point for a second mark – e.g. if you are asked to describe one feature of something for 2 marks.

Consider this question:

Suggest one reason why the largest earthquakes do not always cause the most damage.
(2 marks)

There are two marks for this question and you have to suggest one reason. So, you get one mark (or point) for a reason (e.g. 'because buildings could have been strengthened') and a second mark for developing the point further (e.g. 'so that they can resist the violent shaking which would damage them').

Extended answers (worth 6 marks or more)

Questions carrying 6 marks or more are **level-marked**. The examiner reads your whole answer and then uses a set of criteria – known as **levels** – to judge its qualities. There are three levels for questions carrying 6, 9 or 12 marks. The highest level earns the most marks.

Consider this question:

Explain the causes of tsunami. **(6 marks)**

Examiners would mark this question using the mark scheme in Figure 3, which consists of three levels. The mark scheme is the same for all 6-mark questions in the AS exams. This question tests knowledge and understanding, so it's an AO1 question. To reach the highest marks, your answer must fit the Level 3 criteria.

Level	Mark	Descriptor
Level 1	1-2	• Demonstrates isolated elements of geographical knowledge and understanding, some of which may be inaccurate or irrelevant. • Understanding addresses a narrow range of geographical ideas, which lack detail.
Level 2	3-4	• Demonstrates geographical knowledge and understanding, which is mostly relevant and may include some inaccuracies. • Understanding addresses a range of geographical ideas, which are not fully detailed and/or developed.
Level 3	5-6	• Demonstrates accurate and relevant geographical knowledge and understanding throughout. • Understanding addresses a broad range of geographical ideas, which are detailed and fully developed.

▲ **Figure 3** *The mark scheme for a 6-mark question. These usually assess AO1 questions.*

9-, 12- and 16-mark answers

9- or 12-mark questions also have three levels in their mark schemes. Again, these are identical for both exam papers. However, the balance of the marks is slightly different: the 12-mark questions are split between 3 marks for AO1 and 9 marks for AO2. To reach the top level, you must:

◆ demonstrate accurate knowledge and understanding throughout (AO1)

◆ apply your knowledge and understanding (AO2)

◆ produce a full interpretation that is relevant and supported by evidence (AO2)

◆ make supported judgements in a balanced and coherent argument (AO2)

In addition to the 9- and 12-mark questions, 16-mark questions assess geographical concepts and synoptic themes (see page 324). These questions are marked using four levels. The 16-mark questions are split 4 marks for AO1 and 12 for AO2. To reach the top level, you must, in addition to the qualities for Level 3:

◆ reach a rational, substantiated conclusion, supported by a balanced and coherent argument (AO2)

◆ synthesize ideas and concepts from across the course of study throughout the answer (AO2)

Command words

Command words – that is, those words which tell you what you must do – are shown in Figures 4 and 5. Notice that different command words are used for shorter and extended questions – although 'Explain' is common to both.

Command word	Definition
Identify/Give/ Name/State	Recall or select one or more pieces of information. This assesses AO1.
Define	State the meaning of a term. This assesses AO1.
Calculate	Produce a numerical answer, showing relevant working. This assesses AO3.
Draw/plot	Create a graphical representation of geographical information. This assesses AO3.
Complete	Create a graphical representation of geographical information by adding detail to a resource that has been provided. This assesses AO3.
Describe	Give an account of the main characteristics of something or the steps in a process. Statements in the response should be developed but do not need to include a justification or reason. This assesses AO1.
Compare	Find the similarities and differences of two elements given in a question. Each response must relate to both elements, and must include a statement of their similarity/difference. This generally assesses AO1 or AO3.
Suggest	For an unfamiliar scenario, provide a reasoned explanation of how or why something may occur. A suggested explanation requires a justification/exemplification of a point that has been identified. This generally assesses AO2.
Explain (less than 6 marks)	Provide a reasoned explanation of how or why something occurs. An explanation requires understanding to be demonstrated through the justification or exemplification of points that have been identified. This assesses AO1 and occasionally AO2.

⬆ **Figure 4** *Command words for shorter questions at AS*

Explain (6 marks)	Provide a reasoned explanation of how or why something occurs. An explanation requires understanding to be demonstrated through the justification or exemplification of points that have been identified. This assesses AO1 and occasionally AO2.
Assess (9 or 12 marks)	Use evidence to determine the relative significance of something. Give balanced consideration to all factors and identify which are the most important. This assesses AO1 (because you have to know something) but mainly AO2 (because you are asked to apply your understanding and make a judgment).
Evaluate (16 marks)	Measure the value or success of something and provide a balanced and substantiated judgment or conclusion. Review information and then bring it together to form a conclusion, drawing on strengths, weaknesses, alternatives and relevant data. This assesses AO1 (because you have to know something), AO2 (because you are asked to apply understanding and make a judgment), and also AO3 (because you may interpret data using skills to help you).

⬆ **Figure 5** *Command words for extended questions at AS*

Marks are allocated to questions with particular command words in a standard way (see Figure 6).

▼ **Figure 6** *Mark allocations to questions with particular command words*

Mark tariff	1	2	3	4	6	8	9	12	16
Define	*								
Identify/State/Name	*	*							
Calculate	*	*							
Complete	*	*							
Draw / Plot		*	*						
Describe		*	*						
Compare		*	*						
Suggest		*	*						
Explain				*	*	*			
Analyse						*			
Assess							*	*	
Evaluate									*

Handy hints for high marks!

It's hard to keep cool under exam pressure. However, these hints should help you perform much better.

1 Dissect the question

Look at the example below. Try to break up questions like this – it will help focus you on what the examiner is looking for.

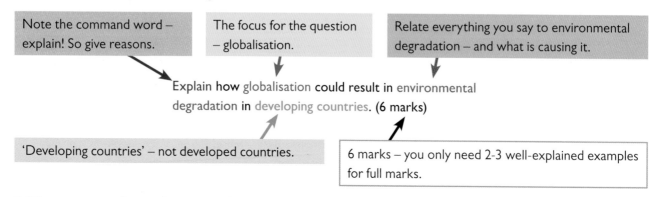

Note the command word – explain! So give reasons.

The focus for the question – globalisation.

Relate everything you say to environmental degradation – and what is causing it.

Explain how globalisation could result in environmental degradation in developing countries. (6 marks)

'Developing countries' – not developed countries.

6 marks – you only need 2-3 well-explained examples for full marks.

2 Choose examples and case studies

Case studies are in-depth examples of particular places, used to illustrate big ideas at localised scales. There are plenty in this book; some examples are a paragraph, while others run to several pages. You should use these to answer the exam questions.

Few questions actually ask for examples. However, you'll get further in answering a question if you quote examples. The following is an example of a question about Regenerating Places (Chapter 5), where good examples could produce a stronger answer:

Explain the impacts of declining rural services upon places. (6 marks)

In this example, an example of a named place – it could be a village you have studied, or a whole county such as Cornwall (see Sections 5.9 and 5.13). The important thing is that named examples make your answer precise – and with only six marks, you could find that three explained examples would earn you all six! But be selective – no question at this level will ever ask you to write everything you know about a case study.

3 Plan answers

Fact – students who plan their answers before writing them usually score higher marks than those who don't. This is because planning:

◆ stops the student from going off-track

◆ prevents 'memory blanks'.

Your plans need not be lengthy – allow 5-10% of the total exam time for planning, e.g. 2 minutes for a 16-mark answer. Also, your plans should not be elaborate – brief notes will keep your answer on track, like the example in Figure 7.

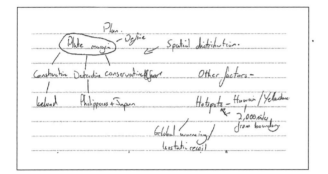

△ **Figure 7** *An example of a student plan for an answer about plate margins in Topic 1*

4 Keep to time

Many students lose track of time in exams. The exam questions are designed to be completed in the time allowed, so the following suggestions may help you to achieve this. BEFORE the exam:

◆ work out how long each section should take to complete

◆ work out 5-10% total planning time, 80-85% writing time, and 10% checking time

◆ practice timed answers – including shorter answers, as well as longer ones.

Using fieldwork and research in the exam

Fieldwork and research skills are a feature of both exam papers at AS. Whichever physical (Chapters 2 or 3) and human topics (Chapters 5 or 6) you have studied, you must use fieldwork and research in the exam.

Some questions will ask you about **familiar fieldwork** – that is, work you have done yourself. Part of the question you have to answer could be something like this:

Explain how one method of analysing your fieldwork data helped you to understand the data better. (4 marks)

In this question, you would identify a method, such as chi-square or Spearman's rank, and explain how the coefficient helped you to understand whether any relationships existed between the data you collected.

Other questions will ask you about **unfamiliar fieldwork** – that is, asking you to analyse a new situation based on your own experience. An example could be:

Explain how fieldwork data could be collected to investigate differences between glacial and fluvio-glacial deposits at different locations. (6 marks)

In this question, you would explain one or two methods of collecting data (e.g. on sediment size and roundness) for about half of the marks. The remaining marks would be awarded if you explained how these data would help to understand any differences between glacial and fluvio-glacial deposits.

Glossary

A

A T Kearney Index – measures how globalised a country has become

ablation zone – where outputs from a glacier exceed inputs

abrasion – the grinding away of bedrock by fragments of rock which may be incorporated in ice. Also known as *corrasion

accumulation zone – where inputs to a glacier exceed outputs

adaptation – strategies designed to prepare for and reduce the impacts of events

Arctic amplification – the phenomenon where the Arctic region is warming twice as fast as the global average

asthenosphere – the part of the mantle, below the *lithosphere, where the rock is semi-molten

attrition – the gradual wearing down of rock particles by impact and *abrasion, leading to a reduced particle size and rounder, smoother stones

B

basal sliding (or slip) – where pressure and friction create meltwater at the base of the glacier, which lubricates the glacier's movement

Benioff zone – the area where friction is created between colliding tectonic plates, resulting in intermediate and deep earthquakes

C

Central Business District (CBD) – the area of a city which is usually the hub of finance, services and retail

centrifugal forces – forces which push people apart, for example changes in employment

centripetal forces – forces which draw people together, for example a strong sense of community

cirque glaciers – see *corrie glaciers

climatic climax community – the final community of species that will be adjusted to the climatic conditions of an area

closed system – there are no inputs or outputs of matter from an external source

compressional flow – when reductions in gradient force a glacier to slow down, causing it to 'pile up' and thicken

concordant coast – where bands of more-resistant and less-resistant rock run parallel to the coast

convection currents – hot, liquid magma currents moving in the *asthenosphere

corrasion – see *abrasion

corrie glaciers – small glaciers occupying hollows on mountains which may feed into *valley glaciers

corrosion – the breaking down of rock by chemical action, often involving the dissolving of alkaline rock by weak acids in seawater

cost-benefit analysis – a process by which the financial, social and environmental costs are weighed up against the benefits of a proposal in terms of social outcomes as well as in terms of profit and loss

crustal fracturing – when energy released during an earthquake causes the Earth's crust to crack

cryosphere – the frozen part of the Earth's hydrological system

cultural diffusion – the spread of one culture to another by various means

cultural erosion – the changing and loss of culture in an area, such as the loss of language and traditional food

cultural fractionization – measures how diverse countries are, by measuring people's attitudes towards, for example, religion, democracy and the law

D

Dalmatian coasts – a type of *concordant coastline formed as a result of a rise in sea level when valleys flooded leaving the tops of the ridges above the surface of the sea as offshore islands

deindustrialisation – the decline of manufacturing industry in an area

dependency ratio – the ratio of dependents (aged under 15 or over 64) to the working-age population (aged 15-64), published as the proportion of dependents per 100 working-age population

deregulation – the reduction in rules which means that any foreign business can set up in the UK

discordant coasts – the geology alternates between bands of more-resistant and less-resistant rock, which run at right angles to the coast

dormitory suburbs – residential areas which are primarily homes for commuters

drift-aligned – where sediment is transferred along the coast by *longshore drift producing a pattern of sediment size and roundness which varies between one location on a beach and another

dynamic equilibrium – where landforms and processes are in a state of balance

E

ecological footprint – a measure of the land area and water reserves that a population needs in order to produce what it consumes (and absorb the waste it generates), using current technology

economies of scale – the ability to reduce costs proportionally by increasing the scale of production

emergent coastline – when a fall in sea level exposes land previously covered by the sea

Enterprise Zones – small areas which offer incentives to attract companies, such as tax discounts and a reduction in planning permission requirements

entrainment – the process by which surface sediment is incorporated into a fluid flow (e.g. air, water or ice) as part of the process of erosion

Environmental Impact Assessment (EIA) – a quantitative means of estimating the environmental changes arising from a proposal

epicentre – the point on the Earth's surface directly above the *focus of an earthquake

equilibrium line – the boundary between the *accumulation zone and the *ablation zone

ethnic enclaves – concentrations of particular communities in an area, such as a high concentration of Asian or Asian British residents with a Pakistani background in East London

eustatic change – when the sea level itself rises or falls, partly as a result of the growth and decay of ice sheets

Export Processing Zones – the term now used in China for *Special Economic Zones (SEZ)

extensional flow – when an increase in gradient causes the ice to flow faster and it become 'stretched' and thinner

F

focus – the point inside the Earth's crust from which the pressure is released when an earthquake occurs

Foreign Direct Investment (FDI) – investment made by an overseas company or organisation into a company or organisation based in another country

G

Gini coefficient/Gini index – fundamentally, these measure the same thing: inequality. They measure how far wealth distribution within a country deviates from perfect equality. The Gini index is shown as a percentage, and the Gini coefficient as a value between 0 and 1 (i.e. the Gini index divided by 100). The larger the number, the greater the concentration of wealth will be amongst only a few. An index/coefficient rating of 0 = perfect equality, and a rating of 100 or 1 = perfect inequality

glacial outburst – when a huge amount of meltwater, that was previously trapped either beneath the ice or as surface lakes, eventually bursts

global homogenisation – the idea that everywhere is becoming the same

glocalisation – when a company re-styles its products to suit local tastes

Gross Domestic Product (GDP) – the same as *Gross National Income, but excluding foreign earnings

Gross National Income (GNI) – the value of goods and services earned by a country (including overseas earnings), formerly known as Gross National Product (GNP)

H

Haff coast – a *concordant coastline which consists of long spits of sands and lagoons

hazard-management cycle – a theoretical model of hazard management as a continuous four-stage cycle involving mitigation, preparation, response and recovery

hazard-response curve – see *Park model

hot spot – points within the middle of a tectonic plate where plumes of hot magma rise and erupt

hub cities – see *world cities

Human Development Index (HDI) – a measure of development which takes into account life expectancy, education and *GDP for every country and converts them into a value between 0 and 1

hydrometeorological hazards – natural hazards caused by climate processes (including droughts, floods, hurricanes and storms)

hyper-urbanisation – rapid *urbanisation

I

imbrication – where glacial deposits are orientated (or aligned), overlapping each other like toppled dominoes

Index of Multiple Deprivation – an overall measure of deprivation which incorporates income, employment, education, health, crime, barriers to housing and services, and living environment

Integrated Coastal Zone Management (ICZM) – a strategy designed to manage complete sections of the coast, rather than individual towns or villages, by bringing together all of those involved in the development, management and use of the coast

Inter-Governmental Organisations – organisations which comprise of two or more countries working together. Examples include the EU and the United Nations

International Monetary Fund (IMF) – a global organisation whose primary role is to maintain international financial stability

interstadials – short-lived warmer periods within a major glacial, associated with ice retreat

intra-plate earthquakes – earthquakes which occur far from plate margins

isostatic change – when the land rises or falls, relative to the sea, often in response to the melting or accumulation of glacial ice

K

knowledge economy – see *new economy. Also associated with the quaternary sector of industry which provides highly specialised jobs that use expertise in fields such as finance, law and IT

KOF Index – measures how globalised a country has become by taking into account international interactions

L

L waves – the slowest seismic waves, which focus all their energy on the Earth's surface

land-use zoning – a process by which local government regulates how land in a community may be used

liquefaction – when the violent shaking during an earthquake causes surface rocks to lose strength and become more liquid than solid

lithosphere – the solid layer, made from the crust and upper mantle, from which tectonic plates are formed

littoral zone – another name for the coastal zone: the boundary between land and sea which stretches out to sea and onto the shore

longshore drift – the movement of sand and shingle along the coast

Lorenz curve – a graph to show and measure any inequality by comparing it to a line of perfect equality

M

market-led regeneration – the improvement of an area which is driven by the potential needs and wants of customers

mass movement – the downward movement of material under the influence of gravity. It includes a wide range of processes such as rockfalls, landslides and *solifluction

Milankovitch cycles – three interacting astronomical cycles in the Earth's orbit around the Sun, believed to affect long-term climatic change

mitigation – action to reduce the impacts of an event

multiple-hazard zone – an area that is at risk from multiple natural hazards such as hurricanes and earthquakes

multiplier effect – when success in one business creates further wealth and spending, boosting the economic development of the local economy as a whole

N

negative feedback – the regulation and reduction of a natural process

negative multiplier – a downward spiral or cycle, where economic conditions produce less spending and less incentive for businesses to invest (therefore reducing opportunities)

neo-liberalism – a belief in the free flows of people, capital, finance and resources. Under neo-liberalism, State interventions in the economy are minimized, while the obligations of the State to provide for the welfare of its citizens are diminished

new economy – where *GDP is earned more through expertise and creativity in services such as finance and media than from the manufacture of goods. Also known as the *knowledge economy

nivation – a range of processes associated with patches of snow

Non-Governmental Organisation (NGO) – a non-profit organisation created by private organisations or people with no participation or representation by any government

O

off-shoring – when a company does work overseas, either itself or using another company

open system – where energy and matter can be lost to and gained from an external source

outsourcing – when work is contracted out to another company (known as *off-shoring when that company is overseas)

P

P waves – the fastest seismic waves which travel through both solids and liquids

palaeomagnetism – the study of past changes in the Earth's magnetic field

paleo-environments – fossil or geologically past environments

Park model (hazard-response curve) – shows how a country or region might respond after a hazard event

periglacial – areas at the edge of permanent ice, characterised by *permafrost and a *tundra climate

permafrost – where a layer of soil, sediment or rock below the ground surface remains almost permanently frozen

pioneer species – the first colonising plants which begin the process of plant succession

polders – land, often reclaimed from the sea, enclosed by embankments

positive feedback – enhances and speeds up processes, promoting rapid change

post-industrial economy – see *new economy

post-production countryside – the situation in rural areas when the 'old economy', consisting mainly of primary sector jobs, has declined

pressure and release (PAR) model – a tool used to work out how vulnerable a country is to hazards

pressure melting – lower down a glacier, where pressure is higher, ice may melt at temperatures below 0°C

pressure melting point – the temperature at which ice is on the verge of melting

Purchasing Power Parity (PPP) – relates average earnings to local prices and what they will buy. This is the spending power within a country, and reflects the local cost of living

Q

quotas – a fixed level indicating the maximum amount of imported goods or persons which a state will allow in

R

re-imaging – how the image of a place is changed and portrayed in the media

regelation – melting and freezing of ice, caused by changes in pressure

remittance payments – income sent home by individuals working elsewhere (usually abroad but can be in urban areas)

rural-urban continuum – the spectrum which moves from a large city or conurbation to countryside areas

S

S waves – seismic waves which only travel through solids and move with a sideways motion

sediment budget – the amount of sediment available within a *sediment cell

sediment cells – a length of coastline and its associated nearshore area within which the movement of coarse sediment (sand and shingle) is largely self contained. There are 11 sediment cells around England and Wales some of which can be divided into sub-cells

Shoreline Management Plan (SMP) – a plan that takes into consideration the risks of coastal processes and attempts to identify sustainable coastal defence and management options

slab pull – when newly formed oceanic crust sinks into the mantle, pulling the rest of the plate further down with it

social clustering – a preference for people to live close to others they associate with, for example communities of similar ethnicity or religion

solar output – the amount of radiation that the sun emits (associated with sunspot activity), which can affect the Earth's temperature

solifluction – a form of *mass movement which is the downhill flow of saturated soil

Special Economic Zones (SEZ) – set up by national governments to offer financial or tax incentives to attract *Foreign Direct Investment, which differ from those incentives normally offered by a country

stadials – short-lived colder periods within a major glacial, associated with ice advance

stakeholders – people who have an interest in a scheme or an area, such as local residents or an environmental group

sub-aerial processes – the processes of weathering and *mass movement

subduction zone – the area in the mantle where a tectonic plate melts

sublimation – the change from the solid state to gas with no intermediate liquid stage

submergent coastline – when a rise in sea level floods a previously exposed coast

subsidies – grants given by governments to increase the profitability of key industries

supraglacial – on top of a glacier

swash-aligned – where sediment moves up and down the beach with little lateral transfer

swell waves – waves originating from the mid-ocean which appear as larger waves amongst smaller, locally-generated waves

T

tariffs – a tax that is paid on goods coming into or going out of a country

terminal groyne syndrome – when higher rates of erosion occur immediately after a set of coastal defences have finished

trade liberalisation – the removal of trade barriers such as *subsidies, *tariffs and *quotas

trade protectionism – the use of methods such as *tariffs and *quotas to attempt to boost a country's exports or reduce its imports

trading blocs – when countries have grouped together to promote free trade between them. The EU is an example of a trading bloc

transform fault – a fault created on a large scale when two plates slide past each other

tundra – *periglacial regions found in the barren plains of northern Canada, Alaska and Siberia where both temperature and rainfall are low

U

urbanisation – the increasing proportion of people living in towns and cities as opposed to the countryside

V

valley glaciers – large masses of ice moving from ice fields or *corries and following river courses

W

World Bank – a global organisation which uses bank deposits placed by the world's wealthiest countries to provide loans for development in other countries

world cities – cities with a major influence, based on: finance, law, political strength, innovation and ICT

World Trade Organisation (WTO) – a global organisation which looks at the rules for how countries trade with each other